Beyond the Biomass

Beyond the Biomass

Compositional and Functional Analysis
of Soil Microbial Communities

Edited by

K. Ritz, J. Dighton and K.E. Giller

JOHN WILEY & SONS

Chichester - New York - Brisbane - Toronto - Singapore

A Co-Publication with the British Society of Soil Science (BSSS)
and Sayce Publishing (United Kingdom)

Published by   John Wiley & Sons
                 Baffins Lane, Chichester
                 West Sussex PO19 1UD, United Kingdom

Other Wiley Editorial Offices

John Wiley & Sons, Inc., 605 Third Avenue,
New York, NY 10158-0012, USA

Jacaranda Wiley Ltd., G.P.O. Box 859, Brisbane,
Queensland 4001, Australia

John Wiley & Sons (Canada) Ltd., 22 Worcester Road,
Rexdale, Ontario M9W 1L1, Canada

John Wiley & Sons (SEA) Pte Ltd., 37 Jalam Pemumpin 05-04,
Block B, Union Industrial Building, Singapore 2057

Co-Publishers

British Society of Soil Science (BSSS), University of Reading,
P.O. Box 233, Reading RG6 2DW, United Kingdom

Sayce Publishing, Hampton Place, 56 Longbrook Street,
Exeter EX4 6AH, United Kingdom

A catalogue record for this book is available from the British Library

ISBN 0471 950 963

Printed by Colorcraft Ltd, North Point, Hong Kong

Contents

Preface

Soil microbes are a fundamentally important component of terrestrial habitats, playing many key roles in ecosystem function. Their primary role is in governing the many nutrient cycling reactions which are essential in the maintenance of soil fertility (for example, through decomposition of plant and animal remains, fixation of atmospheric dinitrogen and formation of symbiotic associations with plants). Microbes are also intimately involved with the genesis and maintenance of soil structure — the architecture of the soil — which delimits the physical environment in which below-ground communities function and plant roots grow. A characteristic feature of soil communities is their complexity, both in terms of numbers of organisms present and their genetic and functional diversity. A cubic centimetre of grassland soil typically contains hundreds of millions of bacteria, tens of thousands of protozoa, hundreds of metres of fungal hyphae, several hundred nematodes, mites and insects, and a myriad of other microbes and larger organisms. In the mid-1970s, the concept of treating such below-ground life collectively, as the *microbial biomass*, emerged, and techniques for measuring this entity were devised. 'Microbial' is actually a misnomer; in practical terms, the biomass constitutes any organism which passes through the sieve (typically 2 mm mesh size) that a researcher uses to prepare soil samples. However, the biomass is comprised predominantly of *micro*organisms.

The soil microbial biomass concept has been vigorously adopted by soil scientists. The ability to quantify microbial pools of nutrient elements and the fluxes of such nutrients between microbial and other compartments of the soil has led to great advances in the analysis of nutrient cycling in soil systems. However, a major drawback of the concept is that it constitutes a 'black box' approach, and as such is highly generalised. It is the view of many soil microbiologists and others that we need to go beyond the biomass concept and acknowledge the diversity of microbial form and function within the total microbial pool.

In March 1993, an international symposium was held at Wye College, University of London, in Kent, UK, to provide a forum for the reporting and discussion of contemporary ideas on characterising complex microbial communities, the functional analysis of such communities and their interactions with other components of the biota, especially in relation to nutrient cycling. The meeting was attended by 170 delegates from 23 countries and comprised 11 keynote papers, 16 oral presentations and 64 posters, interspersed with discussions. The chapters in this book are based on the oral presentations made by researchers at the meeting. The contributions range from overviews of the biomass concept itself, through specific techniques being developed at the forefront of research, to more philosophical pieces. In the final chapter, we attempt to summarise the main themes which emerged during the formal and informal discussion sessions.

The symposium was organised under the auspices of the British Society of Soil Science (BSSS) and the International Society of Soil Science (ISSS). We thank Professor Jim Lynch, Professor Duncan

Greenwood and Dr Martin Wood for their advice, support and assistance over many months during the organisation of the meeting. We also thank the independent referees who so conscientiously reviewed the manuscripts and provided many useful comments on the papers. A book cannot be realised without a publisher, and we are grateful to Kay Sayce for her competence, effort and efficiency in creating the volume. Finally, we acknowledge the support of our host institutions and many colleagues, who, as always, provided much useful discussion.

Karl Ritz (Scottish Crop Research Institute)
John Dighton (Institute of Terrestrial Ecology)
Ken Giller (Wye College, University of London)

Contributors

Antoon Akkermans, Department of Microbiology, Wageningen Agricultural University, Hesselink van Suchtelenweg 4, 6703 CT Wageningen, the Netherlands

Traute-Heidi Anderson, Institut für Bodenbiologie, Forschungsanstalt für Landwirtschaft, Bundesallee 50, D-38116 Braunschweig, Germany

Luigi Badalucco, Dipartimento di Agrobiologia ed Agrochimica, Via S.C. DeLellis, 01100 Viterbo, Italy

Pierre Bottner, Centre d'Ecologie de Fonctionnelle et Evolutive, Centre National de la Recherche Scientifique, Route de Mende, BP 5051, 34033 Montpellier Cedex, France

Alexander Bruckner, Institute of Zoology, University of Agriculture, Gregor-Mendel-Strasse 33, A-1180 Vienna, Austria

Marianne Clarholm, Division for Soil Biology, Swedish University of Agricultural Sciences, EMC Box 7072, S-75007, Uppsala, Sweden

David Coleman, Institute of Ecology, University of Georgia, Athens, Georgia 30602-2603, USA

Marie-Madeleine Coûteaux, Centre d'Ecologie de Fonctionnelle et Evolutive, Centre National de la Recherche Scientifique, Route de Mende, BP 5051, 34033 Montpellier Cedex, France

Frida Lise Daae, Department of Microbiology, University of Bergen, Jahnebakken 5, N-5020 Bergen, Norway

Daniel Deere, Department of Genetics and Microbiology, Life Sciences Building, University of Liverpool, Crown Street, Liverpool L69 3BX, UK

Julian Diaper, Department of Genetics and Microbiology, Life Sciences Building, University of Liverpool, Crown Street, Liverpool L69 3BX, UK

John Dighton, Institute of Terrestrial Ecology, Merlewood Research Station, Windermere Road, Grange-over-Sands, Cumbria LA11 6JU, UK

Clive Edwards, Department of Genetics and Microbiology, Life Sciences Building, University of Liverpool, Crown Street, Liverpool L69 3BX, UK

Jay Garland, The Bionetics Corporation, Mail Code BIO 3, Kennedy Space Centre, Florida 32899, USA

Ken Giller, Wye College, University of London, Wye, Ashford, Kent TN25 5AH, UK

Cécile Gilot, Laboratoire d'Ecologie des Sols Tropicaux, Institut Français de Recherche Scientifique pour le Développement en Coopération (ORSTOM), 72 Route d'Aulnay, 93143 Bondy Cedex, France

Jostein Goksøyr, Department of Microbiology, University of Bergen, Jahnebakken 5, N-5020
    Bergen, Norway

Dittmar Hahn, Department of Soil Biology, Institute of Terrestrial Ecology, Grabenstrasse 3,
    CH-8952 Schlieren, Switzerland

David Harris, Center for Microbial Ecology, Michigan State University, East Lansing,
    Michigan 48824, USA

Peter Harris, Department of Soil Science, University of Reading, London Road, Reading,
    Berkshire RG1 5AQ, UK

Arlene Hilger, Department of Crop and Soil Science, Oregon State University, ALS 3017,
    Corvallis, Oregon 97331-7306, USA

Kerstin Huss-Danell, Department of Plant Physiology, University of Umeå, S-901 87 Umeå,
    Sweden

Christian Kampichler, Institute of Zoology, University of Vienna, Althanstrasse 14, A-1090,
    Vienna, Austria

Ellen Kandeler, Institute for Soil Management, Denisgasse 31-33, A-1200 Wein,
    Austria

Annelise Kjøller, Department of General Microbiology, University of Copenhagen, Salvgade 83H,
    DK-1307, Copenhagen, Denmark

Walter Klingmüller, Department of Genetics, University of Bayreuth, Universitätstrasse 30,
    D-95440 Bayreuth, Germany

Frantisek Kunc, Institute of Microbiology, Academy of Sciences of the Czech Republic, Videnska
    1083, 142 20 Prague 4, Czech Republic

Loretta Landi, Dipartimento di Scienza del Suolo e Nutrizione della Pianta, Università degli Studi,
    Piazzale delle Cascine 28, 50144 Firenze, Italy

Patrick Lavelle, Laboratoire d'Ecologie des Sols Tropicaux, Institut Français de Recherche
    Scientifique pour le Développement en Coopération (ORSTOM), 72 Route d'Aulnay, 93143
    Bondy Cedex, France

SangHoon Lee, Korean Ocean Research and Development Institute, PO Box 29, Ansan, Seoul,
    Korea

Kendall Martin, Department of Crop and Soil Science, Oregon State University, ALS 3017,
    Corvallis, Oregon 97331-7306, USA

Jone Michalsen, Department of Microbiology, University of Bergen, Jahnebakken 5, N-5020
    Bergen, Norway

Aaron Mills, Laboratory of Microbial Ecology, Department of Environmental Sciences, University
    of Virginia, Clark Hall, Charlottesville, Virginia 22901, USA

David Myrold, Department of Crop and Soil Science, Oregon State University, ALS 3017, Corvallis, OR 97331-7306, USA

Paolo Nannipieri, Dipartimento di Scienza del Suolo e Nutrizione della Pianta, Università degli Studi, Piazzale delle Cascine 28, 50144 Firenze, Italy

Roger Pickup, IFE, The Windermere Laboratory, Ambleside, Cumbria LA22 0LP, UK

Jon Porter, Department of Genetics and Microbiology, Life Sciences Building, University of Liverpool, Crown Street, Liverpool L69 3BX, UK

David Powlson, Soil Science Department, Rothamsted Experimental Station, AFRC Institute of Arable Crops Research, Harpenden, Hertfordshire AL5 2JQ, UK

Alan Rayner, School of Biological Sciences, University of Bath, Claverton Down, Bath, Avon BA2 7AY, UK

Karl Ritz, Soil Plant Dynamics Group, Cellular and Environmental Physiology Department, Scottish Crop Research Institute, Invergowrie, Dundee DD2 5DA, UK

Kåre Salte, Statoil, N-4033, Norway

Angela Sessitsch, Soil Science Unit, Food and Agriculture Organisation (FAO)/International Atomic Energy Agency (IAEA) Programme, IAEA Laboratories, A-2444 Seibersdorf, Austria

Neil Smith, Soil Science Department, Rothamsted Experimental Station, AFRC Institute of Arable Crops Research, Harpenden, Hertfordshire AL5 2JQ, UK

Roald Sørheim, Center for Industrial Research, Blindern, N-0314 Oslo 3, Norway

David Stribley, Soil Science Department, Rothamsted Experimental Station, AFRC Institute of Arable Crops Research, Harpenden, Hertfordshire AL5 2JQ, UK

S. Struwe, Department of General Microbiology, University of Copenhagen, Salvgade 83H, DK-1307, Copenhagen, Denmark

Vigdis Torsvik, Department of Microbiology, University of Bergen, Jahnebakken 5, N-5020 Bergen, Norway

Kate Wilson, Center for the Application of Molecular Biology to International Agriculture (CAMBIA), GPO Box 3200, Canberra, ACT 2601, Australia

Anne Winding, Department of Marine Ecology and Microbiology, National Environmental Research Institute, PO Box 358, Frederiksborgvej 399, DK-4000 Roskilde, Denmark

Birgit Winter, Federal Institute for Soil Management, Denisgasse 31-33, A-1200, Vienna, Austria

Peter Young, Department of Biology, University of York, Heslington, York YO1 5DD, UK

Josef Zeyer, Department of Soil Biology, Institute of Terrestrial Ecology, Grabenstrasse 3, CH-8952 Schlieren, Switzerland

Dmitri Zvyagintsev, Department of Soil Biology, Faculty of Soil Science, Moscow State University, Moscow, 119899, Russia

PART I

Review of the microbial biomass concept

Beyond the Biomass
Edited by K. Ritz, J. Dighton and K.E. Giller
© 1994 British Society of Soil Science (BSSS)
A Wiley-Sayce Publication

CHAPTER 1

The soil microbial biomass: Before, beyond and back

D.S. POWLSON

The concept of the soil microbial biomass is simple. It is that, for some purposes, the entire soil microbial population can be treated as a single entity. The concept was proposed by Jenkinson (1966) and a practical method for measuring the size of the biomass was published by Jenkinson and Powlson (1976b). It was an early example of a 'holistic' approach and contrasted sharply with most soil microbial studies in the 1960s, which were concerned with identifying the organism responsible for a specific process and studying its physiology and biochemistry in detail.

This chapter traces the development of the biomass concept and its associated methodologies, and gives examples of practical applications and of fundamental questions that the approach has raised. It is a personal view and is not intended to be a thorough review of the subject; for reviews of various aspects of the subject, the reader is referred to Jenkinson and Ladd (1981), Sparling (1985), Jenkinson (1988), Nannipieri et al. (1990), Smith and Paul (1990), Parkinson and Coleman (1991), Tunlid and White (1992) and Wardle (1992).

CHLOROFORM FUMIGATION AND INCUBATION

The work reported by Jenkinson (1966) was not planned as an attempt to develop a method for measuring the quantity of carbon held in the cells of living microorganisms in soil. Rather, it was a study of the decomposition of ^{14}C-labelled ryegrass in soil from which the biomass concept and methodology was an unforeseen consequence. Those concerned with the planning of research today would do well to remember the limitations of planning and the importance of allowing scientists to follow an idea into uncharted territory.

Jenkinson had allowed ^{14}C-labelled plant material to decompose in different soils for periods of several years. He noted that the percentage of labelled carbon in the CO_2 evolved (9.2%) was much greater than that of labelled carbon in the soil organic carbon (0.043%) (*see* Table 1.1 *overleaf*). Thus, the carbon entering the soil from the labelled plant material was not spread evenly through all fractions of soil organic matter. He then incubated the soils after they had been exposed to a series of treatments

Table 1.1   Labelled and unlabelled carbon evolved by soil during a 10-day incubation at 25°C
               following different treatments and inoculation with fresh soil[a]

Treatment	CO$_2$ evolved (µg C/g soil) labelled	CO$_2$ evolved (µg C/g soil) unlabelled	% labelled C in evolved CO$_2$-C
None	14.7	146	9.2
Air-drying	23.2	195	10.6
Irradiation	33.8	238	12.4
CH$_3$Br vapour	47.5	239	16.6
CHCl$_3$ vapour	49.4	259	16.0
Oven-drying at 80°C	54.0	347	13.6
Oven-drying at 100°C	56.0	493	10.2
Autoclaving at 120°C	58.6	524	10.1

Note: a   Percentage labelled C in soil 0.043
Source: Jenkinson (1966)

known collectively as 'partial sterilisation'. The treatments included drying at various temperatures, followed by rewetting, fumigation with biocidal chemicals (chloroform [CHCl$_3$] or methyl bromide [CH$_3$Br]), gamma-irradiation and autoclaving, followed by inoculation with a small quantity on untreated soil. All treatments increased the CO$_2$ evolved during a 10-day incubation at 25°C, but there were marked differences between the effects of the treatments (see Table 1.1). For example, autoclaving caused the largest increase in CO$_2$ evolution but the percentage of labelled carbon in CO$_2$ differed little from the control. By contrast, fumigation with CHCl$_3$ caused a smaller increase in total CO$_2$ evolution than autoclaving but the greatest increase in the percentage of labelled carbon.

Even in the 1960s, there was a vast literature on the effects of partial sterilisation of soil and a plethora of hypotheses to explain the observations (reviewed by Jenkinson, 1966; Powlson, 1975). Jenkinson (1966) concluded that the extra CO$_2$ evolved (and nitrogen mineralised) came mainly from the decomposition of soil microorganisms which had been the cells killed by the treatment and these constituted a pool of high ^{14}C content. He suggested that in the case of CHCl$_3$ and CH$_3$Br fumigation, this was by far the major source of extra carbon but that other treatments, such as air-drying or autoclaving, also released varying amounts of lightly labelled carbon from non-living fractions of soil organic matter. If this is so, the amount of extra carbon evolved after CHCl$_3$ fumigation, removal of CHCl$_3$ and inoculation with untreated soil should be proportional to the quantity of carbon in the killed cells. This was expressed as:

$$B = \frac{F}{k}$$

where:

B   =   quantity of carbon in the cells of living organisms in the soil (the biomass)
F   =   flush of carbon measured during a 10-day incubation at 25°C, defined as the CO$_2$ evolved from fumigated soil minus CO$_2$ evolved from the unfumigated soil
k   =   fraction of the biomass carbon which was mineralised to CO$_2$ under standard incubation conditions

The expression would now be written:

$$B_C = F_C/k_C$$

to show that it refers to carbon. An equivalent expression for nitrogen would be:

$$B_C = F_N/k_N$$

The k_C factor was originally set at 0.3 on the basis of an experiment in which the decomposition of ^{14}C-labelled *Nitrosomonas europaea* was measured, with and without $CHCl_3$ fumigation. It was later revised to 0.45, the mean value obtained from a range of microorganisms (Jenkinson, 1988).

The assumptions and limitations involved in using the $CHCl_3$ fumigation-incubation (FI) method to estimate soil microbial biomass were set out clearly by Jenkinson, thus:

- the soil should not contain recently added substrate

- $CHCl_3$ gives a near complete kill of the soil microbial population

- $CHCl_3$ does not render non-biomass parts of the soil organic matter decomposable

- $CHCl_3$ leaves neither decomposable nor toxic residues in soil

TESTING THE HYPOTHESIS

In a series of five papers (Jenkinson, 1976; Jenkinson and Powlson, 1976a, b; Jenkinson et al., 1976; Powlson and Jenkinson, 1976), experiments were reported which had tested the hypothesis that $CHCl_3$ selectively renders microbial cells decomposable and, further, that the size of the flush of decomposition caused by $CHCl_3$ fumigation can be used as a measure of the microbial biomass in soil. The strongest evidence came from comparing the flush in eight soils with estimates of biomass obtained from direct microscopy. The soils were dispersed in dilute agar, using a slight modification of the Jones and Mollison technique, films of known thickness prepared and microorganisms stained with phenolic aniline blue (PAB) (Jenkinson et al., 1976). Organisms ranging from less that 1 μm to 20 μm in diameter were observed by bright field microscopy, classified as 'spherical' or 'cylindrical' (hyphae) and divided into size classes. The numbers in each size class were counted and the total 'biovolume' of the soil calculated. Biomass carbon was calculated from biovolume using estimates of the dry matter and carbon content of soil microorganisms. With one exception (an acid soil, pH 3.9), there was a close correlation between biovolume and biomass carbon calculated from the flush of decomposition (*see* Table 1.2 *overleaf*). The key finding was this close correlation between the two totally different ways of estimating microbial biomass. The close numerical correspondence may be fortuitous in view of the assumptions made in calculating biomass carbon from biovolume and the considerable experimental difficulties encountered with direct microscopy. Fluorescent stains rendered organisms much easier to count but did not stain such a wide range of size classes as PAB, so were rejected from this work.

Recently, Vance et al. (1991) repeated the comparison of biovolume and flush for additional soils and confirmed earlier results. They also confirmed an intriguing observation made by Jenkinson et al. (1976) that logarithmically equal volume classes (for example, the volume class between 0.1 μm³ and 1 μm³ and the volume class between 1 μm³ and 10 μm³) contained the same volume of living cells. No convincing explanation for this relationship has yet been proposed.

Table 1.2   Comparison of biomass carbon as calculated from direct microscopy and the fumigation-incubation (FI) method

Soil	Organic C (%)	pH	Biomass C calculated from biovolume[a] (µg C/g soil)	Biomass C calculated from FI method[b] (µg C/g soil)	Ratio of biomass C from biovolume to biomass C from FI
Arable[c]	2.81	7.6	550	547	1.01
Arable[c]	0.93	8.0	190	220	0.86
Deciduous woodland[c]	4.30	7.5	1540	1231	1.25
Arable[c]	2.73	6.4	390	360	1.08
Grassland[c]	9.91	6.3	3200	3711	0.86
Deciduous woodland[c]	2.95	3.9	330	51	6.47
Secondary rainforest[d]	1.46	7.1	430	540	0.80
Cleared forest[d, e]	1.23	6.2	260	282	0.92

Note: a   See Jenkinson et al. (1976) for method of calculation
        b   Calculated using $k_C = 0.45$, not 0.5 as in the original paper
        c   Temperate soils from the UK
        d   Sub-humid tropics, Nigeria
        e   Arable cropping for 2 years after clearing secondary forest, Nigeria
Source: Adapted from Jenkinson et al. (1976)

Other experiments have given supporting, though not unequivocal, evidence for the hypothesis and revealed difficulties if treatments other than the $CHCl_3$ are used to kill the biomass. For example, drying, heating or gamma-radiation (Powlson and Jenkinson, 1976), grinding (Powlson, 1980) or microwave radiation (Puri and Barraclough, 1993) are less selective than fumigation, influencing the decomposability of non-living parts of soil organic matter in addition to killing organisms.

After various tests of the original hypothesis, Jenkinson and Powlson (1976b) proposed a practical method of applying the chloroform FI method to measure soil biomass carbon. The protocol envisaged the use of large soil samples (200-300 g in each replicate) which had been coarsely sieved (6.25 mm). In practice, it has been more common to use smaller samples (25-100 g soil) following 2 mm sieving. Although Lynch and Panting (1980) suggested that sieving soil led to erroneous results, neither Powlson and Jenkinson (1980) nor Ross et al. (1985) found any difference between biomass measured in intact cores and sieved soil.

In addition to causing a flush of CO_2, $CHCl_3$ fumigation also increases the amount of nitrogen mineralised. The ratio (CO_2-C evolved/N mineralised) in fumigated soil is generally about 6, consistent with the C/N ratio of cell cytoplasm, and narrower than for unfumigated soil. The ratio is generally wider following drying and rewetting, irradiation or autoclaving (Powlson and Jenkinson, 1976), consistent with the earlier suggestion that these treatments do more than merely kill biomass. The flush of nitrogen caused by fumigation was therefore regarded as an indication of biomass nitrogen content, although the situation is more complex than for carbon because part of the nitrogen mineralised from killed biomass will be immobilised by the decomposing population. This was investigated by Nicolardot et al. (1984) and Shen et al. (1984) who proposed that biomass nitrogen could be calculated from $B_N = F_N/k_N$, using a k_N factor of 0.68, later revised to 0.57 (Jenkinson, 1988).

ASSUMPTIONS, LIMITATIONS AND CONTROVERSIES

The control

Typically, $CHCl_3$ fumigation doubles CO_2 evolution during a 10-day incubation at 25°C, compared with the unfumigated control soil. Thus, the value of the control has a very significant effect on the calculated flush. By contrast, the increase in nitrogen mineralisation is proportionately greater, so the control value is less critical in calculating F_N. Disturbance as a result of sampling and sieving causes a small additional flush of CO_2, thus altering the control value. It is usual to incubate the soil at 25°C for 3-10 days to allow the effect of disturbance to subside but this means that when biomass is measured it may not represent the situation in the field prior to sampling. Alternatives are to incubate the unfumigated soil for a second 10-day period and use the CO_2 evolved during this period as the control (Jenkinson and Powlson, 1976b) or to deduce the control value from the respiration rate of fumigated soil after the flush has subsided (Chaussod and Nicolardot, 1982).

A key assumption stated by Jenkinson and Powlson (1976b) was that there is a 'background' rate of soil organic matter decomposition that continues unchanged in fumigated soil and that CO_2 evolved from the decomposition of killed biomass is additional to this. Intuitively, this seems somewhat unreasonable as it presupposes that the narrow range of organisms recolonising fumigated soil (fast-growing bacteria; Ridge, 1976) are capable of attacking complex soil organic substances to the same extent as the diverse population of unfumigated soil. However, there is indirect evidence that this is the case because after about 5-7 days the CO_2 evolution rate of fumigated soil subsides to approximately that of unfumigated soil and thereafter remains constant for a long period. Thus, the population present after the flush, which is still much less diverse than in unfumigated untreated soil, is indeed capable of causing the same degree of decomposition. Voroney and Paul (1984) proposed that no control should be subtracted but there is clear evidence that this errs too far in the opposite direction. When a soil containing a heavily ^{14}C-labelled biomass (resulting from previous incubation with ^{14}C-labelled glucose) was fumigated and incubated, the specific activity of the evolved CO_2 declined rapidly to a value similar to that of the unfumigated control (Wu, 1990). This shows that carbon from killed biomass made a diminishing contribution to the total, so there must have been *some* background CO_2 evolution.

In the same experiment (Wu, 1990), portions of fumigated and unfumigated soil were either incubated for 10 days, and CO_2 evolution measured, or extracted with 0.5 M K_2SO_4 and extractable carbon measured. Extraction of 0.5 M K_2SO_4 is the basis of the FI method described below. It was found that the quantities of *extra* carbon (labelled and unlabelled) made extractable to 0.5 M K_2SO_4 by $CHCl_3$ or made decomposable by $CHCl_3$ and subsequently evolved as CO_2 during a 10-day incubation were almost identical if control values were subtracted. The specific activities of the *extra* carbon were also identical. If the CO_2 evolution of the unfumigated soil was not subtracted, then agreement between the two approaches (incubation and extraction) was much poorer. This is strong evidence that a control, based on the carbon mineralised in unfumigated soil, gives a good estimate of 'background' respiration in fumigated soil.

Added substrate

Adding substrate to soil before fumigation causes complications, again in part because of problems with the control. There is evidence that CO_2 from the flush and from the decomposition of complex substrates such as plant material is not additive (Jenkinson and Powlson, 1976a; Sparling et al., 1981; Martens, 1985a). Jenkinson (1966) and Jenkinson and Powlson (1976b) therefore stated that the FI method should not be used in soils containing recent additions of substrate.

The k_C factor

The use of a single k_C factor of 0.45 for all soils is obviously an approximation. Different organisms have different decomposition rates (Jenkinson, 1976), but the range of species present in any given soil is always unknown. Nicolardot et al. (1984) showed that soil clay content has a small effect on k_C. Jenkinson (1988) gathered together measurements of k_C obtained by many different authors and found remarkable agreement on a mean value of 0.45 (range 0.37-0.55) for soils between pH 5.0 and 8.5 incubated at 25°C. However, in more acid soils, k_C was much smaller, ranging between 0.2 and 0.35 for soils with a pH value of about 4.

Acid soils

Studies conducted on two acid soils with pH values of 3.9 and 4.6 (Powlson and Jenkinson, 1976) showed a different response to $CHCl_3$ fumigation, compared with soils with pH values between 5.5 and 8.0. Following inoculation with fresh soil, there was a lag period of several days before respiration increased; the total CO_2 evolution over a 10-day period was sometimes less than that for unfumigated soil. This aberrant behaviour accounted for the lack of agreement between biomass calculated from the flush (which was small or non-existent) and microbial biovolume, which was substantial in the soil with a pH value of 3.9 (see Table 1.2). Indeed, Vance et al. (1987) developed the fumigation-extraction (FE) method for measuring microbial biomass largely because the FI method could not be used in acid soils.

DEVELOPMENTS IN METHODOLOGY

Substrate-induced respiration

Anderson and Domsch (1978) proposed a physiological method for the measurement of soil microbial biomass, based on the way soil respiration responds to the addition of a readily decomposable substrate such as glucose. If an appropriate concentration of glucose was chosen, a short period of constant respiration rate occurred for a few hours after addition but before the onset of rapidly increasing respiration due to microbial growth. This initial rate was interpreted as being proportional to the size of the biomass present when the glucose was added. Anderson and Domsch (1978) demonstrated a correlation between this rate and the size of the flush of decomposition following $CHCl_3$ fumigation. Thus, they were able to calibrate the physiological method known as substrate-induced respiration (SIR) against biomass carbon, as measured by FI.

In addition to establishing SIR as a valuable technique in its own right, the observation of a relationship between SIR and FI was remarkable because the two methods are based on fundamentally different approaches. This is a strong indication that both are providing biologically meaningful information on the soil microbial population and gives credence to the 'holistic' approach.

Adenosine triphosphate and adenylate energy charge

Any chemical which occurs in living cells in soil in a reasonably constant proportion but is absent from non-living parts of the soil could be used as a measure of the size of the soil microbial population. There is now considerable evidence that adenosine triphosphate (ATP) meets these criteria, at least for soils handled in a carefully specified way. Oades and Jenkinson (1979) demonstrated a close linear relationship between soil ATP content and biomass carbon as measured by FI. This was confirmed for a wider range of soils by Jenkinson et al. (1979) (*see* Figure 1.1) and an improved method of measuring ATP in soil extracts was proposed by Tate and Jenkinson (1982).

Figure 1.1   **Relationship between the soil adenosine triphosphate (ATP) content and biomass carbon (B_c) measured by the fumigation-incubation (FI) method**

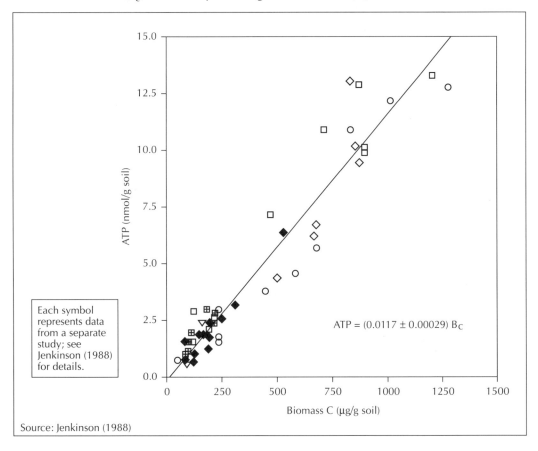

Source: Jenkinson (1988)

ATP from dead microbial cells is rapidly broken down in soil and does not accumulate extracellularly, unlike DNA which is stabilised by sorption on clays. ATP is also present in the cells of plant roots; the results in Figure 1.1 are all taken from studies in which the soils were incubated at 20-25°C for at least a few days before ATP or biomass were measured. In addition to minimising problems of determining the true control value with FI, this allows ATP from plant cells to decompose — an important condition for using ATP as a measure of biomass.

It has become clear that the method used to extract ATP from soil is critical. With strongly acidic reagents such as H_2SO_4 or trichloroacetic acid (TCA), ATPases within cells are inactivated. This is vital, as otherwise ATP and adenosine diphosphate (ADP) are rapidly dephosphorylated to adenosine monophosphate (AMP). With neutral reagents such as $NaHCO_3/CHCl_3$ (Paul and Johnson, 1977), ATPases are not inactivated, so lower concentrations of ATP are inevitably obtained. It is sometimes argued that these lower values represent ATP extracted only from an 'active' section of the biomass (for example, Verstraete et al., 1983) but this is difficult to substantiate. ATP is strongly sorbed on clays and oxide surfaces in soil so that extractants with compounds such as phosphate, paraquat or adenine have been used to reduce sorption by competing for these binding sites. It is essential to measure recovery of ATP, by adding a spike, and thus correct for incomplete extraction (Jenkinson and Oades, 1979). The slope of the regression between biomass carbon (measured, for example, by FI) and ATP represents the concentration of ATP in the biomass. For the Figure 1.1 data, it is 11.7 µmol/g biomass carbon, corresponding to 5.39 µmol ATP/g dry biomass, assuming that the biomass contains 46% carbon. Surprisingly, this is similar to values measured in actively growing microorganisms in culture.

Even more surprising are the values for the adenylate energy charge (AEC) that have been measured in soil extracts. AEC is defined as:

$$AEC = \frac{[ATP] + 0.5[ADP]}{[ATP] + [ADP] + [AMP]}$$

and is a measure of the metabolic energy stored in the adenine nucleotide pool. Theoretically, it can range from 0 (all AMP, low energy) to 1.0 (all ATP, high energy). Brookes et al. (1983) were the first to measure AEC in soil and found a value of 0.85 in a freshly sampled grassland soil. The high values found in soil have led to considerable speculation regarding the survival strategies adopted by microorganisms in soil. Sporulation cannot be the dominant strategy for survival since spores have AECs of less than 0.1. Perhaps reserve substances such as polyhydroxybutyric acid or polyphosphate are laid down shortly after cell division and then used to maintain ATP at a high level. Perhaps resting cells in soil have very efficient mechanisms for minimising their maintenance energy requirements. Perhaps the required energy comes from the slow decomposition of 'inert' soil organic matter.

Measuring the AEC of the whole soil population has raised fundamental questions in microbial ecology and biochemistry. It has shown how studies at the level of a whole population (in this case, the soil microbial biomass) can reveal important properties that might have been missed by a more classical approach in which isolated individuals were examined. Detailed and fundamental studies on the biochemistry and physiology of soil organisms are now needed to elucidate the processes responsible for the observed high AECs; this represents an exciting challenge for soil microbiologists.

A word of caution is required regarding soil AEC values. Reliable measurements of ATP, ADP and AMP are a necessary prerequisite; the use of acidic reagents, to prevent rapid enzymic dephosphorylation of ATP, has generally been regarded as essential. Low AEC values have sometimes been reported (for example, 0.3-0.4 by Martens, 1985b) and have led to controversy; Brookes et al. (1987) attributed such measurements to the decomposition of ATP in the alkaline reagents used by Martens and others. This was based on the finding (Brookes et al., 1987) that ATP concentration and AEC were always lower in a reagent based on $NaHCO_3$ than in an acidic reagent based on TCA, but the total quantity of adenine nucelotides extracted was similar. Recently, Martens (1992) claimed that the lower AEC in a $NaHCO_3$-based reagent was due to this reagent extracting *more* ADP and AMP than an acidic reagent based on H_2SO_4; this controversy has not yet been resolved.

Fumigation-extraction for carbon, nitrogen, phosphorus and sulphur

The original FI method had two important limitations: it could not be applied to acid soils; and it could not be used in the presence of recently added substrate. The FE method overcomes both these limitations and also sidesteps some of the problems concerning the control. In the case of biomass carbon, the FE method is based on the observation that fumigation causes an immediate increase in the amount of organic carbon that can be extracted by aqueous solutions (Jenkinson and Powlson, 1976a; Voroney, 1983). Vance et al. (1987) showed that biomass (B_C) could be calculated from $B_C = 2.64$ where E_C is the quantity of carbon extracted from fumigated soil with 0.5 M K_2SO_4 less that extracted from unfumigated soil. The FE method can be applied to soils over the whole pH range and to those to which substrate has recently been added; this is shown by the close correlations between ATP content and biomass carbon or nitrogen for soils with and without recent straw addition (*see* Figure 1.2). In the original FE method (Vance et al., 1987), organic carbon in solution was measured using dichromate oxidation; analysis is greatly facilitated by the use of an automated instrument designed for measuring dissolved organic carbon in water (Wu et al., 1990).

Figure 1.2   Relationship between the soil adenosine triphosphate (ATP) content and (a) biomass carbon measured by the fumigation-extraction (FE) method, (b) biomass nitrogen measured by FE and (c) ninhydrin nitrogen measured by FE in two soils with or without amendment with wheat straw and incubated for 13 or 35 days at 25°C

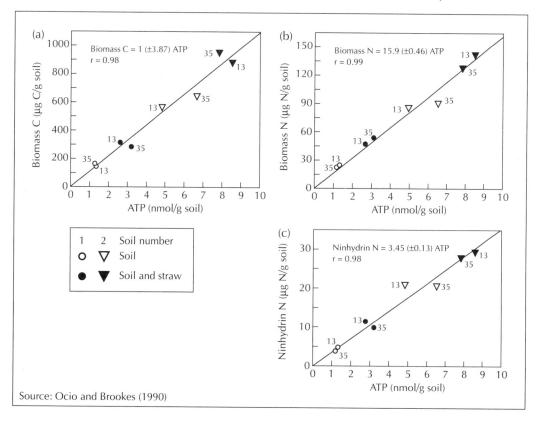

Source: Ocio and Brookes (1990)

Methods for measuring nitrogen (Brookes et al., 1985), phosphorus (Brookes et al., 1982) and sulphur (Saggar et al., 1981; Chapman, 1987) in the soil microbial biomass, all based on variations of the FE method, have been published. In the case of biomass nitrogen, it was shown that the quantity of nitrogen made extractable by $CHCl_3$ in the FE method was strongly correlated with the quantity mineralised during the 10-day incubation in the FI method. There was also strong evidence that the two methods were attacking the same fraction of soil nitrogen, presumably the cytoplasm of the microbial biomass. When FE and FI were applied to soils that had received ^{15}N-labelled fertiliser 1-2 years previously, the ^{15}N enrichments of the nitrogen rendered extractable by FE and that mineralised in FI were equal (Brookes et al., 1985), although much greater than the enrichment of the soil organic nitrogen as a whole.

A valuable variation on the method for measuring biomass nitrogen is based on the increase in ammonium and alpha-amino nitrogen in soil following $CHCl_3$ fumigation (Amato and Ladd, 1988). Although the quantity of nitrogen in these forms is a small fraction of the total amount of organic nitrogen solubilised, it is a remarkably constant proportion and can easily be measured by its reaction with ninhydrin, giving a blue colour. Ninhydrin-reactive nitrogen is closely correlated with other measurements of biomass and is simple and rapid to measure.

In the FI methods for biomass carbon or nitrogen, decomposition of biomass killed during $CHCl_3$ fumigation is mediated by recolonising bacteria during the 10-day incubation. In FE, solubilisation of the constitutents of killed organisms has to occur during the period of fumigation and extraction and appears to depend on enzyme activity within damaged cells, even in the presence of $CHCl_3$. Longer periods of fumigation may be required to allow sufficient time for solubilisation, so the exact period of exposure to $CHCl_3$ is more critical than for FI. For example, Brookes et al. (1982) were initially surprised to find that much of the phosphorus rendered extractable to 0.5 M $NaHCO_3$ by $CHCl_3$ fumigation was inorganic rather than in organic forms. The proportion that was inorganic increased further if fumigation time was increased from 24 to 120 hours, as this allowed more time for phosphatase activity. However, the amount in the inorganic form after 24 hours was both high and a very constant proportion of the total phosphorus released (generally, greater than 85%) in a wide range of soils so that the method proposed for biomass phosphorus determination (Brookes et al., 1982) was based on 24-hour fumigation and measurement of inorganic phosphorus only.

In all the FE methods, the substances which are extracted will be subject to sorption, to an extent which will depend upon the quantity of clays or other active surfaces present. In the biomass phosphorus method (Brookes et al., 1982), a correction for phosphate sorption is made by measuring the recovery of a spike of added inorganic phosphate. This is also done for ATP, ADP and AMP by measuring the recovery of known quantities of added nucleolotides. The close correlations between biomass carbon or nitrogen (as measured by FE without a spike) and ATP (as measured with a spike) suggest that fixation of microbial carbon or nitrogen during extraction is not a major cause of error in the FE method.

Flooded soils

Inubushi et al. (1984) devised a variation of the FI method for measuring biomass nitrogen in flooded soils using anaerobic incubation. The FE methods for carbon and nitrogen, as well as ninhydrin nitrogen, ATP and AEC, have now been tested on waterlogged soils with promising results, but further work is required to see whether the relationships between biomass content and analytical measurement established for aerobic soil also apply to anaerobic soil (Inubushi et al., 1989, 1991).

SOME APPLICATIONS OF BIOMASS MEASUREMENTS

Early warning of changes in soil organic matter content

Many changes in land use or agricultural management lead, in the long term, to a change in soil organic matter content. However, these changes are difficult to measure in the short or medium term because they occur slowly and against the background of a large quantity of organic matter, much of which is relatively alert. The quantity of microbial biomass in soil is much more sensitive to land management than is soil organic matter as a whole, so that biomass measurements can give early warning of changes in total soil organic matter long before they can be detected by classical measurements.

An example was a study of two field experiments in Denmark in which spring barley straw had been either burned or incorporated for 18 years (Powlson et al., 1987). No significant change in total carbon or nitrogen content could be detected but biomass carbon (measured by FI) had increased by about 40%, an easily measurable change. Biomass measurements have also been used to detect changes brought about by differential management of sorghum residues in a Vertisol in sub-tropical Australia (Saffigna et al., 1989), by reduced tillage (Powlson and Jenkinson, 1981; Doran, 1987), by organic farming (Doran et al., 1987) and by clearing tropical forests (Ayanaba et al., 1976).

Impact of residue incorporation on soil nitrogen transformations

It is only since the development of the FE methods that it has been possible to study the short-term impact of crop residue incorporation on carbon and nitrogen transformation. Ocio et al. (1991a) used FE to follow changes in biomass nitrogen and inorganic nitrogen after the incorporation of wheat straw. They found that biomass nitrogen approximately doubled (an increase of 50 kg N/ha) within 7 days of straw incorporation but this was not accompanied by a corresponding decrease in soil inorganic nitrogen, implying that the nitrogen entering the biomass came from other sources. Subsequent work (in which either the straw or soil inorganic nitrogen were labelled with ^{15}N) confirmed that the main source of nitrogen was the straw itself (Ocio et al., 1991b). This unexpected result, which requires further confirmation in a wider range of situations, implies that straw incorporation may be less beneficial than previously thought as a means of reducing the quantity of nitrate in soil at risk to leaching. The experiments also showed that the FE method for biomass nitrogen was sufficiently robust to follow short-term changes under field conditions.

Effects of pollutants on soil microbial populations and activities

Brookes and McGrath (1984) measured biomass carbon in field experiment plots which had received either inorganic fertilisers or organic amendments including sewage-sludge (relatively high in heavy metals, including copper, zinc and cadmium) and farmyard manure (low in metals). The organic amendments had been applied over a 20-year period but the last application had been made 20 years before the soils were sampled for biomass measurements. The total inputs of carbon were very similar from farmyard manure and sewage-sludge, and total soil organic carbon contents were almost equal. However, biomass in the high-metal plots was about half that in the low-metal plots. Later experiments showed than the decrease in biomass in the metal-contaminated soil was due *partly* to lower inputs of carbon from plant roots and *partly* to the smaller proportion of the substrate carbon retained in the biomass — that is, there was greater respiration/unit of input (or biomass) in the metal-contaminated soil (Chander and Brookes, 1991a). Brookes (1993) has suggested that biomass carbon expressed as a percentage of total soil organic carbon could be used as an indicator of stress due to soil contamination; this index could be applied in 'real life' situations where no true control soil is available.

Adverse effects of heavy metals on biomass have since been detected at other sites (for example, Chander and Brookes, 1991b) and studies have progressed to examine impacts on specific microbial processes. Biological nitrogen fixation, both by cyanobacteria (Brookes et al., 1986) and the *Rhizobium*-legume symbiosis (McGrath et al., 1988), is particularly sensitive to interference by heavy metals. This is an example of how biomass measurements have revealed an unrecognised effect of pollutants and opened the way for more detailed studies. It is not the decrease in biomass content *per se* that is necessarily serious — indeed, carbon and nitrogen mineralisation are often unaffected at these levels of metal contamination. Rather, a decrease in biomass content provides an indication of potentially adverse biological effects that can be followed up by studies of specific processes.

Off-target effects of pesticides

Early work (Jenkinson and Powlson, 1970) showed that broad spectrum soil fumigants such as methyl bromide or formaldehyde, used to control soil-borne pests and diseases in the field, caused a decrease in microbial biomass and recovery was slow. Even after 3-5 years, biomass was still smaller than in soil that had never been fumigated. More selective pesticides would be expected to have much less drastic effects on the biomass as a whole, but a measurement of the effect, perhaps in combination with a suitable measure of microbial activity such as respiration or dehydrogenase activity, may provide a useful initial screening for new pesticides.

Use of biomass values in mathematical models of soil organic matter turnover

The soil microbial biomass is one of the few fractions of soil organic matter that is biologically meaningful as well as measurable. Not surprisingly, it features as one of the pools in several models of organic matter turnover (Jenkinson, 1990). Measurements of biomass can be used in parameterising or validating models using either the total contents of carbon, nitrogen, phosphorus or sulphur or the quantity of an isotopic label. Jenkinson and Parry (1989) modelled the flux of total and ^{15}N-labelled nitrogen through the soil microbial biomass over the 4 years after a single application of ^{15}N-labelled fertiliser to winter wheat.

A carbon model which includes a biomass component has been used in reverse to calculate carbon inputs to soil in different ecosystems. For soils in which organic carbon is at a steady state value this is a novel way of calculating net primary production. The model is run for the appropriate soil and climatic conditions and the carbon input varied until the measured soil carbon content is matched. A biomass carbon measurement provides an independent test of the simulation (Jenkinson et al., 1992).

SOME WARNINGS ON METHODOLOGY AND INTERPRETATION

The various techniques for measuring biomass have proved very powerful in probing microbial activity and organic matter turnover in soil but they do have limitations. All of the variations on FI,

FE and SIR are bioassays, depending on the activity of soil enzymes or a recolonising microbial population. Bioassays cannot be expected to be as robust or reproducible as chemical measurements. All types of biological variability can influence the measurement and will add to the usual considerations of spatial variability. It is therefore wise to use more than one independent method wherever possible and to investigate unexpected results in depth before accepting them at face value. One of the inherent strengths of the biomass assays now available is the observation of many internal correlations between, for example, ATP, biomass carbon or nitrogen and ninhydrin-reactive nitrogen. Any deviation from this behaviour may be an indication that an error may have occured or, at least, that further investigation is required.

It is absolutely essential to observe the limitations and exclusions stated in the published methods. Small changes in sample handling or experimental conditions can have unexpected effects. Most of the current methods were originally developed using near-neutral aerobic agricultural soils. When extending outside this range (for example, to soils that are periodically waterlogged, soils with a very high or low pH, highly organic soils or litter layers) the method should be checked carefully (see, for example, Ross and Tate, 1993). There is always scope for improving methods but changes should be investigated thoroughly and the results from the modified and original method compared and reported in refereed publications. It is not good enough to assume that a small modification will not interfere with some other aspect of the method, especially with a bioassay — it is unscientific to deny one's peers the opportunity of critical examination.

A clear example of the inappropriate use of the FI technique is the study reported by Ingham et al. (1991). Estimates of biomass carbon from FI and direct microscopy were compared using soil or litter (sample depth not stated) from three forest sites. Only sketchy information was given regarding the characteristics of the litter and soil but it appears that the material studied was highly organic, comprising fresh litter and copious amounts of fungal hyphae. The authors stated that, in some samples, fungal rhizomorph material accounted for 30-50% of sample dry weight. This clearly contravenes the need to exclude samples containing fresh substrate when using the FI technique, as stated in the original papers (Jenkinson, 1966; Jenkinson and Powlson, 1976b). The very poor agreement between FI and direct microscopy in this particular type of material was totally predictable. However, on the basis of this work, the authors drew the general conclusion regarding biomass estimates that '...direct estimates may be more useful than fumigation estimates.' In fact, their work represents an all too frequent rediscovery of a well-known limitation of the FI method. It also seems likely that the litter and soil used by Ingham et al. (1991) would have been acid, although the authors failed to report its pH. If it is, this is another reason for not using FI in this situation.

Some puzzling results have been reported even where the biomass techniques have been used in apparently appropriate situations. For example, Ritz et al. (1992) made a range of biological measurements on soil in a field experiment in which potatoes were being grown. Soil samples were taken throughout the crop growing period from within the ridges containing the potato plants. Surprisingly, no increase in biomass was detected using FI, FE or SIR in a treatment in which barley straw (at 1200 kg C/ha) had been added prior to planting the potatoes. Assuming a biomass growth efficiency of 20%, an increase of biomass carbon of about 100 µg/g soil might have been expected and this should have been easily measurable. Ritz et al. (1992) also noted that the addition of straw had no effect on soil respiration or on dehydrogenase activity. This may imply *either* that some unknown factor was inhibiting straw decomposition *or* that the straw had not been adequately mixed with the soil. The latter may have occurred as it appears that straw was applied to the soil surface before the planting ridges were formed. Perhaps most of the straw was excluded from the ridges from which soil samples were taken.

Ritz and Robinson (1988) used FI to follow temporal changes in biomass carbon and nitrogen, in the field, during the growth of spring barley and, again, obtained puzzling results. Biomass carbon doubled (from about 300 µg C/g soil to over 500 µg C/g soil) within 50 days of planting spring barley but this was unlikely to have been caused by rhizodeposition because the plants were still very small at this stage. Changes in biomass carbon and nitrogen did not necessarily coincide: biomass carbon peaked before biomass nitrogen and then gradually declined, whereas biomass nitrogen appeared to increase late in the growing season. It is not clear whether these results reflect problems of obtaining a representative soil sample under field conditions, the limitations of the FI method (as opposed to FE) especially in the presence of growing roots or genuine changes in biomass C/N ratios which are, as yet, unexplained. Certainly, these data conflict with the close C/N ratio in the biomass obtained by Jenkinson (1988) from published results from independent workers. Patra et al. (1990) also used FI to follow temporal changes in biomass carbon and nitrogen in soils under winter wheat or grass. They highlighted the difficulties associated with variations in soil moisture throughout the year, a particular problem with FI but probably not with FE. They found that temporal changes were remarkably small and gave theoretical reasons for not expecting large fluctuations in biomass during the year.

CONCLUSION

Current interest in soil microbiology is, in part, a result of the attention brought to the subject by the advent of techniques for studying the microbial biomass as an entity. This provided a new intellectual framework for the subject and made it experimentally accessible to soil scientists who were not specialist microbiologists. New developments in the biological sciences now point to another significant step forward in soil microbiology. Approaches include the whole range of techniques based on DNA technology, studies on the genetic diversity of soil microorganisms, an understanding of how genetic information may be transferred between microorganisms (or between microorganisms and plants), and investigations into the survival of introduced microorganisms and into how native microorganisms survive in soil for long periods in the absence of fresh substrates. Chemical markers are being developed to distinguish different classes of organisms in soil.

   Soil microorganisms and their activities are at the heart of several vitally important processes such as carbon sequestration and CO_2 release, the formation and destruction of trace greenhouse gases and many transformations within the nitrogen cycle. They can also be damaged by pollutants. The subject is as exciting as ever and the widest possible range of approaches and techniques is needed to tackle the problems which are both scientifically interesting and of vital practical importance for agriculture and environmental protection. Both holistic and detailed approaches are needed, and soil scientists must choose that which is appropriate for a specific investigation — no one approach has a monopoly of answers or of scientific credibility. Reports that biomass measurements have been superceded by later developments are premature!

Acknowledgements

The author wishes to thank Professor D.S. Jenkinson and Dr P.C. Brookes for their help and constructive criticism during the preparation of this chapter.

References

Amato, M. and Ladd, J.N. 1988. Assay for microbial biomass based on ninhydrin-reactive nitrogen in extracts of fumigated soils. *Soil Biology and Biochemistry* 20: 107-14.

Anderson, J.P.E. and Domsch, K.H. 1978. A physiological method for the quantitative measurement of microbial biomass in soils. *Soil Biology and Biochemistry* 10: 215-21.

Ayanaba, A., Tuckwell, S.B. and Jenkinson, D.S. 1976. The effects of clearing and cropping on the organic reserves and biomass of tropical forest soil. *Soil Biology and Biochemistry* 8: 519-25.

Brookes, P.C. 1993. The potential of microbiological properties as indicators in soil pollution monitoring. In Schulin, R., Desaules, A., Webster, R. and von Steiger, B. (eds) *Soil Monitoring: Early Detection and Surveying of Soil Contamination and Degradation. Proc. Soil Monitoring Workshop, Monte Verità, Ascona.* Basel, Switzerland: Birkhaüser Verlag.

Brookes, P.C. and McGrath, S.P. 1984. The effects of metal toxicity on the soil microbial biomass. *J. Soil Science* 35: 341-46.

Brookes, P.C., Powlson, D.S. and Jenkinson, D.S. 1982. Measurement of microbial biomass phosphorus in soil. *Soil Biology and Biochemistry* 14: 319-29.

Brookes, P.C., Tate, K.R. and Jenkinson, D.S. 1983. The adenylate energy charge of the soil microbial biomass. *Soil Biology and Biochemistry* 15: 9-16.

Brookes, P.C., Landman, A., Pruden, G. and Jenkinson, D.S. 1985. Chloroform fumigation and the release of soil nitrogen: A rapid direct extraction method to measure microbial biomass nitrogen in soil. *Soil Biology and Biochemistry* 17: 837-42.

Brookes, P.C., McGrath, S.P. and Heijnen, C. 1986. Metal residues in soils previously treated with sewage-sludge and their effects on growth and nitrogen fixation by blue-green algae. *Soil Biology and Biochemistry* 18: 345-54.

Brookes, P.C., Newcombe, A.D. and Jenkinson, D.S. 1987. Adenylate energy charge measurements in soil. *Soil Biology and Biochemistry* 19: 211-17.

Chander, K. and Brookes, P.C. 1991a. Plant inputs of carbon to metal-contaminated and non-contaminated soil and effects on the synthesis of soil microbial biomass. *Soil Biology and Biochemistry* 23: 1169-77.

Chander, K. and Brookes, P.C. 1991b. Effects of heavy metals from past applications of sewage-sludge on microbial biomass and organic matter accumulation in a sandy loam and silty loam UK soil. *Soil Biology and Biochemistry* 23: 927-32.

Chapman, S.J. 1987. Microbial sulphur in some Scottish soils. *Soil Biology and Biochemistry* 19: 301-05.

Chaussod, R. and Nicolardot, B. 1982. Mesure de la biomasse microbienne dans les sols cultivés. I. Approche cimétique et estimation simplifiée du carbon facilement minéralisable. *Revue d'Ecologie et de Biologie du Sol* 19: 501-12.

Doran, J.W. 1987. Microbial biomass and mineralisable nitrogen distributions in no-tillage and plowed soils. *Biology and Fertility of Soils* 5: 68-75.

Doran, J.W., Fraser, D.G., Culik, M.N. and Liebhart, W.C. 1987. Influence of alternative and conventional agricultural management on soil microbial processes and nitrogen availability. *American J. Alternative Agriculture* 2: 99-106.

Ingham, E.R., Griffiths, R.P., Cromack, K. and Entry, J.A. 1991. Comparison of direct vs fumigation-incubation microbial biomass estimates from ectomycorrhizal mat and non-mat soils. *Soil Biology and Biochemistry* 23: 465-71.

Inubushi, K., Wada, H. and Takai, Y. 1984. Easily decomposable organic matter in paddy soil. IV. Relationship between reduction process and organic matter decomposition. *Soil Science and Plant Nutrition* 30:189-98.

Inubushi, K., Brookes, P.C. and Jenkinson, D.S. 1989. Adenosine 5'-triphosphate and adenylate energy charge in waterlogged soil. *Soil Biology and Biochemistry* 21: 733-39.

Inubushi, K., Brookes, P.C. and Jenkinson, D.S. 1991. Soil microbial biomass C, N and ninhydrin-N in aerobic and anaerobic soils measured by the fumigation-extraction method. *Soil Biology and Biochemistry* 23: 737-41.

Jenkinson, D.S. 1966. Studies on the decomposition of plant material in soil. II. Partial sterilisation of soil and the soil biomass. *J. Soil Science* 17: 280-302.

Jenkinson, D.S. 1976. The effects of biocidal treatments on metabolism in soil. IV. The decomposition of fumigated organisms in soil. *Soil Biology and Biochemistry* 8: 203-08.

Jenkinson, D.S. 1988. Determination of microbial biomass carbon and nitrogen in soil. In Wilson, J.R. (ed) *Advances in Nitrogen Cycling in Agricultural Ecosystems.* Wallingford, UK: CAB International.

Jenkinson, D.S. 1990. The turnover of organic carbon and nitrogen in soil. *Philosophical Transactions of the Royal Society, London* B329: 361-68.

Jenkinson, D.S. and Powlson, D.S. 1970. Residual effects of soil fumigation on soil respiration and mineralisation. *Soil Biology and Biochemistry* 2: 99-108.

Jenkinson, D.S. and Powlson, D.S. 1976a. The effects of biocidal treatments on metabolism in soil. I. Fumigation with chloroform. *Soil Biology and Biochemistry* 8: 167-77.

Jenksinson, D.S. and Powlson, D.S. 1976b. The effects of biocidal treatments on metabolism in soil. V. A method for measuring soil biomass. *Soil Biology and Biochemistry* 8: 209-13.

Jenkinson, D.S. and Oades, J.M. 1979. A method for measuring adenosine triphosphate in soil. *Soil Biology and Biochemistry* 11: 193-99.

Jenkinson, D.S. and Ladd, J.N. 1981. Microbial biomass in soil: Measurement and turnover. In Paul, E.A. and Ladd, J.N. (eds) *Soil Biochemistry* (Vol. 5). New York, USA: Marcel Dekker.

Jenkinson, D.S. and Parry, L.C. 1989. The nitrogen cycle in the Broadbalk wheat experiment: A model for the turnover of nitrogen through the soil microbial biomass. *Soil Biology and Biochemistry* 21: 535-41.

Jenkinson, D.S., Powlson, D.S. and Wedderburn, F.W.M. 1976. The effects of biocidal treatments on metabolism in soil. III. The relationship between soil biovolume, measured by optical microscopy, and the flush of decomposition caused by fumigation. *Soil Biology and Biochemistry* 8: 189-202.

Jenkinson, D.S., Davidson, S.A. and Powlson, D.S. 1979. Adenosine triphosphate and microbial biomass in soil. *Soil Biology and Biochemistry* 11: 521-27.

Jenkinson, D.S., Harkness, D.D., Vance, E.D., Adams, D.E. and Harrison, A.F. 1992. Calculating net primary production and annual input of organic matter to soil from the amount and radiocarbon content of soil organic matter. *Soil Biology and Biochemistry* 24: 295-308.

Lynch, J.M. and Panting, L.M. 1980. Cultivation and the soil biomass. *Soil Biology and Biochemistry* 12: 29-33.

Martens, R. 1985a. Limitations in the application of the fumigation technique for biomass estimations in amended soils. *Soil Biology and Biochemistry* 17: 57-63.

Martens, R. 1985b. Estimation of the adenylate energy charge in unamended and amended agricultural soils. *Soil Biology and Biochemistry* 17: 765-72.

Martens, R. 1992. A comparison of soil adenine nucleotide measurements by HPLC and enzymic analysis. *Soil Biology and Biochemistry* 24: 639-45.

McGrath, S.P., Brookes, P.C. and Giller, K.E. 1988. Effects of potentially toxic metals in soil derived from past applications of sewage-sludge on nitrogen fixation by *Trifolium repens* L. *Soil Biology and Biochemistry* 20: 415-24.

Nannipieri, P., Grego, S. and Ceccanti, B. 1990. Ecological significance of the biological activity in soil. In Bollag, J.M. and Stotzky, G. (eds) *Soil Biochemistry* (Vol. 6). New York, USA: Marcel Dekker.

Nicolardot, B., Chaussod, R. and Catroux, G. 1984. Décomposition de corps microbiens dans des sols fumigés au chloroforme: Effets du type de sol et de microorganisme. *Soil Biology and Biochemistry* 16: 453-58.

Oades, J.M. and Jenkinson, D.S. 1979. Adenosine triphosphate content of the soil microbial biomass. *Soil Biology and Biochemistry* 11: 201-04.

Ocio, J.A. and Brookes, P.C. 1990. An evaluation of methods for measuring the microbial biomass in soils following recent additions of wheat straw and the characterisation of the biomass that develops. *Soil Biology and Biochemistry* 22: 685-94.

Ocio, J.A., Brookes, P.C. and Jenkinson, D.S. 1991a. Field incorporation of straw and its effects on soil microbial biomass and soil inorganic N. *Soil Biology and Biochemistry* 23: 171-76.

Ocio, J.A., Martinez, J. and Brookes, P.C. 1991b. Contribution of straw-derived N to total microbial biomass N following incorporation of cereal straw to soil. *Soil Biology and Biochemistry* 23: 655-59.

Parkinson, D. and Coleman, D.C. 1991. Microbial communities, activity and biomass. *Agriculture, Ecosystems and Environment* 34: 3-33.

Patra, D.D., Brookes, P.C., Coleman, K. and Jenkinson, D.S. 1990. Seasonal changes of soil microbial biomass in an arable and a grassland soil which have been under uniform management for many years. *Soil Biology and Biochemistry* 22: 739-42.

Paul, E.A. and Johnson, R.L. 1977. Microscopic counting and adenosine 5'-triphosphate measurement in determining microbial growth in soils. *Applied and Environmental Microbiology* 34: 263-69.

Powlson, D.S. 1975. Effects of biocidal treatments on soil organisms. In Walker, N. (ed) *Soil Microbiology*. London, UK: Butterworths.

Powlson, D.S. 1980. The effects of grinding on microbial and non-microbial organic matter in soil. *J. Soil Science* 31: 77-85.

Powlson, D.S. and Jenkinson, D.S. 1976. The effects of biocidal treatments on metabolism in soil. II. Gamma irradiation, autoclaving, air-drying and fumigation. *Soil Biology and Biochemistry* 8: 179-88.

Powlson, D.S. and Jenkinson, D.S. 1980. Measurement of microbial biomass in intact soil cores and in sieved soil. *Soil Biology and Biochemistry* 12: 579-81.

Powlson, D.S. and Jenkinson, D.S. 1981. A comparison of the organic matter, biomass, adenosine triphosphate and mineralisation nitrogen contents of ploughed and direct-drilled soils. *J. Agricultural Science (Cambridge)* 97: 713-21.

Powlson, D.S., Brookes, P.C. and Christensen, B.T. 1987. Measurement of soil microbial biomass provides an early indication of changes in total organic matter due to straw incorporation. *Soil Biology and Biochemistry* 19: 159-64.

Puri, G. and Barraclough, D. 1993. Comparison of 2450 MHz microwave radiation and chloroform fumigation-extraction to estimate soil microbial biomass nitrogen using ^{15}N-labelling. *Soil Biology and Biochemistry* 25: 521-22.

Ridge, E.H. 1976. Studies on soil fumigation. II. Effects on bacteria. *Soil Biology and Biochemistry* 8: 249-53.

Ritz, K. and Robinson, D. 1988. Temporal variations in soil microbial biomass C and N under a spring barley crop. *Soil Biology and Biochemistry* 20: 625-30.

Ritz, K., Griffiths, B.S. and Wheatley, R.E. 1992. Soil microbial biomass and activity under a potato crop fertilised with N with and without C. *Biology and Fertility of Soils* 12: 265-71.

Ross, D.J. and Tate, K.R. 1993. Microbial C and N in litter and soil of a southern beech (*Nothofagus*) forest: Comparison of measurement procedures. *Soil Biology and Biochemistry* 25: 467-76.

Ross, D.J., Speir, T.W., Tate, K.R. and Orchard, V.A. 1985. Effects of sieving on estimations of microbial biomass, and carbon and nitrogen mineralisation, in soil under pasture. *Australian J. Soil Research* 23: 319-24.

Saffigna, P.G., Powlson, D.S., Brookes, P.C. and Thomas, G.A. 1989. Influence of sorghum residues and tillage on soil organic matter and soil microbial biomass in an Australian Vertisol. *Soil Biology and Biochemistry* 21: 759-65.

Saggar, S., Bettany, J.R. and Stewart, J.W.B. 1981. Measurement of microbial sulphur in soil. *Soil Biology and Biochemistry* 13: 493-98.

Shen, S.M., Pruden, G. and Jenkinson, D.S. 1984. Mineralisation and immobilisation of nitrogen in fumigated soil and the measurement of microbial biomass nitrogen. *Soil Biology and Biochemistry* 16: 437-44.

Smith, J.L. and Paul, E.A. 1990. The significance of soil biomass estimates. In Bollag, J.M. and Stotzky, G. (eds) *Soil Biochemistry* (Vol. 6). New York, USA: Marcel Dekker.

Sparling, G.P. 1985. The soil biomass. In Vaughan, D. and Malcom, R.E. (eds) *Soil Organic Matter and Biological Activity*. Dordrecht, Netherlands: Nijhoff/Junk.

Sparling, G.P., Ord, B.G. and Vaughan, D. 1981. Changes in microbial biomass and activity in soils amended with phenolic acids. *Soil Biology and Biochemistry* 13: 455-60.

Tate, K.R. and Jenkinson, D.S. 1982. Adenosine triphosphate measurements in soil: An improved method. *Soil Biology and Biochemistry* 14: 331-35.

Tunlid, A. and White, D.C. 1992. Biochemical analysis of biomass, community structure, nutritional status, and metabolic activity of microbial communities in soil. In Stotzky, G. and Bollag, J.M (eds) *Soil Biochemistry* (Vol. 7). New York, USA: Marcel Dekker.

Vance, E.D., Brookes, P.C. and Jenkinson, D.S. 1987. An extraction method for measuring soil microbial biomass C. *Soil Biology and Biochemistry* 19: 703-07.

Vance, E.D., Brookes, P.C. and Jenkinson, D.S. 1991. Confirmation of a direct relationship between the size of non-hyphal organisms and their contribution to soil biovolume. *Soil Biology and Biochemistry* 23: 1097-98.

Verstraete, W., van de Werf, H., Kucnerowicz, F., Ilaiwi, M., Verstreten, L.M. J. and Vlassak, K. 1983. Specific measurement of soil microbial ATP. *Soil Biology and Biochemistry* 15: 391-96.

Voroney, R.P. 1983. Decomposition of crop residues. PhD thesis, University of Saskatchewan, Canada.

Voroney, R.P. and Paul, E.A. 1984. Determination of k_C and k_N *in situ* for calibration of the chloroform fumigation-incubation method. *Soil Biology and Biochemistry* 16: 9-14.

Wardle, D.A. 1992. A comparative assessment of factors which influence microbial biomass carbon and nitrogen levels in soil. *Biological Review* 67: 321-58.

Wu, J. 1990. The turnover of organic C in soil. PhD thesis, University of Reading, UK.

Wu, J., Joergensen, R.G., Pommerening, B., Chaussod, R. and Brookes, P.C. 1990. Measurement of soil microbial biomass C by fumigation-extraction — an automated procedure. *Soil Biology and Biochemistry* 22: 1167-69.

PART II

Characterisation of soil
microbial communities

Beyond the Biomass
Edited by K. Ritz, J. Dighton and K.E. Giller
© 1994 British Society of Soil Science (BSSS)
A Wiley-Sayce Publication

CHAPTER 2

Methods for the analysis of soil microbial communities

F. Kunc

The soil microflora is of great importance in the biosphere. Being responsible for processes of mass and energy transformations, it plays a major role in soil fertility and ecosystem functioning. In the past few years, increasing attention has been paid to the methodological approaches for the study of microbial communities in soils. Authors of several reviews have collected and critically evaluated a considerable number of papers documenting the rapid development in this area (Karl, 1980; Jenkinson and Ladd, 1981; Ford and Olson, 1988; Jenkinson, 1988; Trevors and van Elsas, 1989; Nannipieri et al., 1990; Smith and Paul, 1990; Jackman et al., 1992; Kozhevin, 1992; Sayler et al., 1992; Tunlid and White, 1992; Ward et al., 1992).

The aim of this chapter is to provide a short survey of the main methodological tools which soil microbiologists and ecologists now have at their disposal to study soil biomass quantity, its composition, potential and real activities. The chapter discusses the significance of data on soil microbial communities. Methods dealing with soil micro- and mesofauna are not discussed here.

SOIL BIOMASS QUANTITY

To recognise the quantity of the living part of the soil organic matter is the starting point for any consideration of its role, activity and function in the environment. Methods developed for this purpose do not distinguish various parts of microbial community; rather, they treat it as an undifferentiated whole. These methods are based on measuring of some common properties of soil microbial inhabitants which can be related to their total amount:

- The cells can be counted and measured (using fluorescence or electron, phase contrast or bright field microscopy; in thin sections, dispersed soil suspension and thin films, membrane filters), specific staining can be used, and cell volume and carbon content calculated.

- Some constituents of living cells, such as adenosine triphosphate (ATP), phospholipids and DNA, can be extracted and analysed.

- The decomposition and mineralisation of microbial cells killed by fumigation can be measured by carbon dioxide production (using chloroform fumigation methods).

- The respiration rate after the addition of an excess of substrate can be estimated (using the respiratory response method).

- The heat produced by the biomass can be determined (using microcalorimetry).

All these methods have been described in detail and critically reviewed, compared and evaluated by many authors, including Jenkinson and Powlson (1976), Anderson and Domsch (1978), Karl (1980), Jenkinson and Ladd (1981), Sparling (1983), Jenkinson (1988), Smith and Paul (1990), Sayler et al. (1992) and Tunlid and White (1992). The suitable use or combination of methods can also provide data on the elemental content (such as carbon, nitrogen, sulphur and phosphorus) of the biomass.

None of these methods enables direct study of the microbial community in the microhabitat; all of them have specific limitations and, to be used, have to be based on certain assumptions. The difficulties of estimating 'conversion factors' using the Jenkinson and Powlson (1976) fumigation method or the extrapolation of cell biomass from cell constituent determination may be mentioned here as the examples. The accordance between different methods for measuring microbial biomass also depends upon conditions in the particular soil being studied and reliable analysis cannot be based on one approach alone. In no cases are plate counts and the most probable numbers (MPN) of microorganisms suitable for evaluating the general amount of soil biomass; no cultivation medium covers the nutritional requirements of all microbial species.

In this group of methods, more information needs to be accumulated on the recognition of the conditions limiting the use of a particular approach (as has been done in case of the fumigation method) and on the compatibility of different methods.

COMPOSITION AND ACTIVITY OF SOIL MICROFLORA

The soil microbial community is a complex phenomenon, and its constituent parts may be classified in different ways. The taxonomical composition of the particular biomass component, its metabolic abilities and its biochemical and genetical characteristics can serve as an example.

Current taxonomical methods can be used to specify and count the species present after their isolation by shape, size, genetic apparatus and metabolic properties. Specific staining methods can be used to show some other features. Manuals on determinative microbiology are being completed using 'fingerprint' techniques and computerised diagnostical approaches. The diversity of the microbial community, based on a survey of species involved, may be evaluated (Rogosa et al., 1986).

Soil microbial inhabitants are able to conduct almost all known metabolic reactions. Therefore, a large number of methods have been developed to estimate microbial activities in the soil in terms of general or particular aspects of metabolism. Nannipieri et al. (1990) tried to classify microbial activities by measuring common processes such as respiration (that is, oxygen consumption or carbon dioxide production), dehydrogenase activity, ATP content, adenylate energy charge, rates of RNA and DNA synthesis and heat production), as well as specific processes such as those involving nutrient transformation activities, degradation and the formation of biomarkers. Isotope dilution or double labelling techniques may be involved in some approaches. Different physiological groups or groups differing in nutritional requirements as well as in catabolic and synthetic processes can be distinguished among soil microbes. It is important to emphasise that the metabolic rate or the intensity of any microbial process should be evaluated and directly related to microbial biomass size and not to the total weight or volume of soil. If the abiotic part of soil is included in the calculation, misleading conclusions concerning the evaluation of the process may be obtained.

Individual parts of soil microflora differ in their content of specific cell components. The estimation of such 'signatures' may provide valuable insight into community structure, its nutritional status and microbial activity. Thus, among cell wall components, muramic acid is characteristic for bacteria, and glucosamine for fungi. D-alanine or diaminopimelic acid can be monitored in some cases. Similarly, from cell membrane lipids, ergosterol indicates the presence of fungi, the specific chemical structure of phospholid fatty acids indicates the presence of specific actinomycetes, fungi and methane-oxidising or sulphate-reducing bacteria, and the presence of respiratory quinones can differentiate between aerobic and anaerobic microorganisms. The accumulation of typical lipid storage polymers also indicates specific microbial groups and their nutritional status. The incorporation of labelled precursors into phospholipids can indicate the synthetic activity of microbiota. A group of biochemical methods based on the analyses of these compounds was surveyed by Tunlid and White (1992).

There may be some difficulties in linking the measurement of signature components with the microbial composition because of their different distribution throughout the different microbial genera. The formation of antibodies can also be considered as a very specific type of signature. The use of antibody labelling methods is of great importance in the study of the dynamics of soil microbial populations and of the fate of introduced strains in soil (Kozhevin, 1992).

The rapid development of molecular techniques has made it possible to apply them in the structural and functional analysis of microbial communities (Ford and Olson, 1988; Trevors and van Elsas, 1989; Jackman et al., 1992; Ward et al., 1992). These methods represent a new level of information in microbial ecology. Sayler et al. (1992) recently published an overview of the use of molecular techniques in the quantitative and qualitative analysis of soil microflora. The identification and enumeration of individual microbial species (including genetically engineered microorganisms) within a natural microbial community can be achieved with specificity by using nucleic acid hybridisation technology. The specificity is based on the nucleic acid or gene sequence. The long list of techniques includes methods for DNA and RNA extraction from soil, preparation and labelling of gene probes, nucleic acid transfer and hybridisation, and hybridisation protocols and applications. Sayler et al. (1992) state that 'new methods are on the horizon that will contribute to the analysis of the activity of specific populations and genes in environmentally significant processes' and predict that 'within a decade they will be commonplace in their application and as routine as conventional microbial cultivation and enumeration.'

EXPRESSION OF MICROBIAL ABILITIES IN THE SOIL ENVIRONMENT

The qualitative estimation of a soil microbial community provides us with its genotypic characterisation and biological potential. However, we need to know how genetically encoded abilities might be expressed under specific conditions of the soil environment, or, in other words, the phenotypic demonstration of the microflora, its fate and efficacy. The quantity of microorganisms and their composition and activities reflect the fluctuation of biotic and abiotic factors such as the presence of oxygen, water, nutrients, organic and inorganic substances, sources of energy, clay minerals, oxidation-reduction or adsorption conditions, and positive or negative mutual relationships between organisms (microbes, plants and animals). These factors may act at different levels of the biotic system (molecular, subcellular, cellular, population, community or ecosystem). Phenomena such as diversity, dominance and homeostasis in the ecosystem can be understood in the context of such interactions. Typical examples are the ecophysiological studies of genetic transfer within soil microflora, the investigations into the effect of inhibitors and the determination of critical rate limiting step (Oremland and Capone, 1988; Nannipieri et al., 1990; Gammack et al., 1992; Veal et al., 1992).

The experimental approach for studying the behaviour of the soil microbial community has to fit the complicated net of relations and take account of the dynamic and open nature of soil ecosystems. The heterocontinuous-flow cultivation method is an example of an experimental laboratory approach used for the study of the microbial community and its above-mentioned features in soil samples (Macura and Malek, 1958; Macura, 1961; Kunc, 1988). The soil column can be considered as an open micro-ecosystem with the corresponding substrate and energy inputs and outputs, in which a dynamic equilibrium between the environment and the biotic phase can be reached. The method can be used in many different ways, and the natural open environment can be simulated. Having all the qualities of a dynamic and vectorial approach of ecophysiological research, the method is well suited for basic research in microbial ecology. Generally, open experimental systems of continuous cultivation have attracted growing interest amongst microbial ecologists in the past few years. It is likely that a full appreciation of the possibilities offered by continuous cultures in this area will accelerate ecological research considerably, as was the case in biochemistry and genetics.

SIGNIFICANCE OF SOIL MICROBIAL BIOMASS DATA

Once we know the quantity, composition, potential and real abilities of soil microflora, we can use this knowledge in many ways. Fundamental and applied scientific questions can be answered. The significance of information on soil microbial biomass has been reviewed by several authors, including Jenkinson and Ladd (1981), Nikitin and Kunc (1988), Nannipieri et al. (1990), Anderson and Gray (1991), Haider (1992) and Kozhevin (1992). The following important areas need to be emphasised:

- An understanding of the mechanisms controlling the composition and function of microbial communities and their population dynamics in soil allows the possibility of enhancing the growth and activity of desirable microorganisms or of suppressing the undesirable ones (such as decomposers of xenobiotics, plant pathogens, nitrifiers and genetically engineered microorganisms). One should bear in mind that the microbial community in soil, as in all ecosystems, may behave identically in terms of function, despite being composed of diverse species.

- Data on microbial biomass are necessary for studying the flux of energy and material through the soil population, the contribution of biomass to global cycles of carbon, nitrogen, phosphorus, sulphur and other elements, the turnover of biomass and soil organic matter, and the role of microbes as a reservoir of nutrients. Appropriate agricultural management to increase plant productivity (for example, fertilisation, manuring or tillage) can then be applied.

- The microbial community is a sensitive ecological marker of stress situations in soils caused by pollution, agricultural management practices or the recovery of disturbed areas.

- Mathematical models developed to characterise and predict microbial community dynamics, turnover of biomass, nutrient cycling and any metabolic process can be based on exact experimental data; temporal and spatial patterns of structural microbial systems can also be investigated (Stout et al., 1981; Smith, 1982; Odham et al., 1986; Prosser, 1990; Lynch, 1991; Wimpenny, 1992).

CONCLUSION

This chapter has provided a short overview of available methods for analysing the quantity, composition and activities of soil microbial communities. The significance of such methods for basic and applied research as well as for their consequences in environmental and agricultural practices has led to growing interest amongst scientists. To understand the soil microflora, various ideas and facts from different perspectives need to be combined. The investigation of soil biomass is a complex discipline requiring close cooperation between researchers in various disciplines, including soil microbiology, biochemistry, molecular biology and genetics, soil science, ecology and agronomy. The soil should be treated as a dynamic system, at different levels of complexity according to its subsets. The application of open experimental tools is desirable. The results obtained using one method should be compared and verified using another. With the progress made in our ability to quantify biomass, our models no longer need to be based on intuitive ideas about processes. The rapid development of new approaches in this area has resulted from advances in molecular biology and analytical chemistry. We should be aware that 'the new results may challenge the traditional views in microbial ecology and may require us to adopt new ways of thinking' (Ward et al., 1992).

References

Anderson, J.P.E. and Domsch, K. 1978. A physiological method for the quantitative measurement of microbial biomass in soils. *Soil Biology and Biochemistry* 10: 215-21.

Anderson, T.-H. and Gray, T.R.G. 1991. The influence of soil organic carbon on microbial growth and survival. In Wilson, W.S. (ed) *Advances in Soil Organic Matter Research. The Impact on Agriculture and the Environment*. Cambridge, UK: Royal Society of Chemistry.

Ford, S. and Olson, B.H. 1988. Methods for detecting genetically engineered microorganisms in the environment. In Marshall, K.C. (ed) *Advances in Microbial Ecology* (Vol. 10). New York, USA: Plenum Press.

Gammack, S.M., Paterson, E., Kemp, J.S., Cresser, M.S. and Killham, K. 1992. Factors affecting the movement of microorganisms in soils. In Stotzky, G. and Bollag, J.-M. (eds) *Soil Biochemistry* (Vol. 7). New York, USA: Marcel Dekker.

Haider, K. 1992. Problems related to the humification processes in soils of temperate climates. In Stotzky, G. and Bollag, J.-M. (eds) *Soil Biochemistry* (Vol. 7). New York, USA: Marcel Dekker.

Jackman, S.C., Lee, H. and Trevors, J.T. 1992. Survival, detection and containment of bacteria. *Microbial Releases* 1: 125-54.

Jenkinson, D.S. 1988. Determination of microbial biomass carbon and nitrogen in soil. In Wilson, J.R. (ed) *Advances in Nitrogen Cycling in Agricultural Ecosystems*. Wallingford, UK: CAB International.

Jenkinson, D.S. and Powlson, D.S. 1976. The effects of biocidal treatments on metabolism in soil. V. A method for measuring soil biomass. *Soil Biology and Biochemistry* 8: 209-13.

Jenkinson, D.S. and Ladd, J.N. 1981. Microbial biomass in soil: Measurement and turnover. In Paul, E.A. and Ladd, J.N. (eds) *Soil Biochemistry* (Vol. 5). New York, USA: Marcel Dekker.

Karl, D.M. 1980. Cellular nucleotide measurements and applications in microbial ecology. *Microbiological Reviews* 44: 739-98.

Kozhevin, P.A. 1992. The dynamics of microbial population in the soil. *Bulletin of Moscow University, Ser.17., Soil Science* 2: 39-56. (in Russian).

Kunc, F. 1988. Three decades of heterocontinuous flow cultivation method in soil microbiology. In Kyslik, P., Dawes, E.A., Krumphanzl, V. and Novak, M. (eds) *Continuous Culture*. London, UK: Academic Press.

Lynch, J.M. 1991. Sources and fate of soil organic matter. In Wilson, W.S. (ed) *Advances in Soil Organic Matter Research. The Impact on Agriculture and the Environment*. Cambridge, UK: Royal Society of Chemistry.

Macura, J. 1961. Continuous-flow method in soil microbiology. I. Apparatus. *Folia Microbiologica* 6: 328-34.

Macura, J. and Malek, I. 1958. Continuous-flow method for the study of microbiological processes in soil samples. *Nature* 182: 1796-97.

Nannipieri, P., Grego, S. and Ceccanti, B. 1990. Ecological significance of the biological activity in soil. In Bollag, J.-M. and Stotzky, G. (eds) *Soil Biochemistry* (Vol. 6). New York, USA: Marcel Dekker.

Nikitin, D.I. and Kunc, F. 1988. Structure of microbial soil associations and some mechanisms of their autoregulation. In Vancura, V. and Kunc, F. (eds) *Soil Microbial Associations. Control of Structures and Functions*. Amsterdam, Netherlands: Elsevier.

Odham, G., Tunlid, A., Valeur, A., Sundin, P. and White, D.C. 1986. Model systems for studies of microbial dynamics at exuding surfaces such as rhizosphere. *Applied and Environmental Microbiology* 52: 191-96.

Oremland, R.S. and Capone, D.G. 1988. Use of 'specific' inhibitors in biogeochemistry and microbial ecology. In Marshall, K.C. (ed) *Advances in Microbial Ecology* (Vol. 10). New York, USA: Plenum Press.

Prosser, J.I. 1990. Mathematical modeling of nitrification processes. In Marshall, K.C. (ed) *Advances in Microbial Ecology* (Vol. 11). New York, USA: Plenum Press.

Rogosa, M., Krichevsky, M.I. and Colwell, R.R. 1986. *Coding Microbiological Data for Computers*. New York, USA: Springer-Verlag.

Sayler, G.S., Nikbakht, K., Fleming, J.T. and Packard, J. 1992. Applications of molecular techniques to soil biochemistry. In Stotzky, G. and Bollag, J.-M. (eds) *Soil Biochemistry* (Vol. 7). New York, USA: Marcel Dekker.

Smith, J.L. and Paul, E.A. 1990. The significance of soil microbial biomass estimations. In Bollag, J.-M. and Stotzky, G. (eds) *Soil Biochemistry* (Vol. 6). New York, USA: Marcel Dekker.

Smith, O.L. 1982. *Soil Microbiology. A Model of Decomposition and Nutrient Cycling*. Boca Raton, USA: CRC Press.

Sparling, G.P. 1983. Estimation of microbial biomass and activity in soil using microcalorimetry. *J. Soil Science* 34: 381-90.

Stout, J.D., Goh, K.M. and Rafter, T.A. 1981. Chemistry and turnover of naturally occurring resistant organic compounds in soil. In Paul, E.A. and Ladd, J.N. (eds) *Soil Biochemistry* (Vol. 5). New York, USA: Marcel Dekker.

Trevors, J.T. and van Elsas, J.D. 1989. A review of selected methods in environmental microbial genetics. *Canadian J. Microbiology* 35: 895-902.

Tunlid, A. and White, D.C. 1992. Biochemical analysis of biomass, community structure, nutritional status, and metabolic activity of microbial communities in soil. In Stotzky, G. and Bollag, J.-M. (eds) *Soil Biochemistry* (Vol. 7). New York, USA: Marcel Dekker.

Veal, D.A., Stokes, H.W. and Daggard, G. 1992. Genetic exchange in natural microbial communities. In Marshall, K.C. (ed) *Advances in Microbial Ecology* (Vol.12). New York, USA: Plenum Press.

Ward, D.M., Bateson, M.M., Weller, R. and Ruff-Roberts, A.L. 1992. Ribosomal RNA analysis of microorganisms as they occur in nature. In Marshall, K.C. (ed) *Advances in Microbial Ecology* (Vol. 12). New York, USA: Plenum Press.

Wimpenny, J.W.T. 1992. Microbial systems. Patterns in time and space. In Marshall, K.C. (ed) *Advances in Microbial Ecology* (Vol. 12). New York, USA: Plenum Press.

Beyond the Biomass
Edited by K. Ritz, J. Dighton and K.E. Giller
© 1994 British Soil Science Society (BSSS)
A Wiley-Sayce Publication

CHAPTER 3

Vertical distribution of microbial communities in soils

D.G. ZVYAGINTSEV

Several studies of microbial communities in different soil types have shown that dominant and typical species of microorganisms (fungi, yeasts, streptomycetes and bacilli) differ between soils (Mishustin, 1984; Mirchink, 1988; Bab'eva and Zenova, 1989). In general, comparisons of soil types were made in horizon A_1. This chapter summarises some results obtained in a comprehensive programme based at the Department of Soil Biology, Moscow State University. The aim of the work is to examine the vertical distribution of microbial communities in various terrestrial ecosystems (tundra, forest, steppe, desert and peat bog). Studies were conducted on the vertical distribution of microorganisms not only in soil but also in all vertical elements of the ecosystem, including plants and subsoil layers.

MATERIALS AND METHODS

The number and biomass of bacteria were determined using Acridine Orange staining and fluorescence microscopy. The fungi were stained with Calcofluor W® in order to determine total hyphal length and numbers of fungal spores and yeast cells. The biomass of fungi and actinomycetes was calculated using following equation (Moscow State University, 1991):

$$P = 3.14 \times R^2 \times h \times d \times 20\%/100\%$$

where:

P	=	mean biomass of fungal (actinomycete) mycelium
R	=	hyphae radius
h	=	total hyphae length
d	=	density of living hyphae (usually $d = 1$)
20%	=	proportion of dry fungal (actinomycete) biomass

Fungi and yeasts were normally identified at species level and bacteria at genus level, using standard identification procedures. Only dominant genera and species were identified; rare species were not identified.

RESULTS

The experiment showed that ecologically dominant forms differed in the vertical horizons (Dobrovol'skaya et al., 1991; Zenova et al., 1991; Zvyaginstev et al., 1991, 1992 a,b, 1993).

Quantity and quality

Different ecosystems have different vertical stratifications of microorganisms, and there are differences in the quality and depth of horizons. In our studies, tundra and desert had the greatest density of microorganisms in the phylloplane; in forest and steppe the greatest density was in litter, while in peat

Figure 3.1   **Distribution of bacteria (millions/g soil) within the layers of peat-forest ecosystems**

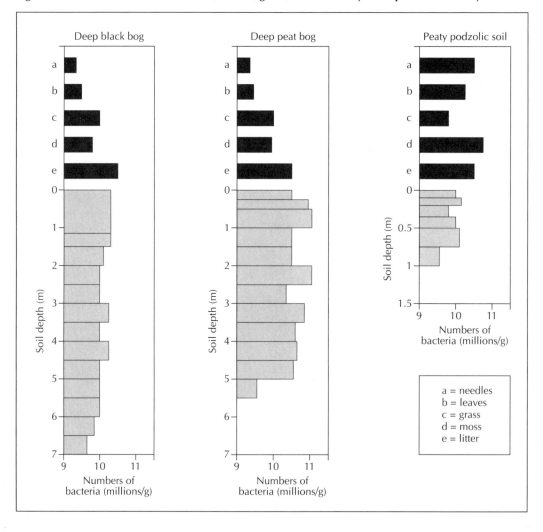

bog it is in the peat layers. For sod-podzols it was found that there were very large numbers of microorganisms in subsoil layers of sediment rocks. In 20 m of subsoil, bacterial biomass was found to be equal to that of the topsoil profile (Zvyagintsev, 1987). Thus, half of the total bacterial biomass was in soil itself and half in subsoil. The subsoil biomass is very important for the purification of soil water which is infiltrated through the rocks and carries organic compounds, nitrates, pesticides and other pollutants. Total microbial biomass was greater in topsoil than in subsoil because the upper layers of soil contained more fungal biomass. There were only very few fungi in subsoil horizons.

Direct microscopic observation showed that, in upper soil horizons, fungal biomass, mycelium, spores and yeast cells usually formed 90-95% of the biomass. Bacterial biomass accounted for 5-10%, of which 25% was actinomycete mycelium. In contrast, bacteria in deep layers were represented only by coryneforms, Gram-negative bacteria and oligotrophs. This was also true for subsoil sediment rocks and especially for permafrost. In permafrost horizons, bacteria remain viable for millions of years (the temperature of the rocks remains constant at -12°) (Zvyagintsev, 1992). The quantity of microorganisms in soil usually decreases considerably with depth and in the lower layers it is reduced by 10-100 times. In our investigations, however, we found an exception to this. In peat and black bog, the number of bacteria, determined using the dilution plate and direct microscopic methods, did not decrease significantly in the vertical profile (see Figures 3.1 and 3.2 and Figure 3.3 overleaf). The number of bacteria was greatest in litter layers; it was lower in peat layers but stayed at a constant level to a depth of 5-7 m (that is, up to the whole depth of the peat). It reached between 10^{10} and 10^{11} cells/g of dry substrate. In contrast, the number of bacteria in leaves, needles, grass and moss were 10 times lower (see Figure 3.1). The length of fungi and actinomycete hyphae, determined using microscopic

Figure 3.2 **Distribution of bacteria, actinomycetes, fungi and yeasts (millions/g soil) wihtin the profile of a peat bog**

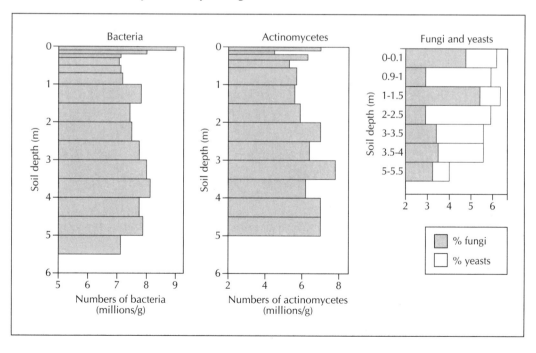

Figure 3.3   Distribution of micromycetes mycelium and actinomycetes mycelium (millions/g soil) within the profile of soils

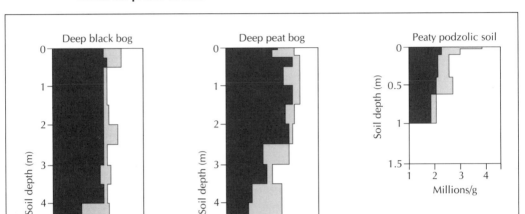

methods, decreased with depth in an ordinary zonal soil. The peat and black bogs contained thousands of metres per gram and this remained constant in all the peat layers. The total fungal biomass contained as much as 40-400 t/ha of dry fungal biomass when calculated per 7 m depth of the peat. Such extremely high fungal biomass appeared to result from the peculiar conditions for microbial conservation in the peat. It should be stressed that luminescent microscopy with Calcofluor staining gives a much higher fungal mycelium length than other methods. The biomass of spores was also estimated.

The length of actinomycete mycelium is less than that of fungi but can be thousands of metres per gram. Actinomycetes were distributed evenly throughout all layers of the peat. Although actinomycete hyphae were thinner than fungal hyphae (0.4 and 4 μm, respectively), there is good reason to believe that their biomass is thousands of times lower than fungal biomass (0.08-0.12 t/ha).

The problem of microbial quality and biodiversity in different layers of soil represents another and more difficult step in understanding the distribution of microbial communities in soils. Three decades ago this problem was studied in detail to obtain the distribution pattern of dominant genera and species in different horizons of the forest litter (Kendrick and Burger, 1962).

Fungi

There were specific dominant forms of fungi in the phylloplane, but different plants generally had the same epiphytic fungi. Fungi are resistant to light due to melanins and carotinoids. Their spores can germinate at low moisture content. Epiphytic fungi continued to germinate and grow even after the leaves had fallen, forming an upper layer of the litter (up to 6 months). In the F-layer (after 6 months) hydrolytic fungi developed. Below this layer, lignin decomposers occurred with oligotrophic fungi, which were also located more deeply.

The number of fungal species increased from plant surface to litter to soil. The major decomposition activity of plant residues in the forest took place in the litter. In mineral horizons of the soil, fungal diversity was greater but many species were in a dormant state. The fungal community changed with depth. Copiotrophs were followed by hydrolytics and the latter by oligotrophs. The vertical species distribution of fungi is presented in Figure 3.4.

Figure 3.4   Distribution of dominant fungi (%) within the layers of a forest ecosystem

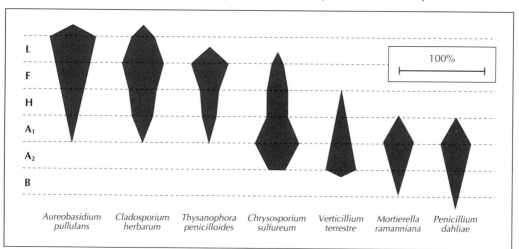

Yeasts

Epiphytic yeasts usually produce carotinoid pigments; they are capsulated and very resistant to drying because of their ability to form chlamydospores. The yeasts usually form ballistospores that are air dispersed. The dominant yeasts were *Cryptococcus laurentii, C. albidus* and *Sporobolomyces roseus* (*see* Figure 3.5 *overleaf*).

Yeasts inhabiting the litter were characterised by high hydrolytic activity and all of them had a mycelial growth stage: *Trichosporon pullulans, Cystafilobasidium capitatum, Candida humicola, C. curvata* and *C. podzolica*. In the mineral soil horizons, oligotrophic capsule-forming yeasts occurred. They were not pigmented and were often able to accumulate lipids.

Bacteria

The vertical distribution of bacteria is shown in Figure 3.6 (*overleaf*). Species of *Pseudomonas, Flavobacterium* and *Erwinia* dominated on plants. These bacteria typically had a high content of carotinoids, R-strategists and copiotrophs. Bacilli and actinomycetes represented a minor component of the community in this habitat.

Figure 3.5   Distribution of dominant yeasts (%) within the layers of a forest ecosystem

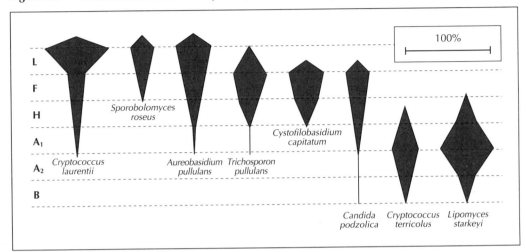

Figure 3.6   Distribution of dominant bacteria (%) within the layers of a forest ecosystem

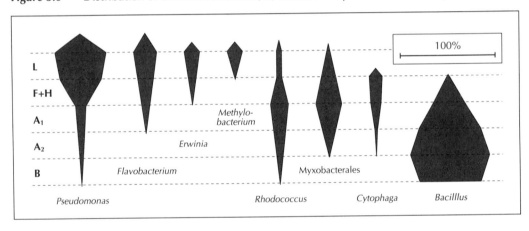

The composition of dominant genera and species of bacteria were the same in the phylloplane and L-layer of the litter. In the F- and H-layers, representatives of *Bacillus*, *Cytophaga*, *Streptomyces* and Myxobacterales appeared. All these belonged to hydrolytics and L-strategists. The maximal diversity of bacteria in the litter was estimated using the dilution plate method. It is possible to view litter as an ecological niche which is inhabited by phylloplane and soil colonists, as well as by native litter species.

In the soil horizons, there was a reduction in bacteria numbers and taxonomic variety. The dominant groups in the bacterial complex were: *Streptomyces*, *Rhodococcus*, *Bacillus*, *Nocardia* and oligotrophs (*Caulobacter*, *Metallogenium*, *Pedomicrobium*, *Seliberia* and *Prosthecomicrobium*). These groups are K-strategists. The greatest variety of bacteria was observed in the A₁ horizon, but it was impossible to isolate them using traditional dilution plate techniques because they have different nutritional requirements and grow slowly. The ability of these bacteria to survive and reproduce in the A₁ horizon has been attributed to the microzonality and heterogeneity of this horizon (Zvyagintsev, 1987).

Thus, the functional groups, ecological strategy and taxonomic structure of dominant bacteria change in vertical layers of terrestrial ecosystems. The domination by different bacterial genera in soil horizons can be clearly demonstrated by determining their frequency of occurrence as a percentage of the total number of samples studied (*see* Figure 3.7).

Figure 3.7   **Frequency of dominance (%) of different taxonomic groups of bacteria in the Bukhara region, Kyzylkum desert, Uzbekistan**

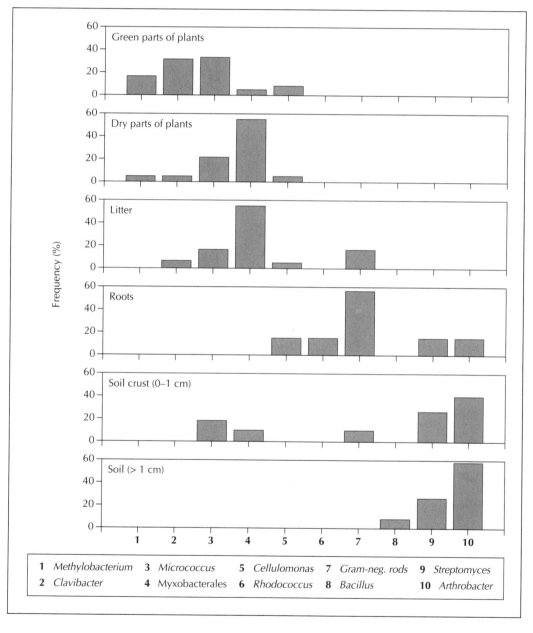

When studying the decomposition of plant residues it is important to realise that bacteria are only 'helpers' of fungi. In terrestrial ecosystems, organic matter is decayed mainly by fungi. Bacteria are connected with fungi and dependent upon them.

Actinomycetes

Actinomycetes are mainly mycelial organisms, and they represent a small but distinct taxonomic group of bacteria. They have a specific ecological status. Mycelial organisation is such that their ecology differs from other bacteria, just as the ecology of mycelial fungi is known to differ from that of single cell yeasts.

In the phylloplane, there were typically few actinomycetes. *Actinoplanes* rarely occurred here. In the litter, actinomycetes of genus *Streptomyces* predominated. However, they reached maximum density in the A_1 horizon (*see* Figure 3.8). Other actinomycetes occurred sparsely. In the L-layer, the following genera were dominant: *Streptosporangium*, *Micromonospora* and *Nocardia*. The genus *Streptoverticillium* was dominant in the H-layer.

Figure 3.8   Distribution of dominant actinomycetes (%) within the layers of a forest ecosystem

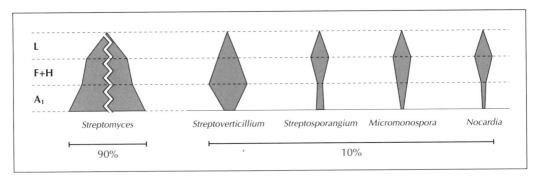

As a rule, soil microbiologists restrict themselves to the study of the genus *Streptomyces*. It appears that the investigation reported here of the distribution of actinomycete genera in a vertical profile is the first of its kind. The investigation is an attempt to study the general laws of the distribution of dominant microorganisms in a vertical profile. However, if we want more detail we should look at the specific soil zones. In this case, we can see a more complex situation in terms of the mesozonality of every vertical layer. Examples are different species of plants, different parts of plants, root zones, mycorrhizae, surface of fungal mycelium, digestional tract of invertebrates, protozoa and iron-manganese concretions.

CONCLUSION

The study outlined here shows that dominant genera and species of microorganisms (fungi, yeasts, bacteria and actinomycetes) differ in vertical layers of terrestrial ecosystems. The main patterns of spatial distribution that were observed in the study were: a correlation between spatial location of microorganisms, their functions, ecological strategy and taxonomic composition; the simultaneous presence of discrete and continuous distribution of genera and species of microorganisms in different horizons; and an extremely high number and biomass of microorganisms in all layers of peat bog and black bog, with little vertical differentiation.

Acknowledgements

The author would like to thank his colleagues Drs T.G. Mirchink, I.P. Bab'eva, T.G. Dobrovol'skaya, G.M. Zenova and L.M. Polyanskaya for their valuable contribution to the work described in this chapter.

References

Bab'eva, I.P. and Zenova, G.M. 1989. *Soil Biology*. Moscow, Russia: Moscow State University. (in Russian).

Dobrovol'skaya, T.G., Polyanskaya, L.M., Golovchenko, A.V., Smagina, M.V. and Zvyagintsev, D.G. 1991. Microbial pool in peaty soils. *Pochvovedenie* 7: 69-77. (in Russian).

Kendrick, W.B. and Burger, A. 1962. Biological aspects of the decay of *Pinus sylvestris* litter. *Nova Hedwig* 4: 313-42.

Mirchink, T.G. 1988. *Soil Mycology*. Moscow, Russia: Moscow State University. (in Russian).

Mishustin, E.N. (ed) 1984. Soil organisms as the component of biogeocenosis. *Nauka (Moscow)*. (in Russian).

Moscow State University. 1991. *Manual of Methods for Microbiology and Biochemistry of Soil*. Moscow, Russia: Moscow State University. (in Russian).

Zenova, G.M., Shirokikh, I.G., Lysak, L.V. and Zvyagintsev, D.G. 1991. Mesophylic and thermotolerant actinomycetes in the recultivated peatland of the southern taiga. *Pochvovedenie* 12: 54-61. (in Russian).

Beyond the Biomass
Edited by K. Ritz, J. Dighton and K.E. Giller
© 1994 British Society of Soil Science (BSSS)
A Wiley-Sayce Publication

CHAPTER 4

Use of DNA analysis to determine the diversity of microbial communities

V. Torsvik, J. Goksøyr, F.L. Daae, R. Sørheim, J. Michalsen and K. Salte

Diversity measurements based on the phenotypical characterisation of isolated strains cover only a small fraction of the total information in a microbial community, because only the fraction (0.1-1%) of the bacteria which can grow and form visible colonies on solid media can be analysed. In addition, the phenotypic tests selected account for a small amount of the total information in the DNA. In a study of diurnal fluctuations in river bacteria, Holder-Franklin et al. (1981) compared numerical taxonomy with DNA homology and found that in many cases there was a coincidence between phenetic clusters and genetic relatedness but that some clusters contained more than one species and even more than one genus.

We have determined the DNA diversity in bacterial populations from soil (Torsvik et al., 1990a) and compared genetic and phenetic diversity of populations exposed to stress in the form of increased temperature. We have also studied the complexity of DNA isolated directly from the bacterial community in natural soil (Torsvik et al., 1990b) and compared it with the DNA complexity of a population of isolated bacteria from the same soil. The DNA analysis shows that the genetic diversity of the total microbial community in the soil was about 200 times higher than the diversity of the population of isolated bacteria, and may comprise thousands of species.

ISOLATION OF DNA

Measurements of DNA complexity by reassociation require highly purified DNA, free from eukaryotic DNA, humic material or other impurities. We have developed a method for isolating DNA with high purity directly from bacteria in the soil. The bacteria are first separated from the soil by a fractionated centrifugation procedure (Faegri et al., 1977). The fractionation yield is estimated from the microscopic count, with the number of bacteria in the bacterial fraction as a percentage of the sum of bacteria in the bacterial fraction plus the sediment after the final low-speed centrifugation. Investigators using this counting method have reported yields of 30-35% (Holben et al., 1988; Steffan et al., 1988). In the organic soil being used in this study, the bacterial yield is 60-65% and the bacterial fraction is substantially free from fungi and other eukaryotic organisms. There are no indications that the fractionation procedure is biased. Our assumption that all the bacterial types are represented in the bacterial fraction is supported by the observation that the yields based on plate count and microscopic count are almost identical (Faegri et al., 1977).

DNA is isolated from the bacterial fraction after removing extracellular DNA and some humic material with hexametaphosphate. The bacteria are lysed with lysozyme and sodium dodecyl sulphate, giving a lysis efficiency of 90-95%. DNA is then extracted from the cells and purified on a hydro-xyapatite column (Bio-Gel HT, Bio-Rad Laboratories) (Torsvik, 1980; Torsvik et al., 1990b). The DNA purification causes loss; the highest loss occurs during centifugation, when cell debris and some humic material are removed (30%), and during the hydroxyapatite purification (50%) (Torsvik, 1980). DNA purification does not cause bias; if bias exists, it would lead to a reduction in the DNA complexity.

When starting with 100 g of wet soil (32 g dry weight) with 4.8×10^{11} bacteria, the DNA yield is 3.5-5.0 µg/g wet soil. Assuming an average DNA amount per bacterial cell of 5×10^{-15} g (Bak et al., 1970), the theoretical yield is 24 µg/g wet soil. Thus, 15-20% of bacterial DNA is recovered from the soil.

DETERMINATION OF GENETIC DIVERSITY

Genetic diversity is a measure of the number and frequency distribution of genetically different bacteria in a bacterial population or community. It can be determined from the complexity of DNA extracted from the mixture of bacteria. DNA complexity is defined as the total length of different DNA sequences measured in number of base pairs (bp) in a defined amount of DNA. This is an ideal quantity and is equal to the genome size for haploid cells with single-copy DNA. The kinetic DNA complexity is a measurable quantity calculated from the reassociation rate of sheared and melted (single-stranded) DNA in solution (Britten and Kohne, 1968). The reassociation of homologous single-stranded DNA follows second-order reaction kinetics where the reaction rate is proportional to the square of the concentration of homologous DNA strands. The fraction of reassociated DNA is expressed as a function of $C_O t$ (C_O is the molar concentration of nucleotides in single-stranded DNA at the beginning of the reassociation, and t is the time in seconds). The reaction rate constant depends upon the relative concentration of complementary DNA sequences and is proportional to the reciprocal of $C_O t$ for half reassociation ($1/C_O t_{1/2}$). Under defined conditions (for example, cation concentration, temperature, DNA fragment length and concentration) $C_O t_{1/2}$ is thus proportional to the complexity of the DNA.

We normally measure the reassociation of sheared (French Press at 20 000 psi) and melted DNA spectrophotometrically in 4-6 x standard saline citrate (SSC) with 30% dimethylsulphoxide (DMSO) added (Torsvik et al., 1990a), using the *Escherichia coli* genome as a reference.

The reassociation curves for DNA from mixtures of bacteria have flatter slopes than that corresponding to an ideal second-order reaction. This may arise from an uneven distribution of the different DNA fractions in the mixture, which gives several second-order reactions with different rate constants. In this case, $C_O t_{1/2}$ has no precise kinetic meaning, but it still can be used as a parameter for expressing the complexity of DNA from mixed bacterial populations and communities.

The genetic diversity of a bacterial community can be expressed by using $C_O t_{1/2}$ under defined conditions as a diversity index. The concept of genetic diversity as we use it corresponds to that used in information theory, where diversity is a measure of the total amount of information in a system (that is, a community) and the distribution of this information (the amount of information found in a few individuals, and the amount found in a large number of individuals). The diversity index $C_O t_{1/2}$ is analogous to the Shannon Weaver index, which is based on numeric taxonomy clusters or species diversities (Atlas, 1984).

APPLICATIONS OF THE DNA REASSOCIATION METHOD

We have used the DNA reassociation method to study the effect of external stress factors on the diversity of bacterial communities, and to compare the diversity of cultured bacteria with diversity of the total bacterial population in soil.

Temperature stress and diversity

In a model study, the diversity indices based on genetic and phenetic measures were compared, and the usefulness of these indices in demonstrating the effect of an external stress factor on the bacterial community was tested. Therefore, only populations of culturable bacteria from the soil were included in the study.

Our model system was a beech forest soil in western Norway (Sørheim et al., 1989). The microscopic count (acridine orange stained) was 1.5×10^{10} and the plate count (Thornton's medium with 10% soil extract; Thornton, 1922) was 4.2×10^7/g dry soil. The soil was stored at 4°C and 30°C for 3 months. By the end of this time the total counts had decreased to 2.5×10^9/g dry soil for the soil stored at 4°C, and to 3.5×10^9/g dry soil for the soil stored at 30°C. The plate counts of the two soils were 2.2×10^7/g dry soil and 5.5×10^7/g dry soil, respectively.

For phenotypic and genotypic testing, 80 strains were selected randomly from the isolates from each soil. The isolates were derived from colonies selected at random from plates used for plate counts. For phenotypic testing, a set of 26 physiological (API 20B and API OF systems; API System S.A., France) and morphological tests were implemented. The proximity between strains was calculated (simple matching coefficient) and they were clustered (complete link, furthest distance method; Sørheim et al., 1989). The dendrograms of the 4°C and 30°C strains (*see* Figures 4.1 and 4.2 *overleaf*) show that the structure of the two populations was entirely different. When a similarity level of 80% was used to define the biotypes, the two populations contained 35 biotypes (*see* Table 4.1). The 4°C population contained 33 biotypes with few strains in each. The 30°C population contained 12 biotypes, two of which were dominant, and the most abundant contained 61% of all the strains in the population.

The phenetic diversity was also expressed as cumulative difference, as described by Torsvik et al. (1990a). This was done by correlating the number of strains in populations of increasing size with the

Table 4.1   Numbers of biotypes, cumulative difference, Shannon index (H' logarithmic base), equitability (J = H'/H'$_{max}$) and population diversity index ($C_0t_{1/2}$) for the 4°C and 30°C populations separately and combined

Population	4°C	30°C	4°C + 30°C
Number of strains	80	80	160
Biotypes	33	12	35
Cumulative difference	121	28	133
H'	4.79	2.11	4.07
J	0.95	0.59	0.79
$C_0t_{1/2}$ [a]	34.7	5.8	22.4

Note: a   In 4 x standard saline citrate (SSC), 30% dimethylsulphoxide (DMSO)

Figures 4.1   Cluster analysis (complete link, furthest neighbour) of the bacterial population isolated from soil stored at 4°C

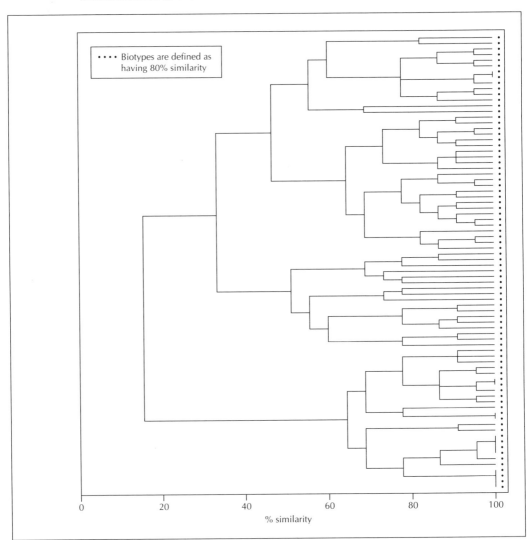

number of different phenotypes or differences in test results. The strains were placed in a randomly chosen sequence and for each new strain selected, the phenotypic test results were compared with those for all previously selected strains. The smallest differences thus obtained represented new phenetic information in the new strain. The sum of the differences was plotted against the number of strains added. An error margin of 2 was included to account for misreading of test results. The cumulative differences for the 4°C and 30°C strains are shown in Figure 4.3 (*overleaf*).

The increase in genetic diversity with an increasing number of strains was determined by DNA reassociation. DNA was isolated using the method described by Marmur et al. (1963), and to reduce the work involved the isolation was performed from groups of 10 strains lumped together. The strain

Figure 4.2   Cluster analysis (complete link, furthest neighbour) of the bacterial population isolated from soil stored at 30°C

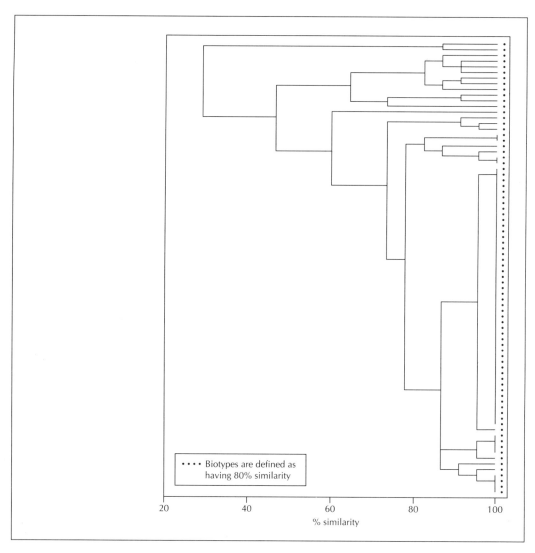

•••• Biotypes are defined as having 80% similarity

20   40   60   80   100

% similarity

sequence was the same as that used for determining cumulative difference. DNA from group 1 was reassociated first, and the reaction rate expressed as $C_0t_{1/2}$ was determined. Equal amounts of DNA from group 1 and group 2 were then mixed and reassociated, and $C_0t_{1/2}$ for this mixture was determined. This process was repeated until all the groups had been included in the reassociation mixture. The $C_0t_{1/2}$ for each reassociation mixture was plotted against the number of strains added (*see* Figure 4.4 *overleaf*). The amount of genetic information in the 4°C and 30°C populations was very different. Figure 4.5 (*overleaf*) shows the reassociation curve for the mixture of DNA from the 4°C and 30°C populations combined (160 strains), compared with the curves for DNA from these two populations individually.

Figure 4.3   Cumulative differences for bacterial populations isolated from soil stored at 4°C and 30°C

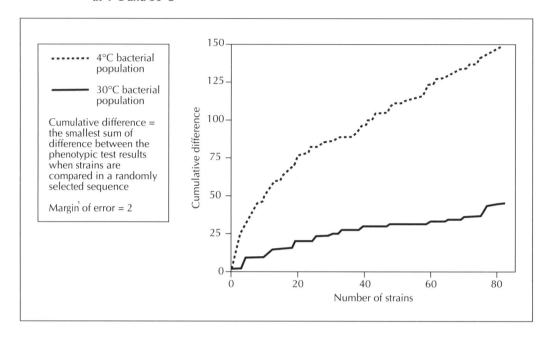

Figure 4.4   The diversity index ($C_0t_{1/2}$) values with increasing numbers of bacterial strains isolated from soil stored at 4°C and 30°C

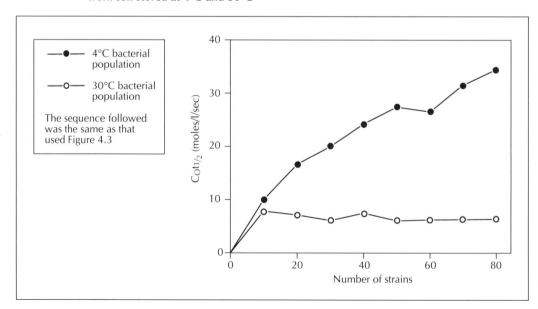

There seems to be a general relationship between DNA diversity and phenotypic diversity in this model system. Under the experimental conditions, the average $C_0t_{1/2}$ for soil bacteria was about 1.4 moles/l x sec, some 60% higher than $C_0t_{1/2}$ for *E. coli* DNA (Torsvik et al., 1990a). In the 4°C population containing 33 biotypes, $C_0t_{1/2}$ for all the 80 strains was 35 moles/l x sec. For the 30°C population containing 12 biotypes, $C_0t_{1/2}$ was 6 moles/l x sec. The curves for cumulative difference and genetic diversity had the same overall appearance, and indicated that the reduction in phenetic and genetic diversity was due to the higher temperature. However, there is a difference in the two diversity measurements, most evident in the curves for the 30°C population. The cumulative difference increased with increasing numbers of strains, but the $C_0t_{1/2}$ levelled off at 10 strains. The reason is that the cumulative difference measures only the amount of information in the population, while the genetic diversity also takes into account how this information is distributed within the population.

Figure 4.5 Reassociation curves at 51°C in 4 x SSC, 30% DMSO for all bacterial strains from soil stored at 4°C and 30°C, separately and combined

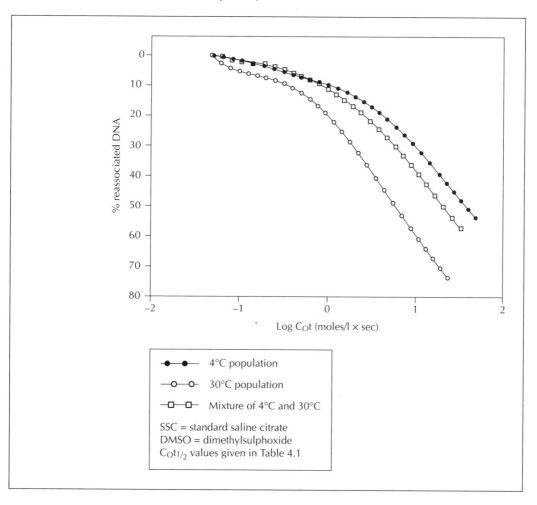

When a mixture of DNA from the two populations was reassociated, the $C_0t_{1/2}$ fell between the $C_0t_{1/2}$ for the two populations reassociated separately. The mixture of the two populations has more biotypes and a higher cumulative difference (*see* Table 4.1), but this was counterbalanced by the increasing number of identical isolates from the 30°C population, increasing the concentrations of homologous DNA in the mixture relative to that of the 4°C population. The $C_0t_{1/2}$ values change in the same manner as the Shannon Weaver index or Equitability (*see* Table 4.1), which also measures both the amount and the distribution of information in the populations.

Diversity of the total bacterial community

The genetic diversity of the total bacterial community in the soil was compared with the genetic diversity of a population of isolated bacteria. The experiments have been described by Torsvik et al. (1990a, b). About 200 strains were randomly chosen from plates of a standard plating medium, and isolated. This population contained 41 biotypes at 80% phenotypic similarity. Figure 4.6 shows the reassociation of DNA from a mixture of the isolated strains, compared with the reassociation of DNA derived directly from the bacterial fraction of the same soil.

The $C_0t_{1/2}$ of DNA from the mixture of the isolated bacterial strains was 28, corresponding to a DNA complexity of 1.4×10^8 bp. The DNA complexity seemed to have reached its maximum value at approximately 90 strains (Torsvik et al., 1990a). The $C_0t_{1/2}$ of DNA isolated from the total bacterial fraction of the same soil was 4700, corresponding to 2.7×10^{10} bp (*see* Table 4.2). Thus, the genetic diversity of the total bacterial population was about 200 times higher than that of the isolated bacteria.

According to a proposal put forward by the committee on the reconciliation of approaches to bacterial systematics (Wayne et al., 1987), strains with at least 70% DNA homology can be defined as belonging to the same species. The maximum value for the number of species would therefore be 3.3 times the number of genomes with no homology. The isolated bacteria therefore consist of a maximum of 66 species with 70% homology, while the maximum number of species in the total bacterial community is approximately 13 000. This indicates that the bacterial types that can be isolated using the standard plating technique are only a small fraction of the soil bacterial population.

Our findings are supported by analysis of 16S ribosomal RNA (rRNA) from environmental samples (Giovannoni et al., 1990; Ward et al., 1990; Fuhrman et al., 1992). It has been found that the 16S rRNA isolated directly from environmental samples does not correspond to that from isolated organisms, and the conclusion drawn was that there may be high genetic variability within natural microbial communities.

It may be difficult to understand that up to 10 000 bacterial species can be harboured in 100 g of soil. The total number of bacteria in 100 g wet soil was about 5×10^{11}. If we assume an even species distribution, each species would consist of about 5×10^7 individuals. Using the immunofluorescence technique, it has been demonstrated that the population size of three specific bacteria in an organic soil ranged from 10^6 to 10^7 cells/g of soil (Sjåstad, 1979). The average plate counts for the three bacteria were about 0.5% of the immunofluorescence counts. Even if 99.5% of the cells were in a non-culturable state, this is an ample population size for species sustainability. Rare species present in frequencies far below the average still have a fairly high population size.

How can so many species have evolved? The high complexity of DNA from soil bacteria may reflect a vast amount of genotypically separate clones in soil, most of them probably unknown. Environments such as soil may contain a large number of different ecological niches which can sustain a great variability of the bacterial flora utilising a combination of substrates under different physico-chemical conditions.

Figure 4.6 Reassociation of DNA from the soil bacterial fraction, 206 isolated bacteria from the same soil and *Escherichia coli* in 4 x SSC and 30% DMSO

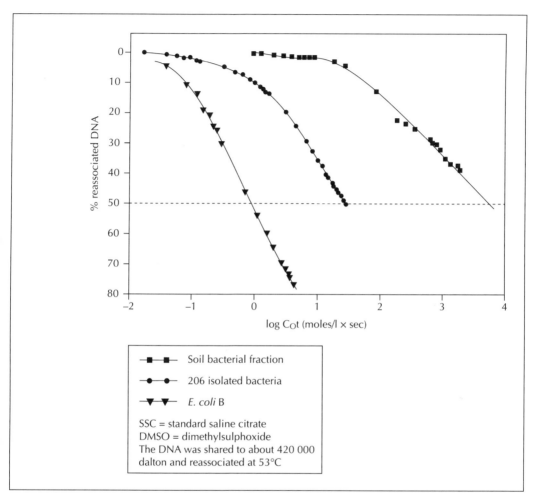

Table 4.2 Mole % G + C, reassociation kinetics ($C_0t_{1/2}$; moles/l x sec at half reassociation), complexity in base pairs (bp) and numbers of standard soil bacterial genomes (6.8×10^6 bp)

DNA source	Mole % G + C	$C_0t_{1/2}$	Complexity (bp)	Standard genomes
Escherichia coli	50.7	0.85[a]	4.1×10^6	0.6
E. coli	50.7	0.72[b]	4.1×10^6	0.6
206 isolates	59-65	28	1.4×10^8	20.6
Bacterial fraction	58.5	4700[b]	2.7×10^{10}	4000

Note: a In 4 x standard saline citrate (SSC), 30% dimethylsulphoxide (DMSO)
            b In 6 x standard saline citrate (SSC), 30% dimethylsulphoxide (DMSO)

CONCLUSION

For populations of isolated bacteria, there is good general agreement between phenetic diversity indices based on a limited number of phenotypic characters and genetic diversity based on DNA complexity, measured as reassociation rates and expressed as $C_0t_{1/2}$. The genetic diversity expressed as $C_0t_{1/2}$ is a good parameter for revealing the differences between a highly diverse population and one in which diversity has collapsed. The genetic diversity can be determined from the complexity of DNA isolated directly from the bacteria in soil, and is currently the only method for assessing the diversity of the total bacterial community. The diversity of this community in soil is very much higher than expected from work with cultured bacteria. Most of the bacteria in these environments are unknown.

References

Atlas, R.M. 1984. Diversity of microbial communities. In Marshall, K.C. (ed) *Advances in Microbial Ecology* (Vol. 7). New York, USA: Plenum Press.
Bak, A.L., Christiansen, C. and Stenderup, A. 1970. Bacterial genome sizes determination by DNA renaturation studies. *J. General Microbiology* 64: 377-80.
Britten, R.J. and Kohne, D.E. 1968. Repeated sequences in DNA. *Science* 161: 529-40.
Faegri, A., Torsvik, V.L. and Goksøyr, J. 1977. Bacterial and fungal activities in soil: Separation of bacteria and fungi by a rapid fractionated centrifugation technique. *Soil Biology and Biochemistry* 9: 105-12.
Fuhrman, J.A., Callum, K.M. and Davis, A.A. 1992. Novel major archaebacterial group from marine plankton. *Nature* 356: 148-49.
Giovannoni, S.J., Britschgi, T.B., Moyer, C.L. and Field, K.G. 1990. Genetic diversity in Sargasso Sea bacterioplankton. *Nature* 345: 60-62.
Holben, W.E., Jansson, J.K., Chelm. B.K. and Tiedje, J.M. 1988. DNA probe method for the detection of specific microorganisms in the soil bacterial community. *Applied and Environmental Microbiology* 54: 703-11.
Holder-Franklin, M.A., Thorpe, A. and Cormier, C.J. 1981. Comparison of numerical taxonomy and DNA-DNA hybridisation in diurnal studies of river bacteria. *Canadian J. Microbiology* 27: 1165-84.
Marmur, J., Rownd, R. and Schildkraut, C.L. 1963. Denaturation and renaturation of deoxyribonucleic acid. In Davidson, J.N. and Conn, W.E. (eds) *Progress in Nucleic Acid Research* (Vol. 1). New York, USA/ London, UK: Academic Press.
Sjåstad, K. 1979. Quantification of soil bacteria by immunotechnique. PhD thesis, Bergen University, Norway.
Steffan, R.J., Goksøyr, J., Bej, A.K. and Atlas, R.M. 1988. Recovery of DNA from soil and sediments. *Applied and Environmental Microbiology* 54: 2908-15.
Sørheim, R., Torsvik, V.L. and Goksøyr, J. 1989. Phenotypic divergences between populations of soil bacteria isolated on different media. *Microbial Ecology* 17: 181-92.
Thornton, H.G. 1922. On the development of a standardised agar medium for counting soil bacteria, with special regard to the repression of spreading colonies. *Annals of Applied Biology* 9: 241-74.
Torsvik, V.L. 1980. Isolation of bacterial DNA from soil. *Soil Biology and Biochemistry* 12: 15-21.
Torsvik, V., Salte, K., Sørheim, R. and Goksøyr, J. 1990a. Comparison of phenotypic diversity and DNA heterogeneity in a population of soil bacteria. *Applied and Environmental Microbiology* 56: 776-81.
Torsvik, V., Goksøyr, J. and Daae, F.L. 1990b. High diversity in DNA of soil bacteria. *Applied and Environmental Microbiology* 56: 782-87.
Ward, D.M., Weller, R. and Bateson, M.M. 1990. 16S rRNA sequences reveal numerous uncultured micro–organisms in a natural community. *Nature* 345: 63-65.
Wayne, L.G., Brenner, D.J., Colwell, R.R., Grimont, P.A.D., Kandler, O., Krichevsky, M.I., Moore, L.H., Murray, R.G.E., Stackebrandt, E., Starr, M.P. and Truper, H.G. 1987. Report of the ad hoc committee on the reconciliation of approaches to bacterial systematics. *International J. Systematic Bacteriology* 37: 463-64.

Beyond the Biomass
Edited by K. Ritz, J. Dighton and K.E. Giller
© 1994 British Society of Soil Science (BSSS)
A Wiley-Sayce Publication

CHAPTER 5

A new approach to direct extraction of microorganisms from soil

N.C. SMITH and D.P. STRIBLEY

It is axiomatic that only a small fraction of the total number of microorganisms in soil can be cultured. So, to obtain a representative sample of the microbial biomass for work on, for example, chemical analysis, fate of added isotopes and analysis of subsets of the microbial community using molecular biological techniques, it is necessary quantitatively to extract microorganisms from soil using physical methods.

The methods described to date for the extraction of microorganisms from soil involve an initial treatment to disperse inorganic material, and especially to disaggregate clay, followed by a beneficiation step in which organisms are separated from the bulk of clay before final purification by density gradient centrifugation (*see* Table 5.1 *overleaf*). The beneficiation step is critical because it controls the ease with which the whole procedure may be scaled up, determines the representativeness of the biomass sampled and affects the efficiency of the final purification. The methods currently used (centrifugation/ resuspension and elutriation) rely on differences in size and density between inorganic matter and organisms. They are unsatisfactory because centrifugation is slow and difficult to scale up, and elutriation yields impractically large volumes of elutriate from even a small sample of soil (Hopkins et al., 1991b).

It was considered that these problems might be overcome by exploiting other physical differences between organisms and soil inorganic matter. Aqueous two-phase partitioning (A2PP) is a separation technique which depends principally on differences in hydrophobicity of added moities (Albertsson, 1986), although other properties such as charge might be involved (Huddleston and Lyddiatt, 1990; Huddleston et al., 1991). It has been used for applications as diverse as the extraction of organelles from disrupted cells, the isolation of one cell type from a mixture (Albertsson, 1986) and the separation of proteins from cellular debris on an industrial scale (Datar and Rosén, 1986). The success of A2PP in the extraction of microorganisms from organic dust and peat (Ström et al., 1987) prompted the present investigation into its use as an alternative method for extracting non-hyphal microorganisms ($< 20\,\mu m$ diameter) from dispersed soil.

Table 5.1   **Some published methods for extraction of non-mycelial microorganisms from soil**

Dispersion stage	Beneficiation	Final purification	Authors
Waring blender (buffered solution)	Repeated centrifugation and resuspension	None	Faegri et al. (1977)
Top drive macerator or sonication or Colworth stomacher (buffered solutions)	Repeated centrifugation and resuspension	Percoll density gradient	Martin and Macdonald (1981)
Waring blender (buffered/detergent solution)	Repeated centrifugation and resuspension	Ludox density gradient	Bakken (1985)
Ion exchange resin in detergent solution	Elutriation	Percoll density gradient	Macdonald (1986a, b, c and d)
Ion exchange resin in detergent/polyethylene-glycol (PEG) solution	Repeated centrifugation and resuspension	None (cultured)	Herron and Wellington (1990)
Waring blender, then ion exchange resin in detergent solution	Repeated centrifugation and resuspension	Percoll density gradient	Hopkins et al. (1991a)
Ion exchange resin in PEG solution	Repeated centrifugation and resuspension	None (cultured)	Turpin et al. (1993)

MATERIALS AND METHODS

The soil used in this study was a silty-clay loam of the Batcombe series (Avery, 1964) from the nil treatment of the Chemical Reference Plots at Rothamsted, UK. It was first air-dried to approximately 15% (w/w) water content, and then passed through a 2 mm sieve and stored at 4°C.

We retained the technique used by Macdonald (1986a, b, c and d) for the dispersion of soils by ion exchange resin because of its proven efficacy (Hopkins et al., 1991a; Turpin et al., 1993). For A2PP, 6 g of dispersed and filtered (20 μm mesh) soil suspension were added to 10 g of 20% (w/w) dextran solution (average molecular mass 503 000 daltons), 15 g of 40% (w/w) polyethylene glycol (PEG) solution (average molecular mass 8000 daltons) and 69 g of water. This gave a biphasic solution with a top/bottom phase volume ratio of 4:1. The mixture was inverted 30 times and then allowed to settle until a stable interface was formed. Top phase liquid was removed by pipette, the mass removed was recorded and then replaced by an equal mass of equilibrated top phase (Albertsson, 1986) and the partitioning procedure repeated, and so on, for sequential extraction of the bottom phase. All operations were at 4°C.

Microorganisms were counted on black isopore membranes using epifluorescence microscopy after staining by acridine orange (Faegri et al., 1977; Kirchman et al., 1982). Fading was retarded using Citifluor® (Wynn-Williams, 1985). Inorganic matter in soils and top phases was determined from the mass remaining after ignition at 500°C for 18 hours. The frequency distributions of volume of particles were measured using a Mastersizer® E (Malvern Instruments Co.) laser particle size analyser. All water used was first distilled, and then filtered (0.22 μm pore size) and autoclaved.

RESULTS AND DISCUSSION

When the dispersed soil was partitioned by A2PP, inorganic matter was retained much more strongly by the bottom phase than were microorganisms (*see* Table 5.2). After four sequential extractions, 59% of the indigenous organisms present in the dispersed soil added to the system was found in the combined (PEG-rich) top phases, but only 3.8% of the inorganic material was present. Expressed on the basis of the mass of soil used for dispersion, the percentage recovery did not differ significantly, suggesting that few organisms were lost during filtration through the 20 μm mesh. This recovery is similar to that obtainable by differentiated centrifugation/resuspension (Hopkins et al., 1991a). Microorganisms in the top phase may be readily recovered, and the solutes diluted out, by diafiltration in a stirred cell filter of pore size 0.2 μm (data not shown).

Table 5.2   Efficiency of extraction of non-mycelial microorganisms (< 20 μm diameter) from dispersed Rothamsted soil using sequential aqueous two-phase partitioning (PEG/dextran system)

	Step number			
	1	**2**	**3**	**4**
Percentage in top phase of total number of microorganisms added to the system	30.7	12.8	7.5	8.2
Cumulative percentage in top phase	30.7	43.5	51.0	59.2
Percentage in top phase of mass of inorganic matter added to system	3.0	0.5	0.2	0.1

It might be argued that the efficiency of counting of microorganisms using direct microscopy is greater for extracted populations than for dispersed soils because the former are relatively free of obscuring material. This would lead to an over-estimation of the efficiency of extraction. However, the combined microbial counts of top and bottom phases from our A2PP fractionations were always within 10% of the total count from the dispersed soil added to the system.

Table 5.3 (*overleaf*) shows that extraction by A2PP increased the ratio of the number of organisms per unit mass inorganic matter by 15 times, yielding a relatively 'clean' preparation. By contrast, the method used by Hopkins et al. (1991a) increased this ratio only twofold. The inefficiency of centrifugation methods might result from the overlap in sedimentation velocities of the clay particles and organisms (Hopkins et al., 1991a) and/or co-sedimentation (hindered settling) of dense suspensions of clay and organisms (Macdonald, 1986a). These problems do not apply to A2PP; however, like centrifugation it will, of course, extract only those organisms that are fully dissociated from clay.

Most organisms in dispersed soil were present singly but an important fraction was present in agglomerates of various sizes (*see* Table 5.4 *overleaf*). We speculate that the dispersion method used here (Macdonald, 1986a) probably breaks up the large microcolonies found *in situ* in soil (Bone and Balkwill, 1986) and that the flocs observed are those which remain intact because of the presence of a surrounding polysaccharide capsule (Macdonald, 1986a). The populations in the A2PP top phase were similar to dispersed soil in their agglomerate distribution, but those obtained using the method

Table 5.3   Number of microorganisms per unit mass of inorganic matter recovered from dispersed and filtered Rothamsted soil[a], either by repeated centrifugation and resuspension (using the method described by Hopkins et al., 1991a) or by aqueous two-phase partitioning (A2PP)

	Number of microorganisms/mg inorganic matter	Increase in ratio: Number of microorganisms/mg inorganic matter
Dispersed soil (control)[a]	8.5×10^5 (5.1×10^4)[b]	0 (control)
Repeated centrifugation and resuspension	1.5×10^6 (1.2×10^5)	1.8x
A2PP[c]	1.3×10^7 (6.5×10^5)	15x

Note: a   Method described by Macdonald (1986a)
        b   Standard errors of replicate extractions (n = 3) are in parentheses
        c   Combined top phases from four sequential extractions

Table 5.4   Frequency distributions of number of microorganisms per agglomerate in Rothamsted soil after three treatments

Treatment	As single organisms	% of total number of microorganisms in groups of								
		2	3	4	5	6	7	8	9	≥10
Dispersion[a] by ion exchange resin in detergent	83.99	4.04	2.14	2.45	1.20	0.89	0.48	0.63	0.22	3.96
Top phase of aqueous two-phase partitioning (A2PP) treatment[b] of dispersed[a] soil	93.58	2.35	1.13	1.10	0.49	0.21	0.21	0.23	0.042	0.69
Repeated centrifugation and resuspension of dispersed soil (using the method described by Hopkins et al., 1991a)	98.95	0.77	0.17	0	0	0	0	0	0	0.29

Note: a   Method described by Macdonald (1986a)
        b   Combined top phases from four sequential extractions

described by Hopkins et al. (1991a) differed markedly in the absence of agglomerates of 3 to 9. More work is required to determine whether centrifugation disrupts agglomerates or selects for them. These data suggest that A2PP yields a more representative sample of the indigenous microbial population than does centrifugation.

Further work on the representativeness of the population of microorganisms extracted by A2PP is needed to establish the degree of sampling bias. For example, the staining technique could be improved: acridine orange does not allow differentiation of prokaryotes and eukaryotes. However, preliminary work in our laboratory (data not shown) indicates that the size frequency distribution of organisms extracted by A2PP does not differ significantly from that of dispersed soil.

Figure 5.1 shows that the particle size distributions of material in top and bottom phases of the partitioned dispersed soil differed strongly. There was little material with a particle size which was

Figure 5.1 **Frequency distribution of volume in relation to particle diameter in dispersed and filtered Rothamsted soils (a) before and (b) after extraction by aqueous two-phase partitioning (A2PP), measured by a laser particle size analyser**

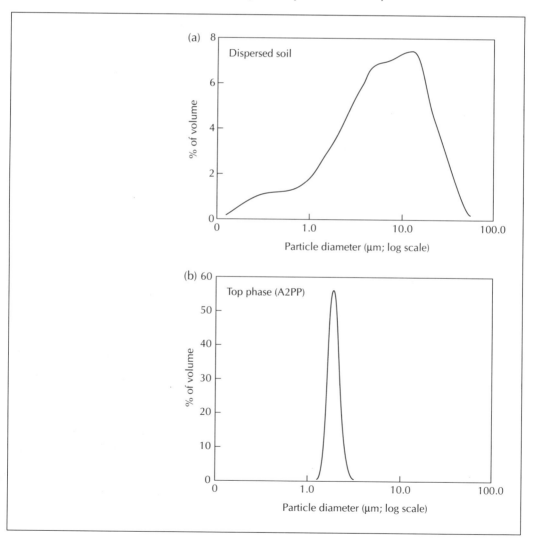

greater than 2 μm in the top phase, suggesting that the silt fraction (2-20 μm) had been retained by the bottom phase.

Future work will investigate the use of countercurrent distribution apparatus (Huddleston and Lyddiatt, 1990; Huddleston et al., 1991) to increase the speed and scale of A2PP as a method for extracting large samples of soil microflora. The use of options which are cheaper than dextran (Skuse et al., 1992) will also be explored.

CONCLUSION

Aqueous two-phase partitioning needs only simple apparatus and can yield cleaner and apparently more representative samples of soil microorganisms than is the case with the much slower technique of repeated centrifugation and resuspension. Its principal advantage is that it may be scaled up readily to prepare large samples of microorganisms for studies on, for example, chemical analyses, cellular physiology and molecular biology.

Acknowledgements

We are grateful to Malvern Instruments Ltd for the particle size analysis of samples and for permission to publish the results.

References

Albertsson, P.A. 1986. *Partition of Cell Particles and Macromolecules*. New York, USA: John Wiley.

Avery, B.W. 1964. The soils and land use of the district around Aylesbury and Hemel Hempstead. *Memoir of the Soil Survey of Great Britain*. London, UK: HMSO.

Bakken, L.R. 1985. Separation and purification of bacteria from soil. *Applied and Environmental Microbiology* 49: 1482-87.

Bone, T.L. and Balkwill, D.L. 1986. Improved flotation technique for microscopy of *in situ* soil and sediment microorganisms. *Applied and Environmental Microbiology* 51: 462-68.

Datar, R. and Rosén, C.-G. 1986. Studies on the removal of *Escherichia coli* cell debris by aqueous two-phase polymer extraction. *J. Biotechnology* 3: 207-19.

Faegri, A., Torsvik, V.L. and Goksøyr, J. 1977. Bacterial and fungal activities in soil: Separation of bacteria and fungi by a rapid fractionated centrifugation technique. *Soil Biology and Biochemistry* 9: 105-12.

Herron, P.R. and Wellington, E.M.H. 1990. A new method for extraction of streptomycete spores from soil and application to the study of lysogeny in sterile amended and nonsterile soil. *Applied and Environmental Microbiology* 56: 1406-12.

Hopkins, D.W., Macnaughton, S.J. and O'Donnell, A.G. 1991a. A dispersion and differential centrifugation technique for representatively sampling microorganisms from soil. *Soil Biology and Biochemistry* 23: 217-25.

Hopkins, D.W., O'Donnell, A.G. and Macnaughton, S.J. 1991b. Evaluation of a dispersion and elutriation technique for sampling microorganisms from soil. *Soil Biology and Biochemistry* 23: 227-32.

Huddleston, J.G. and Lyddiatt, A. 1990. Aqueous two-phase systems in biochemical recovery. *Applied Biochemistry and Biotechnology* 26: 249-79.

Huddleston, J., Veide, A., Köhler, K., Flanagan, J., Enfors, S.-O. and Lyddiatt, A. 1991. The molecular basis of partitioning in aqueous two-phase systems. *Trends in Biotechnology* 9: 381-88.

Kirchman, D., Sigda, J., Kapuscinski, R. and Mitchell, R. 1982. Statistical analysis of the direct count method for enumerating bacteria. *Applied and Environmental Microbiology* 44: 376-82.

Macdonald, R.M. 1986a. Sampling soil microflora: Dispersion of soil by ion exchange and extraction of specific microorganisms from suspension by elutriation. *Soil Biology and Biochemistry* 18: 399-406.

Macdonald, R.M. 1986b. Sampling soil microflora: Optimisation of density gradient centrifugation in Percoll to separate microorganisms from soil suspensions. *Soil Biology and Biochemistry* 18: 407-10.

Macdonald, R.M. 1986c. Sampling soil microflora: Problems in estimating concentration and activity of suspensions of mixed populations of soil microorganisms. *Soil Biology and Biochemistry* 18: 411-16.

Macdonald, R.M. 1986d. Extraction of microorganisms from soil. *Biological Agriculture and Horticulture* 3: 361-65.

Martin, N.J. and Macdonald, R.M. 1981. Separation of non-filamentous microorganisms from soil by density gradient centrifugation in Percoll. *J. Applied Bacteriology* 51: 243-51.

Skuse, D.R., Norris-Jones, R., Yalpani, M. and Brooks, D.E. 1992. Hydroxypropyl cellulose/poly(ethylene glycol)-CO-poly(propylene glycol) aqueous two-phase systems: System characterisation and partition of cells and proteins. *Enzyme and Microbial Technology* 14: 785-89.

Ström, G., Palmgren, U. and Blomquist, G. 1987. Separation of organic dust from microorganism suspensions by partitioning in aqueous polymer two-phase systems. *Applied and Environmental Microbiology* 53: 860-63.

Turpin, P.E., Maycroft, K.A., Rowlands, C.L. and Wellington, E.M.H. 1993. An ion-exchange based extraction method for the detection of salmonellas in soil. *J. Applied Bacteriology* 74: 181-90.

Wynn-Williams, D.D. 1985. Photofading retardant for epifluorescence microscopy in soil micro-ecological studies. *Soil Biology and Biochemistry* 17: 739-46.

Beyond the Biomass
Edited by K. Ritz, J. Dighton and K.E. Giller
© 1994 British Society of Soil Science (BSSS)
A Wiley-Sayce Publication

CHAPTER 6

Analysis of microbial communities by flow cytometry and molecular probes: Identification, culturability and viability

C. EDWARDS, J. DIAPER, J. PORTER, D. DEERE and R. PICKUP

Most natural environments are severely nutrient-limited (Gottschal, 1990), with the result that microorganisms grow very slowly or not at all. The consequences of slow growth are long cell cycle times and an increase in the proportion of energy that must be diverted to maintenance functions. Slow growth also imposes severe stress on the cells so that growth can become unbalanced, with long gaps in the cell cycle during which chromosome replication is absent (Chesboro, 1990). Traditional methods for isolating organisms from natural environments include dilution plating or the most probable number (MPN) method, but these techniques are time consuming and often fail to identify component species. The success of recovery and identification methods will also depend to a certain extent upon the properties and complexity of the natural environment under study. Factors such as population density, temperature, pH, redox and surface sorption characteristics all have a role (O'Donnell and Hopkins, 1993). Amongst natural environments, soil is perhaps the most complex, and a number of studies have attempted to develop methods for extracting viable microorganisms in representative numbers from different soil types (O'Donnell and Hopkins, 1993; Turpin et al., 1993; Wellington et al., 1993). Developments in molecular genetic techniques have recently come to the fore as a potential means of dissecting microbial community structure and diversity. However, these methods rarely indicate whether the target microorganisms are viable and active. This situation is complicated further by the proposal that some bacteria can adopt a viable but non-culturable (VBNC) state, making it difficult to assess their true status in natural environments and rendering them essentially refractory to conventional isolation methods. New methods are therefore required that can rapidly identify and enumerate viable organisms.

A potential way forward is to marry the emerging molecular techniques with more traditional biochemical analyses. Flow cytometry has great potential in this respect. This chapter describes some of the main features of a flow cytometer, its many potential applications and how these methods may be combined with cell sorting to isolate selectively those organisms of interest, even when these organisms are present in diverse and active heterogeneous populations.

FLOW CYTOMETRY PRINCIPLES OF OPERATION

Flow cytometers combine the techniques of microscopy and biochemical methods to analyse thousands of cells per second. Cells are presented singly into an intense and focused light beam by injection into a pressurised stream of fluid (the sheath fluid). As they pass through the excitation beam, light will be scattered and this can be detected by closely aligned photomultiplier tubes. If the cells are stained with appropriate fluorochromes, then providing the incident light beam is of the correct excitation wavelength, each cell will emit fluorescence as it passes through the beam, the intensity of which can be quantified by fluorescence detectors. Most instruments have two light scatter detectors. One measures light scattered at the cell surface which will be a reflection of size; this is known as narrow angle light scatter (NALS). The other measures light that is scattered after passage through the cells and is potentially a reflection of internal structure as refractility; this is known as wide angle light scatter (WALS). The intensity of fluorescence and light scatter is converted into electric signals which can be displayed as event histograms that represent the range of light intensities as a number of channels or electronic bins versus the numbers of particles counted for each light intensity.

The grading of light intensity can be computer-manipulated so that some channels are not monitored, allowing the study of sub-populations in a heterogeneous population of cells. This facility is exploited in more sophisticated instruments for physically sorting a sub-population of interest. This is achieved by converting the sheath fluid into droplets; those containing the particle of interest are electronically charged and are deflected by charged plates into a collection vessel. The criterion used for cell sorting can be size but more usually the instrument is programmed to sort those cells that are fluorescent. More detailed descriptions of flow cytometry and cell sorting, and their applications to microbiology, are available in reviews by Kell et al. (1991), Edwards et al. (1992) and Edwards (1993).

APPLICATIONS OF FLOW CYTOMETRY AND CELL SORTING

Table 6.1 summarises some of the major applications of flow cytometry with respect to the analysis of bacteria. Light scatter is of little use for the analysis of soil organisms unless they can be extracted and the bulk of the particulates removed. One possible way forward would be to fix the cells in soil prior to extraction, after which they could withstand relatively harsh procedures for producing a purified cell suspension. Fixation does not appear to change the light scatter properties of bacteria. However, as demonstrated by Allman et al. (1992), it is difficult to interpret light scatter data. The correlation between size and NALS is frequently poor and WALS rarely reflects changes in internal structure, although these can be followed using electron microscopy. Light scatter is useful, however, for enumerating the total cells present in those samples that give distinctive profiles.

By far the greatest potential of flow cytometry in soil studies involves the use of fluorescent dyes or conjugates. Some of the major stains are listed in Table 6.1. Macromolecular composition can be assessed using Hoechst 33342 for DNA; mithramycin/ethidium bromide staining can also give information regarding chromosome number per cell. Fluorescein isothiocyanate (FITC) is specific for cell protein and has been used to monitor total cell protein during the growth of *Azotobacter vinelandii* (Allman et al., 1990). Numerous dyes are also available, such as propidium iodide for estimating the nucleic acid content of bacteria which will be predominantly ribosomal RNA, the amounts of which are closely coupled with the growth rate and nutritional status of the cell. We have found that flow cytometric measurements of changes in the amounts of these macromolecules correlate well with corresponding values obtained using chemical methods (J. Diaper and C. Edwards, unpubl.).

Table 6.1   Applications of flow cytometry for measuring properties of bacteria

Parameter	Applications	Reference
Light scatter		
Narrow angle (NALS)	Discrimination of bacteria in mixtures	Allman et al. (1992)
Wide angle (WALS)	Monitoring morphological changes in *Azobacter vinelandii*	Allman et al. (1990)
Fluorescence		
Dyes		
Hoescht 33342	Determination of chromosome number; measurement of G-C content	Steen et al. (1990) Sanders et al. (1990)
Fluorescein isothicyanate (FITC)	Protein content of bacteria	Allman et al. (1990)
Propidium iodide	Nucleic acid content	Edwards et al. (1992)
Mithramycin/ethidium bromide	Determination of chromosome number	
Rhodamine 123	Enumeration of viable bacteria	Kaprelyants and Kell (1992); Diaper et al. (1992)
Conjugates		
Fluorescein- , naphthol-coumaryl-linked substrates	Enzyme activity	Kell et al. (1991)
Chemichrome B, carboxy fluorescein	Enumeration of viable bacteria	Pinder et al. (1993)
Antibody-fluorescein (FITC-IgG)	Enumeration and identification of bacteria in soil	Page and Burns (1991)
Rhodamine or fluorescein — 16S rRNA	Species-specific labelling and bacterial identification	Zarda et al. (1991)
Autofluorescence		
Chlorophyll	Discrimination of photosynthetics	Edwards et al. (1992)
F420	Identification and enumeration of methanogens	Edwards (1993)
Cells	Assessment of energy status of cells	Kell et al. (1991)
Cell sorting		
Fluorescence	Isolation of viable *Escherichia coli* from lake water	Porter et al. (unpubl.)

Another facet of the technique is that it enables scientists to conduct multiparametric analysis so that three-dimensional plots of fluorescence, light scatter and cell numbers can be obtained. Such plots can reveal the distribution of a fluorescent dye or probe in cells of different ages and therefore cell cycle analysis is possible. Samples from soil in which there is no meaningful light scatter response can therefore still be analysed, providing that a suitable fluorochrome is available.

Fluorescent conjugated molecules are also available. Perhaps the most exciting are the fluorescent-antibody and 16S ribosomal RNA (rRNA)-fluorochrome conjugates that can be used specifically to label cells which can then be identified and enumerated by flow cytometry. Unfortunately, the one with the most potential is also the most difficult to apply. 16S rRNA molecules theoretically should not only allow the classification of organisms into the domains Archaea, Bacteria and Eucarya (Winker and Woese, 1991) but also contain sufficient variable regions to allow discrimination at species level; the feasibility of this approach has been described by Amann et al. (1990). However, the intensity of the fluorescence signal after hybridisation of fluorescent 16S rRNA oligonucleotide probes is low even in laboratory-grown cultures and hardly detectable in cells from natural environments. This reflects the tendency of rRNA levels to fluctuate in response to nutritional status, as noted earlier. Recently, Zarda et al. (1991) attempted to circumvent the problem by labelling oligonucleotides with digoxigenin (DIG), which, after hybridisation, can be detected using fluorescently labelled antibodies. However, the results were disappointing, with some bacteria exhibiting enhanced fluorescence (but not to the expected degree) and others showing no increase at all.

Some microorganisms produce their own fluorescent molecules and these can be used as a basis for their identification and enumeration. This is especially true for methanogens which contain F420, a deazaflavin which has a specific role in methanogenesis and which is fluorescent. Methanogens cultured in the laboratory are highly fluorescent but we have not tested this property for their flow cytometric detection in soil samples. A potential problem is that light tends to reduce the fluorescence of methanogens, possibly by a photobleaching reaction. Bacterial cells in general produce a measurable amount of fluorescence via the reduced flavin and pyridine nucleotides. This has been studied by Kell et al. (1991) for *Micrococcus luteus*, in which a level of heterogeneity of energy status throughout the cell cycle could be identified.

Viability

The analysis of microorganisms from natural environments is hampered by the inability of traditional microbiological techniques to recover more than 1-10% of the indigenous population. Recently, there has been a growing realisation that some heterotrophic non-differentiating bacteria can adopt a VBNC state, generally as a result of some environmental cue such as temperature or nutrient starvation (Roszak and Colwell, 1987; McKay, 1992). Flow cytometric methods have been developed to enumerate viable bacteria even though they cannot be cultured. Rhodamine 123 (Rh123) has proved to be especially sensitive for the demonstration of viability (Kaprelyants and Kell, 1992) because it is taken up only by cells able to generate a membrane potential (a property of viable cells only). A comparison of cell numbers obtained from colony-forming units (CFUs) with those obtained by flow cytometric enumeration of Rh123-stained cells shows good agreement for laboratory-grown cultures (Diaper et al., 1992). The sensitivity of Rh123 is shown in Figure 6.1. Figure 6.1a shows stained cells of *Salmonella pullorum* analysed in the presence or absence of gramicidin, an agent that abolishes membrane potential. Dead cells are not stained, as shown in Figure 6.1b; the figure shows a mixture of live and formaldehyde-treated *S. pullorum* stained with Rh123. This dye can be used with some success to analyse bacteria extracted from soil. However, the innate background fluorescence due to soil components can make flow cytometric visualisation difficult.

An alternative approach is to employ colourless dye conjugates such as carboxyfluorescein diacetate. These conjugates work by uptake into cells, where they are cleaved intracellularly to release fluorescein (or its derivatives) which is retained only by intact cells; these cells then become fluorescent. Figure 6.2 shows the response of *Staphylococcus aureus* to increasing concentrations of three conjugates with BCEFC-AM, being particularly effective in staining the cells. These dyes may

Figure 6.1   Detection of viable bacteria using flow cytometry, showing fluorescence histograms of (a) *Salmonella pullorum* stained with rhodamine 123 and (b) a mixed culture of viable and formaldehyde-fixed cells of *S. pullorum*

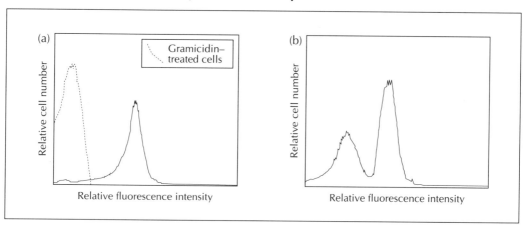

Figure 6.2   Time course for staining viable cells of *Staphylococcus aureus* with fluorescent dye conjugates

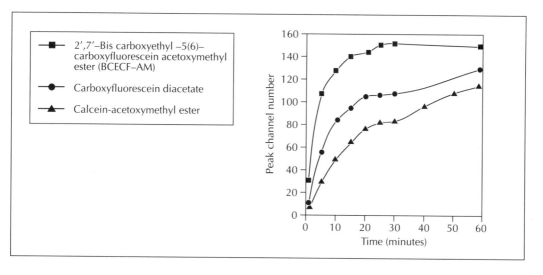

be useful for environmental applications because they fluoresce only inside the cells; non-specific binding is less of a problem than for rhodamine.

Detection

Because of their specificity, fluorescent antibodies have great potential for the flow cytometric detection and enumeration of target bacteria in soil and other environments. This potential is illustrated by the ability of flow cytometry to enumerate and identify *S. aureus* inoculated at different densities into lake water that contained approximately 10^5 bacteria/ml. Immunofluorescent staining with FITC-IgG, which binds to protein A, was employed. Figure 6.3 shows the close agreement between *S. aureus* CFU and FITC-IgG-labelled cells enumerated by flow cytometry. Page and Burns (1991) have confirmed the potential of antibody detection in soil. They used a monoclonal antibody prepared against a *Flavobacterium* sp. for identification and to monitor its numbers by flow cytometry when released into soil. Some problems were encountered in enumeration, which was ascribed to non-specific binding of the antibody to soil particulates.

Figure 6.3   **Relationship between colony-forming units (CFUs) determined on nutrient agar and flow cytometer-determined counts of FITC-IgG stained *Staphylococcus aureus* after inoculation at different densities into non-sterile lake water**

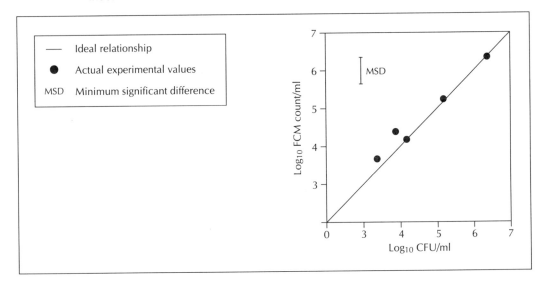

Cell sorting

As far as we know, the use of cell sorting in ecology has not been demonstrated for bacteria. Preliminary experiments have indicated a great deal of promise in using this method for the recovery of viable cells and the isolation of a target sub-population. Figure 6.4 illustrates an experiment in which a mixture of *Escherichia coli* (65%) and *S. aureus* (35%) was prepared in the presence of FITC-labelled IgG, which is bound by protein A on the surface of *S. aureus*. Figure 6.4a shows that the two species are easily discriminated, *S. aureus* (boxed as 1 in Figure 6.4a) being more fluorescent than *E. coli*. The cytometer was then programmed to sort selectively all the particles with fluorescence of sufficient intensity to fall in the area enclosed by box 1. The results after cell sorting are shown in Figure 6.4b; this figure indicates that *S. aureus* has been concentrated from the mixture, so that the proportions are now 95% *S. aureus* and only 5% *E. coli*, as determined by flow cytometry. We have subsequently shown that similar immunofluorescent labelling techniques can be used for selectively enriching viable wild type *E.coli*, using flow cytometric cell sorting of water samples taken near a sewage outlet.

Figure 6.4 Enrichment of flow cytometric cell sorting of immunofluorescently labelled *Staphylococcus aureus* from a mixture of *S. aureus* (35%) and *E. coli* (65%), showing dot plots of (a) the mixture treated with FITC-labelled IgG and (b) the sample sorted for enrichment of *S. aureus*

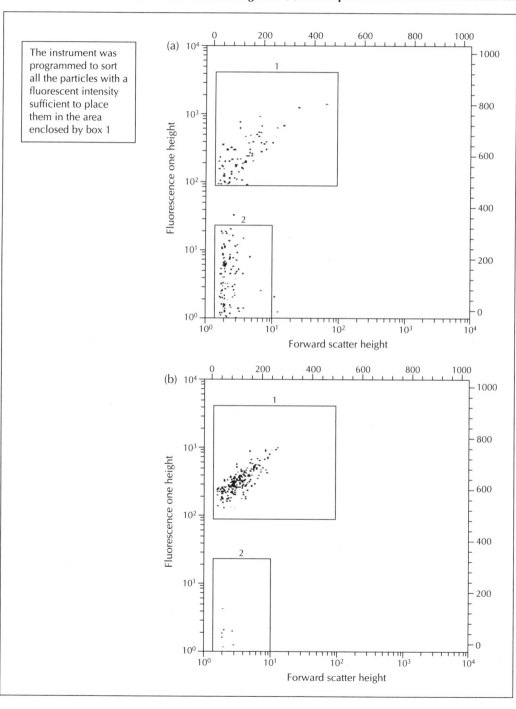

CONCLUSION

In this chapter we have sought to illustrate the potential applications of flow cytometry for analysing bacterial community structure in natural environments. A potential problem is that bacteria need to be relatively free from particulate material before analysis. This is to avoid blockage of fluidic systems of flow cytometers and to ensure effective labelling of cells. However, many of the methods for analysis will work well with fixed cells and these will obviously withstand a much harsher purification protocol. Improved methods for extraction of bacteria from soils are also being developed, as reported recently by Turpin et al. (1993). These authors showed that a Chelex + polyethylene glycol (PEG) protocol was particularly efficient in recovering viable salmonellae from soil.

There is no doubt that flow cytometry holds great promise for applications in natural environments. The major areas for future development include analysis of viability even in the absence of culturability and the use of fluorescent probes, particularly antibodies. Fluorescent oligonucleotides prepared against 16S rRNA sequences also have great promise but will require developmental work to enhance the numbers of copies of the target sequence or improvement of the intensity of fluorescence carried by the probe. The application of cell sorting has the power, in theory at least, to separate and enrich for any target cell or sub-population, providing that these can be made selectively fluorescent.

Acknowledgements

The experimental work described in this chapter was supported by grants from the National Environment and Research Council (NERC), UK and the Ministry of Agriculture, Food and Fisheries (MAFF), UK.

References

Allman, R., Hann, A.C., Phillips, A.P., Martin, K.L. and Lloyd, D. 1990. Growth of *Azotobacter vinelandii* with correlation of coulter cell size, flow cytometric parameters and ultrastructure. *Cytometry* 11: 822-31.

Allman, R., Hann, A.C., Manchee, R. and Lloyd, D. 1992. Characterisation of bacteria by muliparameter flow cytometry. *J. Applied Bacteriology* 73: 438-44.

Amann, R.I., Krumholz, L. and Stahl, D.A. 1990. Fluorescent-oligonucleotide probing of whole cells for determinative, phylogenetic and environmental studies in microbiology. *J. Bacteriology* 172: 762-70.

Chesboro, W., Arbidge, M. and Eiffert, F. 1990. When nutrient limitation places bacteria in the domains of slow growth: Metabolic, morphologic and cell cycle behaviour. *FEMS Microbiology Ecology* 74: 103-20.

Diaper, J.P., Tither, K. and Edwards, C. 1992. Rapid assessment of bacterial viability by flow cytometry. *Applied Microbiology and Biotechnology* 38: 268-72.

Edwards, C. 1993. The significance of *in situ* activity on the efficiency of monitoring methods. In Edwards, C. (ed) *Monitoring Genetically Manipulated Microorganisms in the Environment*. Chichester, UK: John Wiley.

Edwards, C., Porter, J., Saunders, J.R., Diaper, J., Morgan, J.A.W. and Pickup, R.W. 1992. Flow cytometry and microbiology. *SGM Quarterly* 19: 105-08.

Gottschal, J.C. 1990. Phenotypic response to environmental changes. *FEMS Microbiology Ecology* 74: 93-102.

Kaprelyants, A.D. and Kell, D.B. 1992. Rapid assessment of bacterial viability and vitality using rhodamine 123 and flow cytometry. *J. Applied Bacteriology* 72: 410-22.

Kell, D.B., Ryder, M.M., Kaprelyants, A.S. and Westerhoff, H.V. 1991. Quantifying heterogeneity: Flow cytometry of bacterial culture. *Antonie van Leeuwenhoek* 60:145-58.

McKay, A.M. 1992. Viable but non-culturable forms of potentially pathogenic bacteria in waters. *Letters in Applied Microbiology* 14: 129-35.

O'Donnell, A.G. and Hopkins, D.W. 1993. Extraction, detection and identification of genetically engineered microorganisms from soils. In Edwards, C. (ed) *Monitoring Genetically Manipulated Microorganisms in the Environment*. Chichester, UK: John Wiley.

Page, S. and Burns, R.G. 1991. Flow cytometry as a means of enumerating bacteria introduced into soil. *Soil Biology and Biochemistry* 23: 1025-28.

Pinder, A.C., Edwards, C., Clarke, R.G., Diaper, J.P. and Poulter, S.A.G. 1993. *Detection and Enumeration of Viable Bacteria by Flow Cytometry*. SAB Technical Series. Oxford, UK: Blackwell.

Roszak, D.B. and Colwell, R.R. 1987. Survival strategies in the natural environment. *Microbiological Reviews* 51: 365-79.

Sanders, C.A., Yajko, D.H., Kyun, W., Langlois, R.G., Nassos, P.S., Fulwyler, M.J. and Hadley, W.K. 1990. Determination of guanine-plus-cytosine content of bacterial DNA by dual-laser flow cytometry. *J. General Microbiology* 136: 359-65.

Steen, H.B., Skarstad, K. and Boye, E. 1990. DNA measurements of bacteria. *Methods in Cell Biology* 33: 519-26.

Turpin, P.E., Maycroft, K.A., Rowlands, C.L. and Wellington, E.M.H. 1993. Viable but non-culturable salmonellas in soil. *J. Applied Bacteriology* 74: 421-47.

Wellington, E.M.H., Herron, P.R. and Cresswell, N. 1993. Gene transfer in terrestrial environments and the survival of bacterial inoculants in soil. In Edwards, C. (ed) *Monitoring Genetically Manipulated Microorganisms in the Environment*. Chichester, UK: John Wiley.

Winker, S. and Woese, C.R. 1991. A definition of the domains Archaea, Bacteria and Eucarya in terms of small subunit ribosomal-RNA characteristics. *Systematic and Applied Microbiology* 14: 305-10.

Zarda, B., Amann, R., Wallner, G. and Schleifer, K.H. 1991. Identification of single bacterial cells using digoxigenin-labelled rRNA-targeted oligonucleotides. *J. General Microbiology* 137: 2823-30.

Beyond the Biomass
Edited by K. Ritz, J. Dighton and K.E. Giller
© 1994 British Society of Soil Science (BSSS)
A Wiley-Sayce Publication

CHAPTER 7

Physiological analysis of microbial communities in soil: Applications and limitations

T.-H. ANDERSON

Historically, physiological quotients have been used as experimental tools for microbial ecological studies in batch or chemostat cultures. Their application to soil microbial ecology is relatively new. The development of techniques for quantifying the microbial biomass of soils (Jenkinson and Powlson, 1976; Anderson and Domsch, 1978) was required before it became possible to measure rate constants of the performance of microbial communities. In this context, major studies have been conducted on the use of carbon as an energy source for maintenance and on the impact of environmental conditions (*see* Table 7.1 *overleaf*).

Metabolic quotients could be powerful tools in understanding energy transfer or principles of homeostasis at the microbial community level. These, in turn, could elucidate principles of terrestrial ecosystem development. It is important to know whether higher microbial diversity is coupled with a more efficient use of carbon and nutrients, and what the stress costs are in terms of energy loss.

In this chapter, the applications of physiological quotients are reviewed and some of the pitfalls associated with the use of metabolic rate constants are discussed.

USE OF PHYSIOLOGICAL QUOTIENTS IN SOIL STUDIES

Until recently, the assessment of microbial activity in soils was confined to respiratory or enzymatic measurements based on units of soil weight or volume. However, reactions of microbial communities to a changing environment can be more effectively quantified by determining metabolic rate constants (Anderson and Gray, 1991) because they allow a more direct comparison of the microbial communities.

Physiological quotients were first applied in batch or chemostat cultures to gain insight into the growth characteristics or nutrient use of single cell cultures (Pirt, 1975). It can be assumed that, in principle, factors such as nutrient availability, pH, pO_2 and pCO_2 will control cellular metabolism under various soil conditions (Tempest and Neijssel, 1978) in a similar manner. However, different species compete for the same substrate, which certainly will affect the metabolism of the individual members of a community as well as total community metabolism. Earlier experiments on the carbon uptake, growth rate and maintenance rate of soil cultures have shown that rate constants can be lower

Table 7.1   Examples of studies in soil microbiology in which metabolic quotients have been applied

Field of study	Metabolic quotient[a]	Reference
Maintenance carbon requirement	m, qCO_2	Anderson and Domsch (1985a, b) Chapman and Gray (1986)
Carbon turnover	qCO_2, μ, $K_{mGLUCOSE}$, Y, m, qD, C_{mic}/C_{org}	Anderson and Domsch (1986a, b)
Soil management	qCO_2, qD, V_{max}, C_{mic}/C_{org}	Anderson and Domsch (1986a, 1989a, 1990) Insam et al. (1989) Anderson and Gray (1990, 1991) Santruckova and Straskraba (1991) Beck (1991) Grimm (1992)
Impact of climate and temperature	qCO_2, C_{mic}/C_{org}, qD	Insam (1990) Anderson and Domsch (1985b, 1986a) Wardle and Parkinson (1990) Joergensen et al. (1990)
Impact of soil texture and soil compaction	qCO_2, qD	Anderson and Domsch (1989b) Kaiser et al. (1991)
Impact of heavy metals	qCO_2	Brookes and McGrath (1984) Killham (1985) Nordgren et al. (1988) Fliessbach (1991)
Ecosystems, ecosystem theory	qCO_2, qD, C_{mic}/C_{org}	Insam and Domsch (1988) Insam and Haselwandter (1989) Ding et al. (1992) Anderson and Domsch (1993)
Impact of soil animals	qCO_2	Wolters (1991) Wolters and Joergensen (1991)

Note: a   m = maintenance coefficient; qCO_2 = metabolic quotient or specific respiration rate; μ = specific growth rate; K_m = Michaelis-Menten constant; V_{max} = maximum specific uptake rate; Y = growth yield; qD = specific death rate; C_{mic}/C_{org} = microbial carbon/organic carbon ratio expressed as a percentage of microbial carbon to total organic carbon

by one order of magnitude and more compared with values from *in vitro* studies of cells during exponential growth (Anderson and Domsch, 1985b, 1986a). That is, competition for substrates and carbon-limiting conditions are additional controlling factors on metabolic rate constants not normally expressed under pure culture conditions.

Furthermore, the equations for quantifying any metabolic rate were formulated for batch culture as a closed system and for chemostat culture as an open system with respect to nutrient flow (Pirt, 1975). Both systems will promote characteristic growth behaviour which will be reflected in different metabolic rate constants. In general, soil in a natural environment can be considered as an open system (that is, constituents of the system can enter and leave it). As soon as portions of a soil system are taken into the laboratory, the natural flow of nutrients and gas is interrupted and the sample must be treated as a simple batch culture with definite metabolic reactions, particularly with respect to cell death (because of nutrient limitation and animal feeding). This explains the generally observed fact that the average annual level of microbial biomass in a natural site, with energy and/or nutrient flow in a quasi-steady state, remains almost constant year by year, whereas the same soil taken into the laboratory will rapidly lose microbial biomass over time because of the exhaustion of resources during incubation (Anderson and Domsch, 1986a, 1990). However, metabolic rates produced in closed systems can provide information on how an environment affects a microbial community (Anderson and Gray, 1991). It is assumed that any particular set of environmental conditions results in a characteristic rate constant of activity (*see* Figure 7.1) because the reaction of most cells in a community to nutritional,

Figure 7.1 **Working hypothesis for the application of metabolic quotients in ecosystem development at the synecological level**

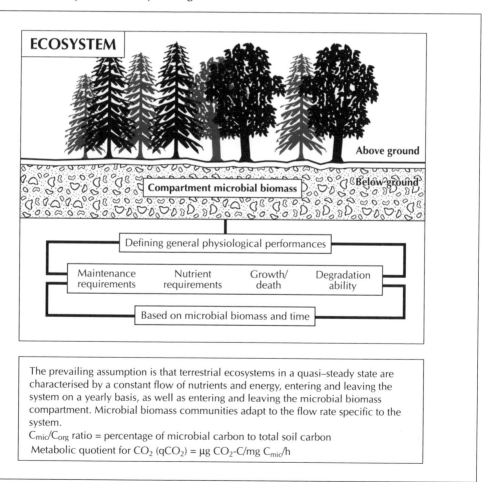

ECOSYSTEM

Above ground

Below ground

Compartment microbial biomass

Defining general physiological performances

Maintenance requirements | Nutrient requirements | Growth/death | Degradation ability

Based on microbial biomass and time

The prevailing assumption is that terrestrial ecosystems in a quasi–steady state are characterised by a constant flow of nutrients and energy, entering and leaving the system on a yearly basis, as well as entering and leaving the microbial biomass compartment. Microbial biomass communities adapt to the flow rate specific to the system.

C_{mic}/C_{org} ratio = percentage of microbial carbon to total soil carbon

Metabolic quotient for CO_2 (qCO_2) = µg CO_2-C/mg C_{mic}/h physical or chemical effects reflects precisely this particular set of conditions. This assumption, however, is valid only to a certain extent. This will be discussed in more detail later in this chapter.

For quantification of microbial performances in soil cultures under laboratory conditions, most of the time equations for metabolic rates (Pirt, 1975) have been applied:

$$x = \frac{1}{q} \frac{dy}{dt}$$

where:

x	=	amount of microbial biomass
q	=	metabolic quotient
y	=	amount of substrate or product
dy/dt	=	metabolic rate

This equation is valid only if performances of the actual (non-growing) biomass are to be characterised.

The percentage of microbial carbon in total soil carbon (C_{mic}/C_{org} ratio) is an additional sensitive parameter which can indicate changes of nutrient availability because the microbial biomass responds more quickly than total C_{org} to any changes (Anderson and Domsch, 1986b, 1989a).

ECOSYSTEM THEORIES

The input/output analysis of nutrient fluxes within ecosystems has been used to understand the dynamics of ecosystem development. Mineralisation of organic substrates and the release of nutrients or elements result from heterotrophic activity of the microbial biomass. This makes the microbial biomass a vital sink and controlling agent in terrestrial ecosystems, a factor which has stimulated extensive research on biomass turnover (for example, Paul and Voroney, 1980; Jenkinson and Ladd, 1981; McGill et al., 1981). The use of metabolic quotients, particularly those which reflect carbon flow through microbial biomass such as the metabolic quotient for CO_2 (qCO_2) and the C_{mic}/C_{org} ratio, could be helpful tools in elucidating annual element or energy fluxes through the microbial biomass.

Table 7.2   Expected trends in ecosystem development extrapolated onto the microbial biomass community

Ecosystem attributes	Developmental stages	Mature stages
Biomass supported/unit energy[a]	Low	High
Species diversity	Low	High
Biochemical diversity	Low	High
Entropy	High	Low
Changed into:		
Respiration/microbial biomass/time (qCO_2)	High	Low
Microbial biomass supported/unit carbon (C_{mic}/C_{org} ratio)	Low	High

Note: a   From Odum (1969)

The view expressed by Odum (1969) on ecosystem development with respect to energy flow implies that ecosystems have phases of successions until a climax is reached, whereby under climax conditions the highest degree of homeostasis or stability is obtained. The communities of successional stages differ in gross production and community respiration. In early stages of succession, for example, the biomass supported per unit energy is low, while in mature stages it is high. This implies that in early stages more energy is used per unit biomass. It would be challenging to test whether this idea can be extrapolated onto microbial community development. For the microbial community, this could be measured by determining community respiration per unit biomass. Furthermore, if community respiration is high, less carbon is available for microbial biomass production, and the C_{mic}/C_{org} ratio should therefore be lower in young systems. In addition, Krebs (1985) cites Margalef who postulated that mature systems differ in degree of species diversity, which causes a higher degree of stability and efficiency (see Table 7.2).

In their study on chronosequences of reclamation sites, Insam and Domsch (1988) used the qCO_2 to follow soil development. They observed a significant decrease of the qCO_2 as the soils aged, a result which agrees with the ecological theory. The same reaction was demonstrated for agricultural plots with a long-term cropping history (Anderson and Domsch, 1990). Insam and Haselwandter (1989) followed the development of the qCO_2 of very young (1 year) to old (~1000 years) moraine soils. Again, old soils showed decreased community respiration. In their study of chronosequences of beech and beech/oak forest stands, Anderson and Domsch (1993) showed that the prevailing soil pH had a tremendous and unexpected influence on the qCO_2 and the C_{mic}/C_{org} ratio (see Table 7.3). Apparently,

Table 7.3 Relationship between soil pH and (a) the metabolic quotient for CO_2 (qCO_2) and (b) the C_{mic}/C_{org} ratio of microbial biomass in forest sites

Type of forest		pH_{KCl} 2-3		pH_{KCl} 5-7	
qCO_2 (mg CO_2-C/mg C_{mic}/h)					
Picea		24.3×10^{-4}	n = 86	17.6×10^{-4}	n = 1
(all sites)	SD[a]	3.5×10^{-4}		SD 3.0×10^{-4}	
Fagus (> 100 years)		23.2×10^{-4}	n = 26	15.1×10^{-4}	n = 24
(> 100 years)	SD	4.0×10^{-4}		SD 3.0×10^{-4}	
Fagus/Quercus		22.0×10^{-4}	n = 53	13.2×10^{-4}	n = 25
(> 100 years)	SD	4.0×10^{-4}		SD 3.4×10^{-4}	
C_{mic}/C_{org} ratio					
Picea		0.29	n = 14	2.0	n = 1
(> 100 years)	SD	0.11			
Fagus		0.78	n = 26	2.0	n = 28
(> 100 years)		0.13		SD 0.5	
Fagus/Quercus		0.82	n = 55	2.6	n = 24
(> 100 years)	SD	0.40		SD 0.5	

Note: a   Standard deviation

the qCO_2 is a sensitive stress indicator, which demonstrates that soil conditions which may influence metabolic quotients must be comparable when analysing ecosystem successions. When these conditions were found in the study of beech soils (A_h horizon) and a very narrow pH range (~ 6.0 to > 7.0) analysed, the data again showed a decrease of the qCO_2 with age of the stand and a concomitant increase of microbial biomass per unit of soil carbon (*see* Figure 7.2). Since site variation in forest soils is high, the values are not statistically significant because only a small number of soils fell into the category of a pH value greater than 6.0. There are also indications that communities from ecosystems of higher diversity (mixed forests) show a lower qCO_2 and a higher C_{mic}/C_{org} ratio, which would agree with Odum's hypothesis (*see* Table 7.3). Similar observations have been made by Ding et al. (1992).

Figure 7.2   **The qCO_2 (μg CO_2-C/mg C_{mic}/h) and the percentage of microbial carbon in total soil carbon (C_{mic}/C_{org}) in beech forest stands of successive ages**

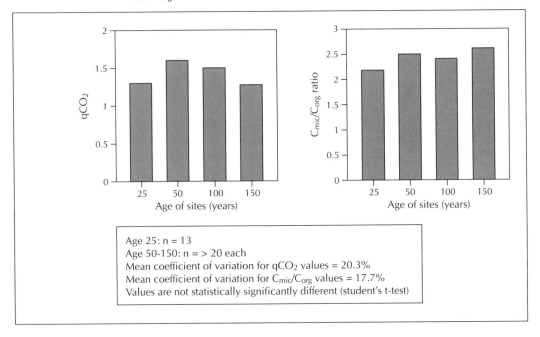

Age 25: n = 13
Age 50-150: n = > 20 each
Mean coefficient of variation for qCO_2 values = 20.3%
Mean coefficient of variation for C_{mic}/C_{org} values = 17.7%
Values are not statistically significantly different (student's t-test)

Odum (1990) discussed a case study of an aquatic ecosystem (freshwater lake) where expected trends related to energetics were not apparent. For example, under stress through acidification, community respiration did not vary significantly. Microbes may be more sensitive indicators of environmental change than higher organisms.

QUANTIFYING ENVIRONMENTAL EFFECTS ON THE MICROBIAL COMMUNITY

The application of metabolic quotients (that is, specific rate constants of activity) can be a meaningful tool in quantifying external factors which may influence soil microorganisms. Changes in performances at the species level which can result from shifts in population structure should be reflected at the community level, which means that total community metabolism will change. The major areas of interest in which metabolic quotients have been applied most frequently are:

- determination of performances of microbial communities under standardised soil conditions

- impact of soil texture and structure

- impact of soil management

- impact of faunal biomass

Understanding the maintenance carbon requirement (m) of microbial communities is of major importance if carbon fluxes from this terrestrial compartment are to be correctly quantified, since it defines the amount of carbon which is unavailable for growth but necessary for cell maintenance functions (Pirt, 1965). Attempts to measure m under soil conditions were made by Anderson and Domsch (1985a, b). While glucose-activated cells showed values comparable with those of actively growing cells known from *in vitro* studies, values obtained from 'natural' dormant biomass were two to three orders of magnitude lower, depending upon the temperature. The maintenance studies under soil conditions were very elaborate, and thus the qCO_2 was used in subsequent work for quantifying catabolic metabolism. However, as pointed out in the original work (Anderson and Domsch, 1985b), m and qCO_2 were not identical. The qCO_2 value reflected a maintenance demand which was at least one order of magnitude higher (under similar incubation temperatures) than m. This problem remains unresolved and needs further research. The conclusion drawn at the time was that a large part of m must be met by endogenously derived carbon sources in mainly dormant soil organisms. This would mean that the endogenous carbon pool needs to be considered to the same extent as the portions of carbon for anabolic and catabolic processes. Cryptic growth (Postgate, 1967) could be a factor in maintenance studies. However, no measurable influence of cryptic growth was observed in the short-term maintenance studies (Anderson and Domsch, 1985b). The maintenance demand of soil microbial communities at a particular time is probably met by both exogenous and endogenous reserve material (Herbert, 1958; Dalenberg and Jager 1981, 1989). These two carbon sources may have different residence times within the soil community. Such factors need to be considered in carbon flux determinations.

Temperature also affects metabolic quotients. For example, when temperatures increase there is an increase in m (Anderson and Domsch, 1985b; Chapman and Gray, 1986), the specific death rate (qD) (Anderson and Domsch, 1985b; Joergensen et al., 1990) and the specific respiration rate (qCO_2) (Anderson and Domsch, 1986a).

Besides temperature and pH, there are indications that the prevailing C_{org} content can affect metabolic quotients. Soils with high C_{org} levels may show a low m value (Anderson and Domsch, 1985b). Also, qD is lower at high C_{org} levels (Anderson and Gray, 1991). It is unclear whether these are metabolic reflections of the total microbial biomass of a soil or are a reflection of shifts at the species level (such as changes in the fungi/bacteria ratio). In their studies on microbial biomass composition, using the selective inhibition technique, Anderson and Domsch (1973, 1975) found a fungi/bacteria ratio of 80/20 in field soils. They later demonstrated that at maximal CO_2 evolution, 1 mg biomass carbon will release 25 µl CO_2/hour, assuming a fungi/bacteria ratio of approximately that order; cells of different ages showed different specific metabolic rates (Anderson and Domsch, 1978) (*see* Table 7.4 *overleaf*). A drastic shift in favour of bacterial biomass could change rate constants. This must be borne in mind in the application of metabolic quotients to comparative studies.

Table 7.4   Specific respiration rate of soil fungi according to the physiological age of mycelia[a]

qCO_2 (µl CO_2/mg C/h)			
Early linear	Middle linear	Late linear	Stationary
42.2	22.8	20.3	17.5

Note: a   Average CO_2 release of 12 soil fungi at 22°C in glucose buffer solution
Source: Anderson and Domsch (1978)

CONCLUSION

The use of metabolic quotients has great potential in improving our understanding of the development of microbial communities and the ecosystems they inhabit. This approach may help to resolve the question of how community structure relates to its function, and to decide if the appearance or disappearance of populations within a community is, in terms of energy, advantageous or not. In combination with other techniques, such as those discussed elsewhere in this book, the use of physiological quotients in comparative analyses of microbial communities under distinct environmental conditions may help us to understand the nature of community development.

    If maximum potential changes in response to an environmental impact, this is a direct indication that the biomass or fractions of it must have changed. Yet, under field conditions the biomass is not static. It undergoes changes, including changes in its physiological status, during the year. It is likely that the qualitative changes of the biomass pool demonstrated above are caused by shifts in the ratio between the active and dormant portion (Nedwell and Gray, 1987). It can be assumed that at times of substrate surplus the biomass will grow above an annual mean, with a subsequent increase of the active

Figure 7.3   Cyclical fluctuations of the metabolic quotient for CO_2 (qCO_2) around a
theoretical mean in relation to the size of the microbial biomass and the shifts in
its active and dormant fractions over one year

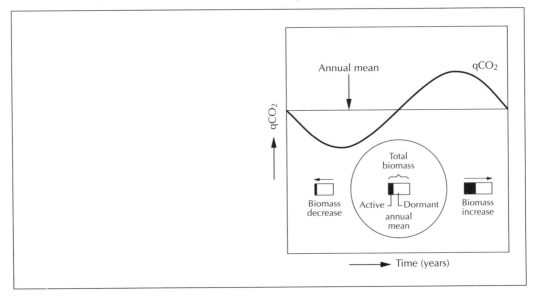

fraction and a higher state of physiological activity per unit cell mass. The reverse would be true if the biomass fell below the annual mean. Figure 7.3 depicts this graphically by using CO_2 evolution as an example. Annual mean values could stand for an ecophysiological constant for comparative purposes. In this context, it is of paramount importance when working with metabolic quotients for comparative purposes that time of site sampling, soil storage, soil handling and preparation must follow the same schedule. Quantification of an annual mean microbial biomass pool, and its corresponding physiological performance under controlled laboratory conditions, can only approximate rates of processes at the natural site but it can help elucidate underlying principles of microbial community behaviour.

References

Anderson, J.P.E. and Domsch, K.H. 1973. Quantification of bacterial and fungal contributions to soil respiration. *Archiv für Mikrobiologie* 93: 113-27.

Anderson, J.P.E. and Domsch, K.H. 1975. Measurement of bacterial and fungal contributions to respiration of selected agricultural and forest soils. *Canadian J. Microbiology* 21: 315-22.

Anderson, J.P.E. and Domsch, K.H. 1978. A physiological method for the quantitative measurement of microbial biomass in soils. *Soil Biology and Biochemistry* 10: 215-21.

Anderson, T.-H. and Domsch, K.H. 1985a. Maintenance carbon requirements of actively metabolising microbial populations under *in situ* conditions. *Soil Biology and Biochemistry* 17: 197-203.

Anderson, T.-H. and Domsch, K.H. 1985b. Determination of ecophysiological maintenance carbon requirements of soil microorganisms in a dormant state. *Biology and Fertility of Soils* 1: 81-89.

Anderson, T.-H. and Domsch, K.H. 1986a. Carbon assimilation and microbial activity in soil. *Zeitschrift für Pflanzenernaehrung und Bodenkunde* 149: 457-68.

Anderson, T.-H. and Domsch, K.H. 1986b. Carbon-link between microbial biomass and soil organic matter. In Megusar, F. and Gantar, M. (eds) *Proc. Fourth International Symposium on Microbial Ecology*. Ljubljana, Yugoslavia: Mladinska Knjiga.

Anderson, T.-H. and Domsch, K.H. 1989a. Ratios of microbial biomass carbon and total organic carbon in arable soils. *Soil Biology and Biochemistry* 21: 471-79.

Anderson, T.-H. and Domsch, K.H. 1989b. Der Einfluss des Bodengefüges auf mikrobielle Stoffwechselleistungen. *Mitteilungen der Deutschen Bodenkundlichen Gesellschaft* 59 (1): 523-28.

Anderson, T.-H. and Domsch, K.H. 1990. Application of eco-physiological quotients (qCO_2 and qD) on microbial biomasses from soils of different cropping histories. *Soil Biology and Biochemistry* 22: 251-55.

Anderson, T.-H. and Domsch, K.H. 1993. The metabolic quotient for CO_2 (qCO_2) as a specific activity parameter to assess the effects of environmental conditions, such as pH, on the microbial biomass of forest soils. *Soil Biology and Biochemistry* 25: 393-95.

Anderson, T.-H. and Gray, T.R.G. 1990. Soil microbial carbon uptake characteristics in relation to soil management. *FEMS Microbiology Ecology* 74: 11-20.

Anderson, T.-H. and Gray, T.R.G. 1991. The influence of soil organic carbon on microbial growth and survival. In Wilson, W.S. (ed) *Advances in Soil Organic Matter Research*. Melksham, UK: Redwood Press.

Beck, T. 1991. Einsatzmöglichkeiten der substratinduzierten Atmungsmessung bei bodenmikrobiologischen Untersuchungen. *Mitteilungen der Deutschen Bodenkundlichen Gesellschaft* 66 (1): 459-62.

Brookes P.C. and McGrath, S.P. 1984. Effects of metal toxicity on the size of the soil microbial biomass. *J. Soil Science* 35: 341-46.

Chapman, S.J. and Gray, T.R.G. 1986. Importance of cryptic growth, yield factors and maintenance energy in models of microbial growth in soil. *Soil Biology and Biochemistry* 18: 1-4.

Dalenberg, J.W. and Jager, G. 1981. Priming effect of small glucose additions to ^{14}C-labelled soil. *Soil Biology and Biochemistry* 13: 283-89.

Dalenberg, J.W. and Jager, G. 1989. Priming effect of some organic additions to ^{14}C-labelled soil. *Soil Biology and Biochemistry* 21: 443-48.

Ding Ming Mao, Yi Wei Min, Liao Lan Yu, Martens, R. and Insam, H. 1992. Effect of afforestation on microbial biomass and activity in soils of tropical China. *Soil Biology and Biochemistry* 24: 865-72.

Fliessbach, A. 1991. Auswirkungen mehrjähriger, abgestufter Klärschlammgaben auf die mikrobielle Biomasse des Bodens und ihre Aktivität. PhD thesis, University of Göttingen, Germany.

Grimm, J. 1992. Einfluss langjährig differenzierter Bewirtschaftung auf mikrobielle biomasse und Aktivität im Boden und ihre Beziehungen zu Standortfaktoren und Ertrag. PhD thesis, Berlin University, Germany.

Herbert, D. 1958. Some principles of continuous culture. In Tunevall, G. (ed) *Recent Progress in Microbiology: VI International Congress for Microbiology, Sweden.* Stockholm, Sweden: Almquist and Wiksell.

Insam, H. 1990. Are the soil microbial biomass and basal respiration governed by the climatic regime? *Soil Biology and Biochemistry* 22: 525-32.

Insam, H. and Domsch, K.H. 1988. Relationship between soil organic carbon and microbial biomass on chronosequences of reclamation sites. *Microbial Ecology* 15: 177-88.

Insam, H. and Haselwandter, K. 1989. Metabolic quotient of the soil microflora in relation to plant succession. *Oecologia* 79: 171-78.

Insam, H., Parkinson, D. and Domsch, K.H. 1989. Influence of macroclimate on soil microbial biomass. *Soil Biology and Biochemistry* 21: 211-21.

Jenkinson, D.S. and Powlson, D.S. 1976. The effect of biocidal treatments on metabolism in soil. V. A method for measuring soil biomass. *Soil Biology and Biochemistry* 8: 189-202.

Jenkinson, D.S. and Ladd, J.N. 1981. Microbial biomass in soil: Measurement and turnover. In Paul, E.A. and Ladd, J.N. (eds) *Soil Biochemistry* (Vol. 5). New York, USA: Marcel Dekker.

Joergensen, R.G., Brookes, P.C. and Jenkinson, D.S. 1990. Survival of the soil microbial biomass at elevated temperatures. *Soil Biology and Biochemistry* 22: 1129-36.

Kaiser, E.-A., Walenzik, G. and Heinemeyer, O. 1991. The influence of soil compaction on decomposition of plant residues and on microbial biomass. In Wilson, W.S. (ed) *Advances in Soil Organic Matter Research*, 253-266. Melksham, UK: Redwood Press.

Killham, K. 1985. A physiological determination of the impact of environmental stress on the activity of microbial biomass. *Environmental Pollution* 38: 283-94.

Krebs, C.J. 1985. *Ecology — The Experimental Analysis of Distribution and Abundance.* New York, USA: Harper and Row.

McGill, W.B., Hunt, W.H., Woddmansee, R.G. and Reuss, J.O. 1981. Phoenix — A model of the dynamics of carbon and nitrogen in grassland soils. In Clark, F.E. and Rosswall, T. (eds) *Terrestrial Nitrogen Cycling.* Stockholm, Sweden: Ecological Bulletin.

Nedwell, D.B. and Gray, T.R.G. 1987. Soil and sediments as matrices for microbial growth. *Symposia of the Society for General Microbiology* 41: 21-54.

Nordgren, A., Bååth, E. and Söderström, B. 1988. Evaluation of soil respiration characteristics to assess heavy metal effects on soil microorganisms using glutamic acid as substrate. *Soil Biology and Biochemistry* 20: 949-54.

Odum, E.P. 1969. The strategy of ecosystem development. *Science* 164: 262-70.

Odum, E.P. 1990. Field tests of ecosystem-level hypotheses. *Trends in Ecology and Evolution* 5: 204-05.

Paul, E.A. and Voroney, R.P. 1980. Nutrient and energy flows through soil microbial biomass. In Ellwood, D.C., Hedger, J.N., Latham, M.J., Lynch, J.M. and Slater, J.H. (eds) *Contemporary Microbial Ecology.* London, UK: Academic Press.

Pirt, S.J. 1965. The maintenance energy of bacteria in growing cultures. In *Proc. Royal Society of London (Series B)* 163: 224-31.

Pirt, S.J. 1975. *Principles of Microbe and Cell Cultivation.* Oxford, UK: Blackwell.

Postgate, S.J. 1967. Viability measurements and the survival of microbes under minimum stress. *Advances in Microbial Physiology* 1: 2-23.

Santruckova, H. and Straskraba, M. 1991. On the relationship between specific respiration activity and microbial biomass in soils. *Soil Biology and Biochemistry* 23: 525-32.

Tempest, D.W. and Neijssel, O.M. 1978. Eco-physiological aspects of microbial growth in aerobic nutrient-limited environments. *Advances in Microbial Ecology* 2: 105-53.

Wardle, D.A. and Parkinson, D. 1990. Interactions between microclimate variables and the soil microbial biomass. *Biology and Fertility of Soils* 9: 273-80.

Wolters, V. 1991. Biological processes in two beech forest soils treated with simulated acid rain: A laboratory experiment with *Isotoma tigrina* (Insecta, Collembola). *Soil Biology and Biochemsitry* 23: 381-90.

Wolters, V. and Joergensen R.G. 1991. Microbial carbon turnover in beech forest soils at different stages of acidification. *Soil Biology and Biochemistry* 23: 897-902.

Beyond the Biomass
Edited by K. Ritz, J. Dighton and K.E. Giller
© 1994 British Society of Soil Science (BSSS)
A Wiley-Sayce Publication

CHAPTER 8

A community-level physiological approach for studying microbial communities

J.L. GARLAND and A.L. MILLS

Microbial ecology lacks a clear paradigm regarding the relationship between structure and function in microbial communities. It is often unclear if functional changes in microbial communities are the result of physiological responses or structural shifts within the community. This shortcoming is due to the paucity of studies which have intensively analysed both the structure and function of microbial communities. In fact, two distinct approaches in microbial ecology have evolved. Some workers prefer to concentrate solely on the microbial processes that contribute to ecosystem function. For example, in many studies samples of the ecosystem of interest are amended with a radio-labelled compound at tracer level and the rate of a particular reaction is determined without regard to the dynamics of microbial community structure. Conversely, many workers will determine the effect of a compound or a process on the community by isolating as many organisms from the affected site as possible, identifying the isolates and publishing lists of species with little regard for the dynamics of ecosystem function. Obviously, these are the extremes in approach in microbial ecology, but successful work that reconciles these extremes has not been reported. Many researchers recognise that the integration of these two approaches is necessary in order to understand the behaviour of microbial communities.

The greatest impediment to conducting integrated studies is the lack of effective methods for evaluating microbial community structure. Isolate-based techniques offer a limited, biased view of communities due to the selective nature of cultural media and the non-culturable status of many microorganisms. Even if a 'perfect' isolation medium was developed, the labour-intensive nature of the isolate-based approach would still severely limit the spatial and temporal intensity of sampling and the number of individuals analysed per sample. Recent evidence indicates that the genotypic diversity of environmental samples is so great that characterisation of individuals, even with unbiased techniques, would capture only a small percentage of existing information (Torsvik et al., 1990; Torsvik, *Chapter 4, this volume*).

We report here on the development and application of a community-level approach to the study of microbial communities. The approach does not rely on the isolation of organisms, but attempts to examine the distribution of physiological characteristics in samples of intact communities. The ability to distinguish among samples from different habitats and along spatial gradients within habitats based on patterns of physiological abilities has been previously reported (Garland and Mills, 1991). Recently, we have explored temporal trends within a specific type of habitat (the rhizosphere of hydroponically grown plants) in order to further examine the effectivenenss of this approach. The ability to detect the effects of inoculation, and its significance in community-level assay, are also discussed. Finally, we consider how structural and functional information may be concurrently evaluated using this approach.

MATERIALS AND METHODS

Sole-carbon-source utilisation of a range of substrates for microbial growth was used for community-level physiological tests in these studies. We employed Biolog® GN microplates (Biolog, Inc.) in a rapid assay of the utilisation of a large number of carbon sources. A description of the system and a list of specific substrates is given in Bochner (1989) and Garland and Mills (1991).

The community-level approach is based on the direct inoculation of whole environmental samples in the Biolog microplates, and the subsequent development of colour production in wells (absorbance at 590 nm) as a result of microbial respiration. Earlier research (Garland and Mills 1991) has indicated that distinctive patterns among samples arise from differences in the relative degree of colour production in wells and not from differences in which of the wells change colour. Accordingly, the resolution of the assay is enhanced if colour is quantitatively measured and if plates are read at approximately the same point in colour development. Our current method involves reading plates at 2-hour intervals from the point of initial colour formation, using an automated microplate reader (Biotek® Model EL 320). The patterns of colour production among samples of a given average well colour development (AWCD) are then compared using multivariate statistical techniques. We report the results of principal component analysis (PCA) in this chapter, but we have also used detrended correspondence analysis and clustering algorithms (Pielou, 1984). An AWCD of 0.75 absorbance units was selected because at this point in colour development the majority of wells show positive responses and the most responsive wells are approaching maximal colour development.

The first study evaluated the ability to resolve samples along temporal gradients, based on patterns of colour development. White potato (*Solanum tuberosum* L. cv. Norland) was grown in a large, atmospherically closed plant growth chamber (Corey and Wheeler, 1992). The chamber contained four separate nutrient delivery systems, each supporting 16 plant growth trays approximately 0.25 m^2 in size. Rhizosphere samples were taken from a single tray in each nutrient delivery system 10, 37, 71, and 104 days after planting. Root samples were added to 0.1% sodium pyrophosphate and hand shaken for 2 minutes in the presence of glass beads. The resulting suspension was diluted 20-fold in 0.85% NaCl and directly inoculated into Biolog GN plates.

A subsequent study evaluated the inoculation effects on rhizosphere communities from dwarf wheat (*Triticum aestivum* L. cv. Yecora Roja). Four replicate, benchtop hydroponic systems containing 2 l of nutrient solution and 29 cm^2 of growing area were used in these studies. All systems were decontaminated prior to planting (hardware rinsed with 5% HNO$_3$, and seeds surface-sterilised with 10% bleach). Two of the four systems were inoculated with 1 ml of an enrichment culture. The enrichment was produced by seeding an aqueous wheat extract medium with a sample from a freshwater impoundment on the Kennedy Space Center, and incubating this for 48 hours at 25°C. Samples were frozen at -70°C in glycerol for storage prior to the start of the experiment. Rhizosphere samples were collected and analysed as above at 7, 14, and 21 days after planting.

Samples from an aerobic bioreactor which contained ground, inedible wheat residues were analysed for comparative purposes. The overall volume of the reactor was 8 l and the substrate loading rate was 40 g/l. Average residence time of solids was 8 days, effected by daily replacement of 1/8 of the culture solution. Samples were taken after 7, 14, and 21 days of operation, mixed with an equal volume of 0.1 % sodium pyrophosphate, blended for 30 seconds, allowed to settle to remove large particulates, and diluted in 0.85% NaCl before inoculation into Biolog plates.

RESULTS AND DISCUSSION

Evaluation of the relationships in the patterns of colour production among samples requires multivariate statistical analysis to reduce the multidimensional data into a more readily interpretable form. PCA projects the multivariate data onto a reduced number of dimensions (principal components), which capture a subset of the total variance in the data set (Pielou, 1984). The major relationships among samples can be visualised by examining the relative similarity of their principal component (PC) scores. The cause for the differences among samples can be determined by evaluating the correlation of the original variables to the principal components.

Several important temporal trends were apparent in the white potato rhizosphere data, as shown in Figure 8.1.

Figure 8.1 Results of principal component analysis (PCA) of rhizosphere samples from white potato grow-out

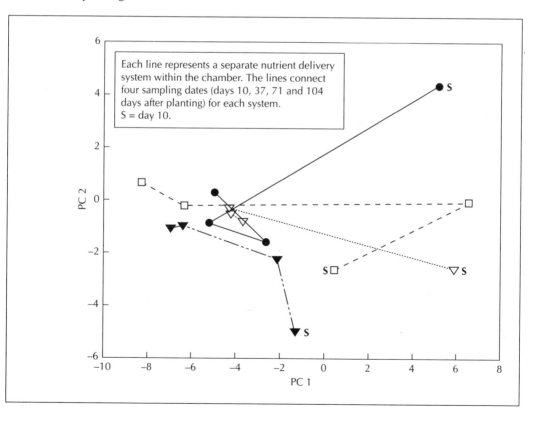

Coordinates for the first two principal components indicate:

- a relatively consistent change in all four systems with time (that is, separation of early from late samples)

- a greater similarity among samples from the separate nutrient delivery systems at the end of the grow-out compared with the beginning

- a decreasing rate of change among samples from the same nutrient delivery system over time

The presence of relatively consistent changes among all four systems over time indicates that the community-level assay can resolve temporal shifts within microbial communities. Both the convergence in the community-level response from the different systems and the decrease in the rate of change within each system over time suggest that the rhizosphere community tends to approach a stable state. In plant communities, stochastic dispersal and colonisation events may cause significant initial differences in community composition among spatially distinct sites, but consistent environmental conditions select for similar communities over time (Peet, 1991). The community-level data from the chamber study could be interpreted as reflecting a similar trend in microbial community composition among four spatially distinct sites (levels) with similar environmental conditions.

Figure 8.2   **Results of principal component analysis (PCA) of replicated 21-day wheat grow-outs in nutrient delivery systems previously inoculated with an enrichment of microbial cells or uninoculated**

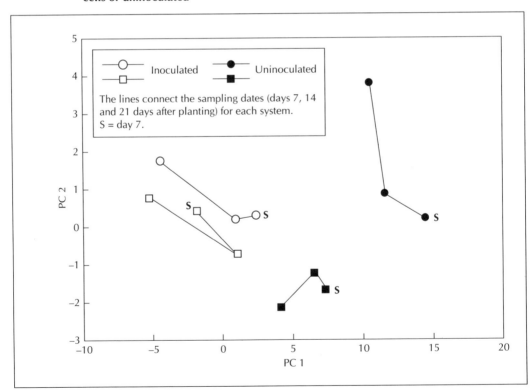

The inoculation study tested the hypothesis that the community-level signal reflects community composition by determining whether inoculation could produce differences in the signals. Inoculated and uninoculated systems showed distinctive responses; PC 1 scores were higher for uninoculated systems (*see* Figure 8.2). The patterns of carbon source utilisation were also more similar between the two inoculated systems than the two uninoculated ones. This can be expressed as a smaller standard deviation in principal component scores among all samples from the inoculated (2.81, 0.73 and 0.54 for PCs 1, 2 and 3, respectively) than from the uninoculated systems (3.34, 2.03 and 1.13, respectively). This trend can also be viewed as the greater separation between scores for the two uninoculated systems in Figure 8.2. These data indicate that the community-level signal distinguishes between inoculation treatments and is less variable among replicate systems receiving an identical enrichment inoculum.

The distinctive patterns in colour production caused by inoculation, in systems with identical functional requirements for microbial communities, indicate that the community-level assay reflects microbial community structure. The assay apparently can resolve temporal trends and the influence of inoculum, suggesting that short- and long-term effects of colonisation factors (natural dispersal or anthropogenic manipulation) on the stability of community structure can be studied using community-level assay. Longer-term, replicated studies in a variety of systems are necessary to verify this finding.

The correlation of carbon-source variables with PCs can be used to evaluate the underlying basis for differences among samples. It may also reflect the selective processes giving rise to a particular community structure. For most data sets we have analysed (Garland and Mills, 1991; Garland, 1992; J.L. Garland and A.L. Mills, unpubl.), PCs did not represent broad classes of compounds (such as amino acids, carboxylic acids and carbohydrates). In most cases, individual compounds from a variety of classes were most strongly associated with the differences among samples. In some data sets, however, more broadly based shifts in carbon-source utilisation have been detected. As shown in Figure 8.3, results of the PCA of bioreactor and rhizosphere samples showed a clear separation of the

Figure 8.3   **Results of principal component analysis (PCA) of samples from a bioreactor fed inedible wheat biomass and from white potato rhizosphere**

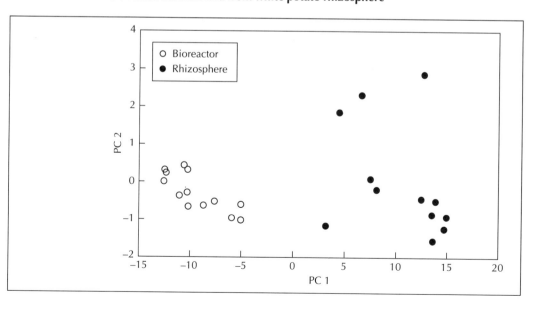

two habitats along the PC 1 axis; that axis explained 48% of the variance in the data (*see* Figure 8.3). Many of the amino acids and carboxylic acids were positively correlated with PC 1 (that is, greater utilisation by rhizosphere samples), while many of the carbohydrates were negatively correlated with PC 1 (that is, greater utilisation by bioreactor samples).

These differences in carbon-source utilisation can also be shown by averaging the absorbance values for all substrates within a given class for each sample (*see* Table 8.1). Bioreactor samples had a higher AWCD response for carbohydrates, but a lower AWCD response for amino acids and carboxylic acids. In the bioreactor samples, the presence of the broad functional shift is probably due to the selection of community members for utilisation of the range of carbohydrate materials found in the plant material. In many natural environments, the spectrum of compounds present, at least intermittently, may be much broader. Shifts in community structure would therefore be related more to individual compounds than to classes of compounds. Shifts in community structure related to broad classes of compounds should be positively related to duration of exposure to a specific class of compounds (that is, continuous culture with a constant feed stream). Further work with laboratory continuous culture systems should help to define these relationships more clearly. Functional shifts may be more readily detected in environmental samples if alternative carbon sources are selected which are known to vary in their availability across slected gradients and to be utilised by fewer types of microorganisms.

Table 8.1   Average well colour development (AWCD) response for different types of substrates in bioreactor and potato rhizosphere samples

| | Substrate type | | |
	Carbohydrate	Carboxylic acid	Amino acid
Bioreactor[a]	1.19[c] (0.08)	0.46 (0.06)	0.52 (0.08)
Rhizosphere[b]	0.85 (0.10)	0.75 (0.07)	0.94 (0.13)

Note: a   Values for 15 samples from an aerobic bioreactor containing inedible wheat residue
        b   Values for 20 white potato rhizosphere samples
        c   Numbers represent mean and standard deviation in absorbance at all wells containing specified substrate type; data are from plates with AWCD of 0.75 absorbance units

CONCLUSION

The data presented in this chapter, particularly those which relate to the effects of inoculation, indicate that the community-level assay can be used as a measure of community structure. Concurrent evaluation of samples using this assay and separate measures of community structure (isolate-based techniques, genetic analysis, and quantification of structural 'markers') would help verify this finding. The rapid nature of the assay may make concurrent evaluation of community structure (based on patterns of colour production) and function (such as growth rates and degradation rates) feasible. Functional information can also be derived directly from evaluating the types of carbon-source utilisation. We would encourage the evaluation of alternative methodologies using the community-level approach. Potential modifications include the use of different carbon sources coupled with the redox dye used in this research, or suitable alternative methods to produce a set of physiological (or structural) responses.

References

Bochner, B. 1989. Breathprints at the microbial level. *ASM News* 55: 536-39

Corey, K.A. and R.M. Wheeler. 1992. Gas exchange in NASA's biomass production chamber. *Bioscience* 42: 503-09

Garland, J.L. 1992. Carbon flux with hydroponically based plant growth systems: Analysis of microbial community structure and function. PhD dissertation, University of Virginia, USA.

Garland, J.L., and A.L. Mills. 1991. Classification and characterisation of heterotrophic microbial communities on the basis of patterns of community-level sole-carbon-source utilisation. *Applied and Environmental Microbiology* 57: 2351-59.

Peet, R.K. 1991. Community structure and function. In Glenn-Lewin, D.C., Peet, R.K. and Veblen, T.T. (eds) *Plant Succession: Theory and Prediction*. London, UK: Chapman and Hall.

Pielou, E.C. 1984. *The Interpretation of Ecological Data*. New York, USA: John Wiley.

Torsvik, V., J. Goksøyr, and F.L. Daae. 1990. High diversity in DNA of soil bacteria. *Applied and Environmental Microbiology* 56: 782-87.

Beyond the Biomass
Edited by K. Ritz, J. Dighton and K.E. Giller
© 1994 British Society of Soil Science (BSSS)
A Wiley-Sayce Publication

CHAPTER 9

Fingerprinting bacterial soil communities using Biolog microtitre plates

A. WINDING

The bacterial community structure in soil microbial ecosystems is crucial when estimating the effects of introduced microorganisms or the effects of heavy metals and xenobiotic chemicals. Dormant bacteria might increase in activity, cell size and number in response to environmental changes. The description of bacterial communities has traditionally been based on the colony-forming units (CFUs), total number of bacteria (AODC), and biochemical testing of bacterial isolates from the soil (Atlas, 1983). Biochemical testing of isolates is rather time consuming, and only a limited number of isolates can be tested. Other approaches include the description of total community enzyme activity (Kanazawa and Filip, 1986), number of bacteria able to degrade specific compounds (Kauri, 1983) and total community growth rate (Albrechtsen and Winding, 1992; Bååth, 1992). A new technique is community structure analysis based on lipid composition (Tunlid and White, 1992), which may differentiate between soil habitats (Bååth et al., 1992; Zelles et al, 1992). A DNA reassociation technique has been used to estimate the number of genotypes in soil (Torsvik et al., 1990a, b) and profiling of communities by 16S ribosomal RNA (rRNA) has been reported (Muyzer et al., 1993). Despite these new methodologies, rapid and easy methods of fingerprinting the metabolic potentials of bacterial communities would be most valuable.

The Biolog® system (Biolog Inc.) consists of a 96-well microtitre plate with 95 different carbon sources and a control well without a carbon source. Each well also contains nutrients, salts, a small amount of peptone and the redox dye tetrazolium violet. The dye is reduced during respiratory activity, and insoluble formazan (violet) accumulates inside the cells (Bochner and Savageau, 1977; Bochner, 1978). Inoculation of one microtitre plate thus results in 95 simultaneous biochemical tests. The system has been used for the characterisation of microbial communities (Garland and Mills, 1991) and has been shown to differentiate microbial communities extracted from freshwater, hydroponic solutions and pre-inoculated soil slurries.

The objective of the study reported here was to investigate metabolic fingerprinting in different bacterial soil communities and to compare it with quantitative and classical methods of describing bacterial communities. Differences between microbial communities in agricultural soils, beech forest soils and meadow soil, and in aggregates and size fractions of the agricultural soils, were investigated.

MATERIALS AND METHODS

Soil samples

Nine soils were collected: a sandy loam agricultural soil (A1); a loamy sand agricultural soil (A2); a meadow soil (M); and six beech forest soils — three from Helsingør (B1), one from Danstrup Hegn (B2), one from St Dyrehaven (B3) and one from Uggeløse (B4) in Denmark. At Helsingør, the samples were taken from three sites on the slope ($B1_{top}$, $B1_{slope}$, and $B1_{bottom}$, respectively). From each of the soils, three replicate cores with a length of about 10 cm and a diameter of 5 cm were sampled from the topsoil and transferred to plastic bags, which were sealed and stored in the dark at 4°C for a maximum period of 1 month. The soil was passed through a sieve (4 mm) before testing.

Bacteria were extracted from the soil by blending 10 g of soil (dry weight) with 200 ml of Winogradsky salt solution (W) (Pochon, 1954) at low speed in a Waring commercial blender (three times/1 minute with intermittent cooling on ice). The coarser soil particles were removed by centrifugation at 2500 rpm for 10 minutes at 4°C. Although some bacteria were left in the pellet, the supernatant was taken to represent the entire bacterial community of the soil and was used for direct cell counts and incubation in the microtitre plates. Triplicates of each soil type were tested.

Soil fractionation

Triplicates of the two agricultural soils were fractionated into macroaggregates (> 250 µm), three size classes of microaggregates (2-20 µm, 20-53 µm and 53-250 µm) and a < 2 µm fraction (Oades, 1984) using a modification of the procedure described by Jocteur-Monrozier et al. (1991).

The modifications included using W throughout, as this salt solution facilitates extraction of a higher number of bacteria from sediment (Albrechtsen, 1989). Initially, the soil was allowed to suck W overnight through sterile gauze, which should increase the stability of the macroaggregates (Kemper and Rosenau, 1986). The soil was rolled with glassbeads for 15 minutes instead of overnight. The macroaggregates were disrupted by sonication at maximum speed, using a Sonifier Cell Disrupter B-12 (Branson Sonic Power Co.) equipped with a microtip, three times for 10 seconds (with 10 second intervals) while kept on ice. Particles 20-53 µm were separated from particles < 20 µm by four centrifugations at 90 g with intermittent resuspension. Particles 2-20 µm were separated from particles < 2 µm by four centrifugations at 2400 g with intermittent resuspension. The fractions were resuspended in 100 ml W, except for the < 2 µm fraction, which was suspended in 600 ml W.

During the entire fractionation procedure, all solutions were kept at 0-4°C to reduce bacterial growth. All sieves had a diameter of 10 cm and were made of stainless steel (Endecotts Ltd.). The fractionation procedure was found to be reproducible in three consecutive fractionations of soil A1 with quantification of dry-matter content, ignition loss, total number of bacteria and CFUs in the five fractions, as determined by chi-squared tests.

Abiotic soil analysis

The dry-matter content of the bulk soil and soil fractions was determined by drying them at 105°C for a minimum of 24 hours. The dried soil fractions were burned at 550°C for a minimum of 4 hours to determine ignition loss. Determination of dry-matter content and ignition loss was not possible for the < 2 µm fraction.

Biolog incubation

Biolog GN microtitre plates, designed for indentifying Gram-negative bacteria, were inoculated with the samples representing the soil. The bulk soil samples were diluted 100 times before inoculation, while the 2-20 µm and 20-53 µm fractions were diluted 100 times and the 53-250 µm and > 250 µm fractions 10 times; the < 2 µm fraction was not diluted after fractionation. The plates were incubated at 15°C in plastic bags to minimise desiccation and agitated at 100 rpm to prevent the cells from settling at the bottom of the wells. The formazan formation in the individual wells of the microtitre plates was measured by a video image analysing system, CREAM® (Kem-En-Tec A/S), designed to read enzyme-linked immunosorbent assay (ELISA) plates. The optical density of the grey level value was digitised and totalled over all wavelengths in each of the 96 wells. The value of the control well was subtracted from the values of the wells with carbon.

To visualise and estimate the relationship between the soil types and five size fractions of the agricultural soils, average linkage cluster analysis followed by a tree diagram (dendrogram) were performed using SAS®/STAT Version 5 (SAS Institute, 1985). The probability of the number of clusters formed was tested using pseudo F statistics, and the probability of any two clusters being separated was tested using pseudo t^2 statistics. Differences between soils and between size fractions, and interactions between soils and size fractions, were tested by two-way analysis of variance (ANOVA) (SAS Institute, 1985). The statistical analyses were performed on the data recorded after 3 days of incubation.

Colony-forming units

The number of CFUs were determined after spread-plating on agar medium containing 0.1 g Tryptic Soy Broth (Difco Laboratories), 15 g Bacto-agar (Difco Laboratories) and 25 mg natamycin as a fungal inhibitor (Pedersen, 1992) per litre of filter-sterilised water and incubation at 15°C for 14-19 days. The dilute medium was chosen because a larger number of soil bacteria will grow on very dilute media (Olsen and Bakken, 1987)

Direct counts

Suspensions of the bacterial communities were fixed with formaldehyde (final concentration 2 %) and stored at 4°C until counting. The total number of bacteria was counted using epifluorescence microscopy (Zeiss Axioplan) after staining with acridine orange. The filters were black Nuclepore® polycarbonate membrane filters (Costar) with a diameter of 25 mm and a pore size of 0.2 µm (Hobbie et al., 1977). A minimum of 200 bacteria were counted per filter.

RESULTS

Bulk soils

The beech forest, meadow and agricultural soils were characterised by the total number of bacteria in the extraction solution used in the Biolog microtitre plates (*see* Table 9.1 *overleaf*). Investigations with pure cultures have shown that cell concentration in the inoculum affects the lag time before colour development in the wells, but has no influence on the total colour intensity or the number of wells with

Table 9.1   Means ± standard deviation (SD) of total number of bacteria (AODC)/ml inoculum in
the Biolog GN microtitre plates and descriptions of the soil types

Soil type	Description	AODC (10^7/ml) ± SD	
Agricultural (A1)	Sandy loam	2.3	0.3
Agricultural (A2)	Loamy sand	3.0	0.9
Beech forest (B1$_{top}$)	Poor humic	7.2	3.0
Beech forest (B1$_{slope}$)	Humic	13.1	0.9
Beech forest (B1$_{bottom}$)	Sand	3.3	0.6
Beech forest (B2)	Humic	7.8	1.2
Beech forest (B3)	Humic mould	9.5	1.2
Beech forest (B4)	Mould	5.6	0.7
Meadow (M)	Loam	3.0	0.7

Figure 9.1   Dendrogram of the metabolic fingerprints of bacterial communities extracted
from nine soils

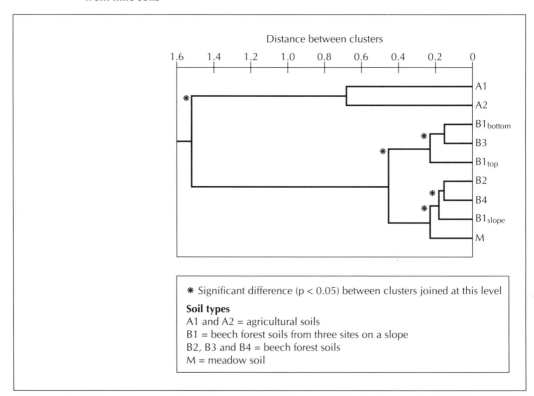

* Significant difference (p < 0.05) between clusters joined at this level

Soil types
A1 and A2 = agricultural soils
B1 = beech forest soils from three sites on a slope
B2, B3 and B4 = beech forest soils
M = meadow soil

colour development within the range of the cell numbers used in the present study (A. Winding and N.B. Hendriksen, unpubl.). The pattern of formazan formation was the basis for average linkage cluster analysis and the construction of a dendrogram (*see* Figure 9.1). From the dendrogram, it appears that habitats with a small degree of disturbance (beech forest soils and meadow soil) clustered together, while the highly disturbed agricultural soils clustered together. However, the distance between the agricultural soils indicated differences between the two communities. The pseudo F statistics suggested four clusters. The pseudo t^2 statistics identified significant differences between five clusters.

Fractionated agricultural soils

The dry-matter contents of the soil fractions are shown in Table 9.2. The dry-matter content showed significant differences between the soils for three fractions, as tested by the t-test for small samples (Bailey, 1981). The ignition loss, which is taken to represent the content of organic matter, showed that the smaller fractions contained relatively more organic material than the larger fractions.

The total number of bacteria extracted in the various fractions differed significantly between the two soils in the 53-250 μm sized fraction (*see* Table 9.2). The number of CFUs differed significantly between the two soils in all the size fractions except the 53-250 μm fraction. Generally, the total and culturable number of bacteria was higher in A1 than in A2. Calculating the culturability of the bacterial populations as the relationship between CFU and AODC showed the culturability to be higher in A1

Table 9.2   Means ± standard deviation (SD) of dry matter, ignition loss, total (AODC) and culturable (CFU) number of bacteria in five size fractions of two agricultural soils[a]

Fraction	Dry matter (mg/g dw ± SD)		Ignition loss (mg/g dw ± SD)		AODC[b] (10^9/g dw ± SD)		CFU[b] (10^8/g dw ± SD)		CFU/AODC (%)
A1 < 2 μm	bd[c]		bd		44	30	43	20	9.8
A2 < 2 μm	bd		bd		9.0	4.5	3.5*	1.8	3.9
A1 2-20 μm	55	4	99	57	17.6	4.9	16.5	4.9	9.4
A2 2-20 μm	35**[d]	4	98	6	28.8	9.9	3.5**	0.4	1.2
A1 20-53 μm	267	14	93	51	3.7	0.3	4.8	0.7	13.1
A2 20-53 μm	208***	5	84	3	3.9	1.5	2.7**	0.1	6.8
A1 53-250 μm	253	31	14	2	1.9	0.4	4.1	4.1	22.3
A2 53-250 μm	191	37	18	4	0.5**	0.1	0.2	0.03	3.2
A1 > 250 μm	426	43	20	4	2.3	1.2	1.9	0.5	8.2
A2 > 250 μm	566*	32	10*	1	0.7	0.1	0.4**	0.03	5.4

Note: a   30 g (dw) of soil was fractionated
       b   In the fraction < 2 μm, the total and culturable number of cells extracted from 30 g dw
       c   Below detection limit
       d   Significant differences between the same size fraction of the two soils: * = $p < 0.02$; ** = $p < 0.01$; *** = $p < 0.001$

for all the fractions (*see* Table 9.2). This estimate of culturability is very crude and depends upon the medium used, the incubation time and the temperature.

The results from the incubation in Biolog microtitre plates were analysed; the dendrogram is shown in Figure 9.2. Three size fractions of A2 (2-20 μm, 20-53 μm and 53-250 μm) clustered together in a group separate from the other fractions. The same three size fractions of A1 also clustered together. The macroaggregates (> 250 μm) and the free and loosely associated bacteria (< 2 μm) of A2 clustered within a small distance. The macroaggregate fraction of A1 (> 250 μm) was clearly distinct from all other fractions, while that of the free and loosely associated bacteria (< 2 μm) in A1 clustered close to the same fraction in A2. Four clusters were suggested by the pseudo F statistics. The pseudo t^2 test identified a significant difference between the five clusters. The same pattern emerged when clustering the five fractions of A1 without the A2 fractions and vice versa.

Figure 9.2   **Dendrogram of the metabolic fingerprints of five size fractions of two agricultural soils (A1 and A2)**

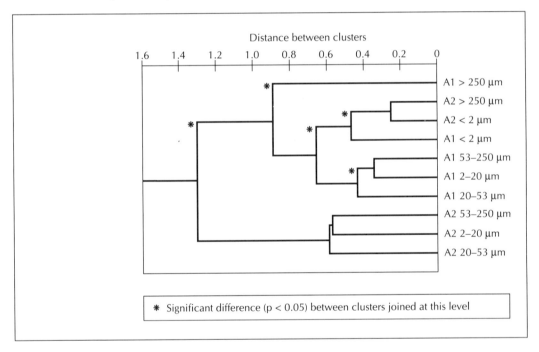

Using two-way ANOVA, the formazan formation was found to be significantly different (p < 0.05) between the two soils for 71 of the 95 tested carbon sources and significantly different between the size fractions for 53 of the 95 tested carbon sources. Interactions between soil and size fraction were significant (p < 0.05) for 10 carbon sources; nine of these sources showed significant difference between the soils and between the size fractions (L-arabinose, D-galactose, D-mannose, D-galactonic acid lactone, bromo succinic acid, L-alanine, glycyl-L-glutamic acid, hydroxy L-proline and L-proline). Evaluation of the actual grey level values for the nine carbon sources showed that the colour formation was always higher for the A2 soil; for this soil, the 2-250 μm size fractions showed the highest formazan formation, while the $< 2\,\mu m$ and $> 250\,\mu m$ fractions showed about the same formazan formation. For the A1 soil, the $< 2\,\mu m$ fraction resulted in the highest formazan formation, while the $> 250\,\mu m$ fraction showed the least formazan formation. This pattern of formazan formation occurred in 45% (A1) and 64% (A2) of the 53 carbon sources, with significant differences between soil fractions, and is in accordance with the cluster analysis (*see* Figure 9.2).

DISCUSSION

The incubation of whole bacterial communities from two agricultural soils and several native forest or meadow soils in Biolog microtitre plates resulted in fingerprints based on the metabolic potentials (*see* Figure 9.1). Because the assay is based on aerobic respiratory activity and growth, strictly anaerobic bacteria and bacteria lacking crucial enzymes of the electron transport chain are excluded. The subsequent cluster analysis depicts the relationship between the communities (*see* Figure 9.1). Differences between disturbed soils (agricultural) and undisturbed soils (beech forest and meadow) were large, and the distance between the two agricultural soils also indicated a difference. This agrees with the findings reported by Garland and Mills (1991), who detected differences between microbial soil communities when fingerprinted using the Biolog assay. In their study, the communities were enriched from three different soil types: a clay loam, a silt loam and a clay.

The quantitative description of the two agricultural soils and their size fractions showed differences in particle size composition and organic matter content. The A2 soil was more sandy than A1. The number and culturability of bacteria were generally higher in A1 than in A2, as shown in Table 9.2. Differences between the size fractions within the same soil were also evident. The organic matter content, the total number of bacteria, the number of CFUs, and the relationship between CFU and AODC all show the A1 soil to be an environment with larger and more active microbial communities in the different microhabitats within the soil than the more sandy A2 soil. The metabolic fingerprints obtained with the Biolog microtitre plates indicated differences between the soil fractions (*see* Figure 9.2). The bacterial communities of the three size fractions of microaggregates (2-20 µm, 20-53 µm and 53-250 µm) exhibited similar metabolic potentials within each soil. This indicates that the bacterial communities are closely related. Bacterial communities in the macroaggregates (> 250 µm) and the free and loosely associated bacteria (< 2 µm) may (A2) or may not (A1) be similar, as shown in Figure 9.2 and indicated by the formazan formation of the nine carbon sources showing significant differences between the soils and size fractions and significant interactions between soils and size fractions. The formazan formation in the wells with these carbon sources contain the most information about the differences between the soils and size fractions. The reason for the pattern of the dendrogram could perhaps be found in the described pattern of formazan formation, and the nine carbon sources could potentially be selected as metabolic indicators of the communities. Bacteria in soil are distributed in different sized pores between and within soil aggregates (Hattori and Hattori, 1976). The free and loosely associated bacteria are likely to interchange with the bacteria of the macroaggregates *in situ*, as pores in the macroaggregates may be considered as part of the outer surface and thus in contact with free or loosely associated bacteria (Hattori, 1988). The dendrogram depicted in Figure 9.2 indicates that the bacterial communities of the microaggregates are different in the two soils and distinct from the bacterial communities of the macroaggregates and the free bacteria. The size fractions of the two agricultural soils showed a closer relationship between the free and loosely associated bacteria of both soils than between this fraction and the microaggregates of both soils. This indicates that the assay can resolve very small spatial differences. The fingerprints of the unfractionated soils were unable to show such a relationship. A more detailed description of the bacterial communities is thus obtained when the bulk soil is size fractionated.

The cluster analysis shows that the Biolog fingerprint of bacterial soil communities is able to discriminate bacterial communities from different soils (*see* Figure 9.1) and microbial communities associated with the different size fractions of agricultural soils (*see* Figure 9.2). The description of bacterial communities is often based on isolations and subsequent tests. This characterisation then forms the database for diversity indices (Atlas, 1983; Atlas et al., 1991; Tate and Mills, 1983) and factor analysis (Sundman, 1970). Apart from being very laborious, isolating and culturing the bacteria for several generations in the laboratory potentially changes the picture of the community considerably. Garland and Mills (1991) included an enrichment procedure with incubation of the communities prior to inoculation into Biolog microtitre plates. Such an enrichment leads to a change in the bacterial species composition and number, as demonstrated by comparing the same soil grown on different media. In the present study, an enrichment took place in the wells during incubation, as bacterial growth in the individual wells is a fact and a necessity for measurable formazan formation with the inoculum densities used (A. Winding and N.B. Hendriksen, unpubl.). Morphologically different bacteria became dominating in the different wells during incubation (data not shown). The bacteria are grown for only a few generations in the microtitre plates and tested during the growth. The metabolic potential of the *in situ* community is thus more reliable when testing the communities immediately after sampling in the field, as possible adaptation to specific substrates during isolation procedures is limited.

In conclusion, the fingerprints of bacterial communities obtained with Biolog microtitre plates discriminate between closely related habitats. In addition, the method is fast and easy to handle, which enables one to describe a large number of bacterial communities within a realistic time period.

Acknowledgements

The research reported in this chapter was financed by the Danish Center for Microbial Ecology. The author wishes to thank B. Hansen, J.C. Pedersen, and J. Sørensen for their critical comments on the manuscript, N.B. Hendriksen for valuable advice during discussions on planning the experiment, C.T. Agger for assistance with the statistical analysis and B.R. Hansen for technical assistance.

References

Albrechtsen, H.-J. 1989. Investigations at Vejen landfill: Microbiology of the groundwater zone. In *Lossepladsprojektet, Report P6-2*. Copenhagen, DK: Danish Environmental Protection Agency. (in Danish).

Albrechtsen, H.-J. and Winding, A. 1992. Microbial biomass and activity in subsurface sediment from Vejen, Denmark. *Microbial Ecology* 23: 303-17.

Atlas, R.M. 1983. Diversity of microbial communities. *Advances in Microbial Ecology* 7: 1-47.

Atlas, R.M., Horowitz, A., Krichevsky, M. and Bej, A.K. 1991. Response of microbial populations to environmental disturbance. *Microbial Ecology* 22: 249-56.

Bailey, N.T.J. 1981. *Statistical Methods in Biology*. London, UK: Hodder and Stoughton.

Bochner, B.R. 1978. Device, composition and method for identifying microorganisms. *United States Patent* No. 4 129 483.

Bochner, B.R, and Savageau, M. 1977. Generalised indicator plate for genetic, metabolic, and taxonomic studies with microorganisms. *Applied and Environmental Microbiology* 33: 434-44.

Bååth, E. 1992. Measurement of heavy metal tolerance of soil bacteria using thymidine incorporation into bacteria extracted after homogenisation-centrifugation. *Soil Biology and Biochemistry* 24: 1167-72.

Bååth, E., Frostegård, A. and Fritze, H. 1992. Soil bacterial biomass, activity, phospholipid fatty acid pattern, and pH tolerance in an area polluted with alkaline dust deposition. *Applied and Environmental Microbiology* 58: 4026-31.

Garland, J.L. and Mills, A.L. 1991. Classification and characterisation of heterotrophic microbial communities on the basis of patterns of community-level sole-carbon-source utilisation. *Applied and Environmental Microbiology* 57: 2351-59.

Hattori, T. 1988. Soil aggregates as microhabitats of microorganisms. *Report of the Institute for Agricultural Research, Tohoku University* 37: 23-36.

Hattori, T. and Hattori, R. 1976. The physical environment in soil microbiology: An attempt to extend principles of microbiology to soil microorganisms. *Critical Reviews of Microbiology* 4: 423-61.

Hobbie, J.E., Daley, R.J. and Jasper, S. 1977. Use of Nucleopore filters for counting bacteria by fluorescence microscopy. *Applied and Environmental Microbiology* 33: 1225-28.

Jocteur-Monrozier, L., Ladd, J.N., Fitzpatrick, R.W., Foster, R.C. and Raupach, M. 1991. Components and microbial biomass content of size fractions in soils of contrasting aggregation. *Geoderma* 49: 37-62.

Kanazawa, S. and Filip, Z. 1986. Distribution of microorganisms, total biomass, and enzyme activities in different particles of brown soil. *Microbial Ecology* 12: 205-15.

Kauri, T. 1983. Fluctuations in physiological groups of bacteria in the horizons of beech forest soil. *Soil Biology and Biochemistry* 15: 45-50.

Kemper, W.D. and Rosenau, R.C. 1986. Aggregate stability and size distribution. In Klute, A. (ed) *Methods of Soil Analysis*. Madison, Wisconsin, USA: American Society of Agronomy, Soil Science Society of America.

Muyzer, G., de Waal, E.C. and Uitterlinden, A.G. 1993. Profiling of complex microbial populations by denaturing gradient gel electrophoresis analysis of polymerase chain reaction-amplified genes coding for 16S rRNA. *Applied and Environmental Microbiology* 59: 695-700.

Oades, J.M. 1984. Soil organic matter and structural stability: Mechanisms and implications for management. *Plant and Soil* 76: 319-37.

Olsen, R.A. and Bakken, L.R. 1987. Viability of soil bacteria: Optimisation of plate-counting technique and comparison between total counts and plate counts within different size groups. *Microbial Ecology* 13: 59-74.

Pedersen, J. 1992. Natamycin as a fungicide in agar media. *Applied and Environmental Microbiology* 58: 1064-66.

Pochon, J. 1954. *Manual Technique d'Analyse Microbiologique du Sol*. Paris, France: Masson et Cie.

SAS Institute. 1985. *SAS® User's Guide: Statistics, Version 5 Edition*. Cary, North Carolina, USA: SAS Institute Inc.

Sundman, V. 1970. Four bacterial soil populations characterised and compared by a factor analytical method. *Canadian J. Microbiology* 16: 455-64.

Tate, R.L. III and Mills, A.L. 1983. Cropping and the diversity and function of bacteria in Pahokee muck. *Soil Biology and Biochemistry* 15: 175-79.

Torsvik, V., Salte, K., Sørheim, R. and Goksøyr, J. 1990a. Comparison of phenotypic diversity and DNA heterogeneity in a population of soil bacteria. *Applied and Environmental Microbiology* 56: 776-81.

Torsvik, V., Goksøyr, J. and Daae, F.L. 1990b. High diversity in DNA of soil bacteria. *Applied and Environmental Microbiology* 56: 782-87.

Tunlid, A. and White, D.C. 1992. Biochemical analysis of biomass, community structure, nutritional status, and metabolic activity of microbial communities in soil. In Stotzky, G. and Bollag, J.-M. (eds) *Soil Biochemistry* (Vol. 7). New York, USA: Marcel Dekker.

Zelles, L., Bai, Q.Y., Beck, T. and Beese, F. 1992. Signature fatty acids in phospholipids and lipopoly–saccharides as indicators of microbial biomass and community structure in agricultural soils. *Soil Biology and Biochemistry* 24: 317-23.

Beyond the Biomass
Edited by K. Ritz, J. Dighton and K.E. Giller
© 1994 British Society of Soil Science (BSSS)
A Wiley-Sayce Publication

CHAPTER 10

Recent advances in the analysis of microbial communities and activity in marine ecosystems

S. LEE

Our knowledge of microbial processes in natural environments has expanded greatly in recent years. It is now clear that microorganisms play indispensable roles in material cycles and energy flow in marine environments (Azam et al., 1983). Despite the growing knowledge, however, little is known about the species and their abundance in natural microbial communities. Unlike other ecological studies that deal with higher organisms, identification has been put aside in microbial ecology. This is primarily because conventional identification of bacteria requires pure cultures, but most natural bacteria are non-culturable (Ferguson et al., 1984; Lee and Fuhrman, 1991a).

The lack of proper methods for identification causes other problems in studies of population dynamics or species interactions. For example, we currently have few methods to study how individual species respond to each other or to environmental changes. Despite the immense diversity of metabolic capabilities of prokaryotes, current techniques for measuring bacterial activities invariably treat a bacterial community as a single entity. Measurements are usually averaged over the entire community, with little attention given to the components and their variation among communities. Responses or interactions of individual populations in a community often have to be ignored because of the method-ological deficiencies. This is why the natural microbial community is often referred to a 'black box'.

Recent advances in molecular techniques and their application to microbial ecology offer new perspectives in opening the black box. Molecular approaches seem so promising that a new term, 'molecular ecology', has been coined. Molecular techniques deal with cellular macromolecules at different levels (for example, nucleic acids, allozymes, lipid, and protein) for different purposes. This chapter introduces the approaches that use nucleic acids for the analysis of natural microbial communities and their activities in marine systems.

APPROACHES BASED ON 16S RIBOSOMAL RNA

Comparisons of the base sequence of 16S ribosomal RNA (rRNA) molecules have revealed that there are portions relatively conserved or more varied at different levels of phylogeny during evolution

(Woese, 1987; Olsen, 1988). Sequence homology can therefore serve as a measure of phylogenetic distances among organisms (Olsen, 1988). Those regions where base sequences are unique to an organism (species) also serve as a suitable 'target' site for molecular probes designed for that taxon (Giovannoni et al., 1988; DeLong et al., 1989).

Single-cell identification by 16S rRNA probes

Using fluorescently labelled oligonucleotides that are complementary to portions of the 16S rRNA sequence, DeLong et al. (1989) demonstrated that single bacterial cells can be identified by microscopy. By selecting appropriate sequences, probes can label cells at the species, genus or higher levels. The method is now frequently used for direct detection or identification of bacterial cells in natural samples without incubations or pure cultures (Amann et al., 1990a; Distel et al., 1991).

This method has contributed to a significant breakthrough, whereby individual cells can be identified without any sample manipulation. This is truly an *in situ* method, with many potential applications to field microbiology. However, the technique needs a few improvements before it can be used routinely for field studies. First, the fluorescence signal results from the probes hybridised to the RNA molecules, and thus cells with very low RNA content may not appear 'labelled' even if they contain target sequences. Natural planktonic bacteria often do not appear labelled, because of their small cell size and slow growth (Lee and Kemp, 1994). For this reason, field studies are currently limited to cells from nutrient-rich environments, such as symbionts (Distel et al., 1991). Second, slight sequence variations (degeneracies) among organisms within a defined taxon can cause minor base mismatches between the probe and the target. This can result in reduced signals (Amann et al., 1990a). Third, to generate a probe, one must have the sequence information of the target organism; it is not trivial, if the organism is not culturable or previously unknown.

For signal enhancement, one may use a probe labelled with multiple fluorochrome molecules; that is, a probe to which multiple reporter molecules are attached via reporter-carrying molecules (DeLong, 1990; Zarda et al., 1991; Amann et al., 1992). Cells require special treatments, often with enzymes, in order for the large carrier molecules to pass through the cell walls. Despite such special treatments, delivering the carrier molecule into the cell was only partially successful (for example, for some Gram-negative bacteria; Amann et al., 1992). Alternatively, one may use multiple probes, each labelled with a fluorochrome molecule, but designed for multiple independent target sites (Amann et al., 1990b; Lee et al., 1994). To use multiple probes, one must have target sites for a defined taxon.

With regard to sequence variations, mixed probes with redundant bases may absorb the variation. However, they may increase the possibility of false-positive signals from closely related, non-target species. For unknown species in natural environments, analyses of 16S rRNA sequences present in nucleic acids directly obtained from environmental samples are expanding our inventory of bacteria.

Genetic diversity of natural communities by 16S rRNA sequence analysis

The problem of non-culturable bacteria and the idea that 16S rRNA sequence is another type of identification have altered the direction of research, from studying organisms themselves to direct analyses of their nucleic acids without culturing. By isolating and sequencing individual 16S rRNA sequences present in environmental samples, one can characterise the sequences (that is, organisms) whether or not the 'host' organism is culturable. Genetic diversity of a community can be inferred from the diversity of organisms recovered from the community in the form of 16S rRNA sequences.

The technique, introduced by Pace et al. (1986), has been modified and updated. New techniques isolate and amplify 16S rRNA sequences either from complementary DNA (cDNA) generated from RNA (Weller and Ward, 1989) or more directly from the 16S rRNA gene present in total genomic DNA extracts (Giovannoni et al., 1990; Britschgi and Giovannoni, 1991; Fuhrman et al., 1993). To avoid time-consuming searches for the 16S rRNA sequences in a nucleic acid sample (Schmidt et al., 1991), some methods use oligonucleotides that hybridise to specific sites in the target molecule. This mole–cule serves as a template, and complementary strands are enzymatically elongated from the oligonucleo–tides, which we call 'primers' instead of probes for this application. A powerful automated procedure, polymerase chain reaction (PCR) technique (Saiki et al., 1988), is now commonly used for sequence amplification (Giovannoni et al., 1990; Britschgi and Giovannoni, 1991; Fuhrman et al., 1993).

Recent studies based on this approach report sequences found from the black box. Sequences filed to date from various marine environments reveal many surprises. There are indications of the common presence of previously 'undescribed' organisms (Giovannoni et al., 1990; Britschgi and Giovannoni, 1991; Fuhrman et al., 1992, 1993). Sequence analyses often show that few sequences (organisms) were identical to the organisms previously isolated from the study sites (Weller and Ward, 1989; Giovannoni et al., 1990; Ward et al., 1990). Identical sequences are often found between oceans, or between different studies, but are rare within a sample (Giovannoni et al., 1990; Britschgi and Giovannoni, 1991; Fuhrman et al., 1993). These indicate high species diversities in marine bacterial communities, the common occurrence of unknown species and the occurrence of the same or near-identical species in different oceans. Interestingly, Archaea (in the Woese et al. [1990] classification), which are characteristically adapted to extreme environments, are found in ocean waters (Fuhrman et al., 1992) as well as in coastal waters of North America (DeLong, 1992).

At present, the analysis of sequence diversity is rather qualitative; that is, sequences recovered from an environmental sample are not a complete and exact replica of the existing species and their abundances. Use of cDNA (DNA copy of RNA) will over-represent metabolically active organisms, while the use of rDNA (rRNA gene) would tend to reflect the 'presence' of organisms. Sequence comparisons alone do not always resolve the phylogeny to the level of species (Stackebrandt, 1985). Sequence does not provide information for the phenotypic characteristics, which can be studied by culturing the organism. As more naturally occurring bacteria are registered to our catalogue, we will get closer to understanding the natural microbial systems and their changes over time and space.

SPECIES COMPOSITION COMPARISON VIA HYBRIDISATION OF COMMUNITY DNA

Developed by Lee and Fuhrman (1990), this method uses total DNA extracted from natural bacterial communities to compare species compositions. Complete cataloguing by the 16S rRNA sequence analysis would allow full-scale comparisons of the community structure. However, with this method, one must spend considerable time and effort on each sample for a community-wide comparison of species-frequency distributions. Exploiting the discriminating power of DNA-DNA hybridisation (total DNA cross-hybridises significantly only between closely related species; Wayne et al., 1987; Lee and Fuhrman, 1990), community DNA hybridisation measures the 'community DNA similarity', an estimate of the fraction of identical or near-identical DNA species present in two communities.

Using this technique, Lee and Fuhrman (1991b) reported that bacterial communities from different ocean basins showed a similarity of only about 10%. Communities at different depths (surface, 100 m, 500 m and 1000 m) at an oceanic (Pacific) sampling location showed similarities from less than 5% to almost 100%; samples from the same depth were 50-95% similar, and samples from shallow depths (surface and 100 m) were very different (a similarity of about 5%) from the deep samples (500 m and 1000 m). Coastal samples taken from one sampling site over a 2-year period showed similarities of higher than 40%, suggesting the year-round presence of residual subpopulations. In a transect study near Bermuda (Lee and Fuhrman, 1991c), comparisons were made between samples collected from the centre of a coral reef lagoon, mid-way to open ocean (Sargasso Sea) and the Sargasso Sea (surface, 100 m and 500 m). There was a directional change of species compositions along the transect as well as along the depth. These studies showed that species compositions of natural bacterial assemblages vary significantly over time and space.

Unlike the 16S rRNA sequence cataloguing, this method does not produce data that can be compiled and used again in the future. The technique is suitable for rapid comparisons of many samples, but data are always relative and interpretations of intermediate similarities are not always clear. Different complexities of community structure may result in asymmetric similarities when hybridised reciprocally; it is for this reason that reciprocal hybridisations are recommended. Concept, data interpretations and limitations have been discussed in more detail by Lee and Fuhrman (1990).

USE OF FLUORESCENT 16S rRNA PROBES TO MEASURE SINGLE-CELL RNA CONTENT AND CELLULAR ACTIVITY

Fluorescence from a cell hybridised with 16S rRNA probes can serve as a quantitative measure of single-cell RNA content (DeLong et al., 1989), because probe molecules hybridise in 1:1 proportion to 16S rRNA molecules. As most cellular RNA is ribosomal, and 16S rRNA is a relatively constant fraction of the ribosomal RNA (Bremer and Dennis, 1987), probe fluorescence should be proportional to the RNA content of cells. DeLong et al. (1989) demonstrated a close relationship between the cellular RNA content and the fluorescence of *Escherichia coli* cells hybridised with fluorescent probes.

Past studies showed strong correlations between the growth rates of cultured, fast-growing bacteria and their cellular RNA contents measured from nucleic acid preparations (Schaechter et al., 1958; Bremer and Dennis, 1987; DeLong et al., 1989; Kemp et al., 1993). The correlation results from the direct role of ribosomes in protein synthesis. Probe-based measurements of single-cell RNA contents should be very useful in studying *in situ* activities of natural bacteria. Potential advantages include:

- there is no need for the sample incubations that often perturb the natural systems (Ferguson et al., 1984; Kroer and Coffin, 1992)

- by using species-specific probes, quantification of RNA content and detection (identification) of the organisms can be done simultaneously

- measurements from individual cells provide new information (such as frequency distributions; Lee and Kemp, 1994); current methods for bacterial activity measurements, which invariably average the activity over the entire assemblage, cannot provide such information

However, natural marine planktonic bacteria, unlike cells from laboratory cultures or nutrient-rich natural environments (DeLong et al., 1989; Distel et al., 1991), often contain insufficient rRNA for a measurable level of fluorescence because of their small size and slow growth (Lee and Kemp, 1994). As noted above, other approaches are under development to increase signal strength. Using five or six different probes targeted to independent sites in a 16S rRNA molecule, Lee et al. (1994) quantitatively evaluated the multiple-probe method for use in field studies. They showed that fluorescence from natural bacteria and cultured marine isolates increased additively with multiple probes.

From 12 natural planktonic bacterial assemblages collected over a 7-month period, Lee and Kemp (1994) observed a close correlation ($r^2 = 0.82$) between cell fluorescence and the RNA content independently determined from nucleic acid preparations. This method, measuring femtogram levels of cellular RNA in single natural cells (average 3-5 fg RNA/cell), revealed distinctive seasonal patterns of fluorescence-frequency distributions and occasional blooms of subpopulations. However, some natural cells, mostly small cocci, were not labelled even with five probes. Labelled fractions increased asymptotically with the number of probes hybridised, from about 20% (one probe) to about 75% (five probes) of the total cell count (Lee and Kemp, 1994).

Although the technique proved useful and sensitive, RNA contents need to be interpreted in more useful terms (for example, growth rate). Kemp et al. (1993) examined the relationship between RNA and growth rate with four marine bacterial isolates that were chemostat-cultured at growth rates spanning the range typical in natural environments. They observed a close relationship within each isolate. However, there was no single universal relationship for the four isolates, probably because of their inherent differences in cell size and physiology. Temperature was also a factor affecting cellular RNA content; from the study of 12 natural samples, Lee and Kemp (1994) observed a great effect of temperature on the RNA content. Further studies are needed for application and interpretation of RNA content measurements in the field (for example, on how to convert RNA content to cell activities, such as growth rate, or on what factors affect the relationships between the two parameters). The method presented in this chapter allows us to begin examining the usefulness of RNA-frequency distributions.

References

Amann, R.I., Krumholz, L. and Stahl, D.A. 1990a. Fluorescent-oligonucleotide probing of whole cells for determinative, phylogenetic and environmental studies in microbiology. *J. Bacteriology* 172: 762-70.

Amann, R.I., Binder, B.J., Olson, R.J., Chisholm, S.W., Devereux, R. and Stahl, D.A. 1990b. Combination of 16S rRNA-targeted oligonucleotide probes with flow cytometry for analysing mixed microbial populations. *Applied and Environmental Microbiology* 56: 1919-25.

Amann, R.I., Zarda, B., Stahl, D.A. and Schleifer, K.-H. 1992. Identification of individual prokaryotic cells by using enzyme-labeled, rRNA-targeted oligonucleotide probes. *Applied and Environmental Microbiology* 58: 3007-11.

Azam, F., Fenchel, T., Gray, J.G., Meyer-Reil, L.A. and Thingstad, T. 1983. The ecological role of water-column microbes in the sea. *Marine Ecology Progress Series* 10: 257-63.

Bremer, H. and Dennis, P.P. 1987. Modulation of chemical composition and other parameters of the cell by growth rate. In Neidhardt, F.C. (ed) Escherichia coli *and* Salmonella typhimurium: *Cellular and Molecular Biology*. Washington DC, USA: American Society for Microbiology.

Britschgi, T. and Giovannoni, S.J. 1991. Phylogenetic analysis of a natural marine bacterioplankton population by rRNA gene cloning and sequencing. *Applied and Environmental Microbiology* 57:1707-13.

DeLong, E.F. 1990. A signal amplification method for *in situ* hybridisations using fluorescently-labeled, rRNA-targeted probes. In Abstracts of 90th Annual Meeting of American Society for Microbiology, 1990, Anaheim, California, USA.

DeLong, E.F. 1992. Archaea in coastal marine environments. *Proc. National Academy of Science, USA.* 89: 5685-89.

DeLong, E.F., Wickham, G.S. and Pace, N.R. 1989. Phylogenetic stains: Ribosomal RNA-based probes for the identification of single cells. *Science* 243: 1360-63.

Distel, D.L., DeLong, E.F. and Waterbury, J.B. 1991. Phylogenetic characterisation and *in situ* localisation of the bacterial symbiont of shipworms (Teredinidae: Bivalvia) by using 16S rRNA sequence analysis and oligodeoxynucleotide probe hybridisation. *Applied and Environmental Microbiology* 57: 2376-82.

Fuhrman, J.A., McCallum, K. and Davis, A.A. 1992. Novel major archaebacterial group from marine plankton. *Nature* 356: 148-49.

Fuhrman, J.A., McCallum, K. and Davis, A.A. 1993. Phylogenetic diversity of subsurface marine microbial communities from the Atlantic and Pacific Oceans. *Applied and Environmental Microbiology* 59: 1294-302.

Ferguson, R.L., Buckley, E.N. and Palumbo, A.V. 1984. Response of marine bacterioplankton to differential filtration and confinement. *Applied and Environmental Microbiology* 47: 49-55.

Giovannoni, S.J., DeLong, E.F., Olsen, G.J. and Pace, N.R. 1988. Phylogenetic group-specific oligodeoxynucleotide probes for identification of single microbial cells. *J. Bacteriology* 170: 720-26.

Giovannoni, S.J., Britschgi, T.B., Moyer, C.L. and Field, K.G. 1990. Genetic diversity in Sargasso Sea bacterioplankton. *Nature* 345: 60-63.

Kemp, P., Lee, S. and LaRoche, J. 1993. Estimating the growth rate of slowly growing marine bacteria from RNA content. *Applied and Environmental Microbiology* 59: 2594-601.

Kroer, N. and Coffin, R.B. 1992. Microbial trophic interactions in aquatic microcosms designed for testing genetically engineered microorganisms: A field comparison. *Microbial Ecology* 23: 143-57.

Lane, D.J., Field, K.G., Olsen, G.J. and Pace, N.R. 1988. Reverse transcriptase sequencing of ribosomal RNA for phylogenetic analysis. *Methods in Enzymology* 167: 138-44.

Lee, S. and Fuhrman, J.A. 1990. DNA hybridisation to compare species compositions of natural bacterioplankton assemblages. *Applied and Environmental Microbiology* 56: 739-46.

Lee, S. and Fuhrman, J.A. 1991a. Species composition shift of confined bacterioplankton studied at the level of community DNA. *Marine Ecology Progress Series* 79: 195-201.

Lee, S. and Fuhrman, J.A. 1991b. Spatial and temporal variation of natural bacterioplankton assemblages studied by total genomic DNA cross-hybridisation. *Limnology and Oceanography* 36: 1277-87.

Lee, S. and Fuhrman, J.A. 1991c. The unique species composition of natural bacterial communities from a Bermuda coral reef lagoon studied by community DNA hybridisation. In Abstracts of 91st Annual Meeting of American Society for Microbiology, 1991, Dallas, Texas, USA.

Lee, S. and Kemp, P. 1994. Single-cell RNA content of natural marine planktonic bacteria measured by hy–bridisation with multiple 16S rRNA-targeted fluorescent probes. *Limnology and Oceanography* (in press).

Lee, S., C. Malone, and Kemp, P. 1994. Use of multiple 16S rRNA-targeted fluorescent probes to increase signal strength and measure cellular RNA from natural planktonic bacteria. *Marine Ecology Progress Series* (in press).

Olsen, G.J. 1988. Phylogenetic analysis using ribosomal RNA sequences. *Methods in Enzymology* 164: 793-812.

Pace, N.R., Stahl, D.A., Lane, D.L. and Olsen, G.J. 1986. The analysis of natural microbial populations by rRNA sequences. *Advances in Microbial Ecology* 9: 1-55.

Saiki, R.K., Gelfand, D.H., Stoffel, S., Scharf, S.J., Higuchi, R., Horn, G.T., Mullis, K.B. and Erlich, H.A. 1988. Primer directed enzymatic amplification of DNA with a thermostable DNA polymerase. *Science* 239: 487-91.

Schaechter, E., Maaloe, O. and Kjeldgaard, N.O. 1958. Dependence on medium and temperature of cell size and chemical composition during growth of *Salmonella typhimurium*. *J. Gen. Microbiology* 19: 592-606.

Schmidt, T.M., DeLong, E.F. and Pace, N.R. 1991. Analysis of a marine picoplankton community by 16S rRNA gene cloning and sequencing. *J. Bacteriology* 173: 4371-78.

Stackebrandt, E. 1985. Phylogeny and phylogenetic classification of prokaryotes. In Schleifer, K.-H. and Stackebrandt, E. (ed) *Evolution of Prokaryotes*. Orlando, Florida, USA: Academic Press.

Ward, D.M., Weller, R. and Bateson, M.M. 1990. 16S rRNA sequences reveal numerous uncultured microorganisms in a natural community. *Nature* 345: 63-65.

Wayne, L.G., Brenner, D.J., Colwell, R.R., Grimont, P.A.D., Kandler, O., Krichevsky, M.I., Moore, L.H., Moore, W.E.C., Murray, R.G.E., Stackebrandt, E., Starr, M.P. and Trüper, H.G. 1987. Report of the ad hoc committee on the reconciliation of approaches to bacterial systematics. *International J. Systematics in Bacteriology* 37: 463-64.

Weller, R. and Ward, D.M. 1989. Selective recovery of 16S rRNA from natural microbial communities in the form of cDNA. *Applied and Environmental Microbiology* 55: 1818-22.

Woese, C.R. 1987. Bacterial evolution. *Microbiological Review* 51: 221-71.

Woese, C.R., Kandler, O. and Wheelis, M.L. 1990. Towards a natural system of organisms: Proposal for the domains Archaea, Bacteria, and Eucarya. *Proc. National Academy of Science, USA* 87: 4576-79.

Zarda, B., Amann, R., Wallner, G. and Schleifer, K.-H. 1991. Identification of single bacterial cells using digoxigenin-labelled, rRNA-targeted oligonucleotides. *J. Gen. Microbiology* 137: 2823-30.

Beyond the Biomass
Edited by K. Ritz, J. Dighton and K.E. Giller
© 1994 British Society of Soil Science (BSSS)
A Wiley-Sayce Publication

CHAPTER 11

Sex and the single cell: The population ecology and genetics of microbes

J.P.W. Young

Microbial ecology has developed as a subject which is somewhat distinct from large-organism ecology, emphasising processes rather than populations. This chapter considers how advances in molecular biology may help to integrate the different approaches. Can microbial species be defined, identified and quantified? How well do processes map onto populations?

The soil microbial biomass has been described as a 'black box'. The image that the term 'biomass' conjures up, though, is that of a shapeless mass of translucent protoplasm oozing indefinitely through the gaps between minerals. Of course, everyone knows that in reality the microbes in soil form a diverse community of individuals, each going about its own business, but the biomass concept arose because it just was not practicable to describe the system at that level of resolution. The strengths of the biomass approach are the ability to summarise the activities of a whole community in just a few numbers, to place this whole 'black box' as a single element into a broader picture of nutrient cycling, and to compare communities without getting bogged down in the details. The corresponding weaknesses stem from a lack of mechanistic understanding of the individual components, making it difficult to predict the evolution of the system under changing environmental conditions and to integrate microbes into the conceptual framework of mainstream population biology.

Population ecologists rely on their ability to distinguish species, to count individuals and to observe their interactions. Population geneticists need to recognise genetic variation within species, and they generally need to examine individuals. How far can these activities be extended from macrobes to microbes? The rapidly growing field known as 'molecular ecology' makes use of the technology of molecular biology to study natural populations, and nucleic acids provide a common currency across organisms of all sizes. In particular, the polymerase chain reaction (PCR) opens up the possibility of studying organisms even if they are small, rare and poorly characterised.

Many of the first fruits of the molecular revolution are described elsewhere in this book; my aim here is to assess the conceptual consequences of treating microbes as species and individuals. The question is posed most sharply for bacteria, as these are phylogenetically, physiologically and genet–ically the furthest removed from the familiar eukaryotic macrobes. I shall concentrate on bacteria, therefore, but no doubt fungi, protists and the rest will spring their own surprises in due course.

DO BACTERIA HAVE SPECIES?

The first important question is whether bacteria have species in the same sense that is familiar to eukaryote population biologists. I have argued before that they probably do (Young, 1989, 1992), so I shall simply restate the main points here.

For sexually reproducing eukaryotes, the species is a taxonomic rank (the only one?) that corresponds to a real biological entity. The individuals in a species form a reproductive community, bound together by genetic recombination into a single gene pool and isolated from other species by barriers to gene exchange. The two essential elements are gene exchange within the species and the barriers between them. Of course, one or other of these elements is sometimes deficient even in eukaryotes; this concept of a biological species then breaks down, which is why taxonomists argue about the status of plants that reproduce asexually, such as blackberries.

It has sometimes been suggested that bacteria do not have good biological species in this sense. There are two strong reasons for this idea: bacteria do not have enough sex; or bacteria have too much sex. The first point is that bacteria reproduce by a process of binary fission, so that recombination is not a necessary part of their life cycle. Indeed, studies of enzyme polymorphism have indicated that most bacterial species have an essentially clonal population structure (Selander et al., 1986); there are strong disequilibria between polymorphic loci, and very similar multilocus genotypes crop up more frequently than would be expected by chance. The contrary argument is that some accessory elements (plasmids and transposable elements) have such broad and overlapping host ranges that they can carry essentially any DNA from any bacterial species to any other. Hence, it is argued, there is a common gene pool into which all bacteria can dip to find a solution to the challenge of the moment. This concept of bacteria as a vast underground genetic market, which sheds new light on the concept of the biomass as a superorganism, has been suggested by various authors (possibly first by Hedges, 1972) and taken to the brink of science fiction paranoia in the book by Sonea and Panisset (1983).

In reality, however, it seems that bacteria are neither strictly clonal nor wantonly promiscuous. Polymorphism studied at the level of DNA sequence has now modified the picture we had obtained from enzyme electrophoresis. When the corresponding region is sequenced from a number of independent isolates there is clear evidence for recombination. For example, the *phoA* gene of *Escherichia coli* supports different phylogenies depending upon the section of the gene examined; two isolates that are closely related in one section may diverge elsewhere (DuBose et al., 1988). There is now similar evidence for various loci in *E. coli*, *Streptococcus* and *Neisseria* (reviewed by Maynard Smith, 1990). The discrepancy between these results and the apparent clonality revealed by enzyme polymorphism may perhaps be resolved by a consideration of the nature of recombination events in bacteria. There are three widely used routes of gene transfer between bacteria (transformation, transduction and conjugation) and they all provide the means to substitute short segments of DNA from the donor for the corresponding segment of the recipient chromosome. This contrasts with the large-scale reassortment of two genomes characteristic of eukaryotic meiosis, and may explain why fine-scale recombination is detected (DNA sequence) while linkage disequilibrium is not disturbed between distant flanking markers (isozymes).

It is true that some plasmids and transposons (notably those carrying antibiotic resistance determinants) do have the ability to move across a remarkably wide range of bacteria, but this does not necessarily invalidate the distinctness of species. Although such mobile elements could carry chromosomal DNA from one species to another, the mismatch repair mechanism ensures that foreign DNA is less likely to become incorporated by homologous recombination than is conspecific DNA (Kondorosi et al., 1980; Baron et al., 1986; Rayssiguier et al., 1989). It seems that this provides an effective barrier allowing the genomes of related species to diverge, while there is still sufficient intraspecific exchange to maintain the coherence of each species (Sharp, 1991). Although mobile genetic elements are common in bacterial populations, most appear to be much more restricted in their distribution than the cosmopolitan resistance determinants, and particular plasmids or insertion sequences may be restricted to a single species or even to certain genotypes (Wheatcroft and Watson, 1988; Young and Wexler, 1988).

Since it seems that bacteria do have gene exchange within species and barriers to gene exchange between species, they could have biological species in the conventional sense. It is not yet clear whether in fact most bacteria do; the critical evidence will come from studies of the distribution of genetic diversity within and between related species.

GENETIC DIVERSITY AND COMMUNITY FUNCTION

If our aim is to understand nutrient cycling through the soil community, is there any need to concern ourselves with species composition or with genetic diversity within species? As long as these remain constant, the community will have predictable properties. The problem, of course, is that real communities are constantly changing in response to environmental factors, and microbial communities can change rapidly and radically. Even if a particular metabolic process continues, there may be a shift in the relative abundance of the species carrying it out. Since, as we shall see later, these different species may be rather distantly related, distinctive features of their enzyme pathways and general physiology may confer quite different properties on the biomass. Furthermore, even identifying the species concerned may not tell the full story, since we know that there is considerable genetic diversity within bacterial species, and local conditions will select local variants that may differ significantly from the type strain.

PHYLOGENY AND PHYSIOLOGY

The most dramatic outcome of the molecular genetic approach to bacterial diversity has been the complete revolution that has taken place during the past decade in our understanding of evolutionary relationships. The key to this revolution has been comparative sequencing of the small subunit ribosomal RNA (16S rRNA) molecule or the genes encoding it. This is a vital structural and functional component of the ribosome, and thus it is universally present and sufficiently conserved to be compared over wide evolutionary distances. The broad picture is presented and discussed in the important review by Woese (1987). Sequence information is now available for well over a thousand bacterial species, so that quite detailed evolutionary trees can be constructed for many important groups of bacteria, using 16S rRNA sequence as a consistent yardstick throughout. The assumption, of course, is that most genes in a bacterium share a common history, and thus the tree constructed for 16S rRNA is in fact a tree for the bacterial species as a whole. Clearly, this will be true only if wide-range lateral transfer between species is not a significant factor for most of the genome. As indicated above, this is probably a reasonable assumption. Where phylogenies can be constructed for other genes, they are usually consistent with that based on 16S, which is reassuring (Amann et al., 1988; Ludwig et al., 1990).

Until the advent of DNA sequence comparisons, bacterial genera were grouped into families on the basis of shared physiological capabilities. Thus, purple non-sulphur phototrophs were placed in the

Rhodospirillaceae, many intracellular animal pathogens in the *Rickettsiaceae* and legume symbionts in the *Rhizobiaceae*. While sequence comparisons have upheld such families, many appear to be heterogeneous, and sometimes the molecular phylogenies imply that organisms with conspicuous physiological differences are much more closely related than those that are superficially similar in physiology. If the 16S phylogeny (*see* Figure 11.1) is to be believed, then the *Rhodospirillaceae*, *Rickettsiaceae*, *Rhizobiaceae* and representatives of other families are thoroughly mixed up. Their distinctive features crop up repeatedly in different parts of the great sequence-based cluster of organisms that is now known as the alpha subdivision of the Proteobacteria (Young, 1992).

Figure 11.1   16S rRNA phylogeny of selected species in the alpha subdivision of the Proteobacteria, indicating physiological properties used in traditional family assignments (branch lengths arbitrary)

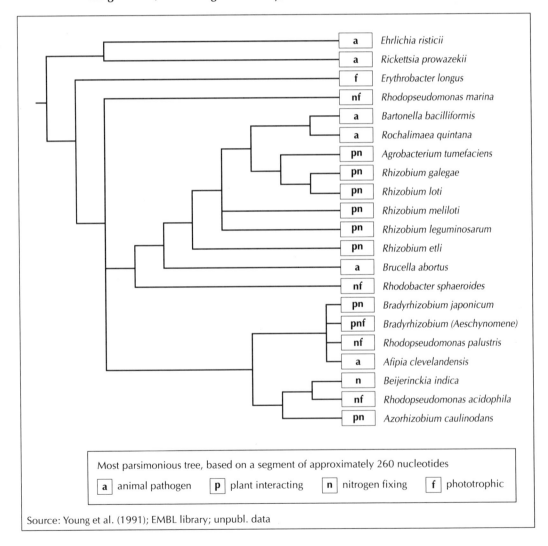

Source: Young et al. (1991); EMBL library; unpubl. data

At first sight, it is puzzling that the 16S rRNA sequence of *Rhodopseudomonas palustris* is more similar to that of the type strain of *Bradyrhizobium japonicum* than are those of many so-called *B. jap–onicum* strains (Young et al. 1991). Are bradyrhizobia really rhodopseudomonads? *B. japonicum* is a symbiont and is not phototrophic, while *R. palustris* is a phototroph that is not symbiotic. A clear distinction, surely? The crucial evidence that lends credence to the molecular phylogeny is provided by an isolate known as BTAi1, which forms nitrogen-fixing nodules on the stems of the legume *Aeschynomene indica*. Not only is this a symbiont, though, it is also pigmented and phototrophic. Thus, it combines the main distinguishing features of the *Rhodospirillaceae* and *Rhizobiaceae*, a problem in terms of classical taxonomy. In the 16S phylogeny, however, its place is clear and consistent; it falls into the *Bradyrhizobium/Rhodopseudomonas* cluster and neatly combines the properties of these two genera. A new piece has recently fallen into this particular puzzle: the 'cat-scratch bacillus' *Afipia* also belongs to this cluster based on 16S sequence (Willems and Collins, 1992; *see* Figure 11.2). As this is a human pathogen and not, as far as we know, symbiotic or phototrophic, the phenotypic versatility of this cluster of organisms appears to be even greater than previously thought.

Figure 11.2   The *Bradyrhizobium/Rhodopseudomonas/Afipia* cluster, showing the number of nucleotide differences in partial 16S rRNA sequences

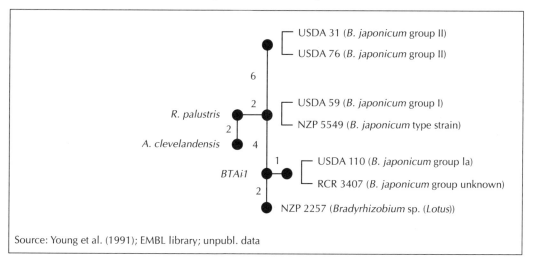

Source: Young et al. (1991); EMBL library; unpubl. data

Our work at the University of York has centred on rhizobia and their relatives, and therefore on the alpha proteobacteria, but the same essential message appears to apply also to bacteria responsible for other important aspects of mineral cycling. The ammonia-oxidising nitrifiers, for example, are split between the gamma and beta proteobacteria, with *Nitrosomonas oceanus* on the one hand and *Nitrosospira* and relatives on the other (Head et al., 1993). The methylotrophs are also phylogenetically diverse (Tsuji et al., 1990).

One result of all this comparative sequence information is that it is relatively easy to design DNA hybridisation probes or polymerase chain reaction (PCR) primers that will serve to detect and identify bacteria from a natural community with a greater or lesser degree of phylogenetic resolution. Examples are discussed elsewhere in this book.

Why does all this phylogenetic excitement matter to the biomass concept? First, parts of the nutrient cycle that have been treated as separate conceptual compartments may be taking place in a single versatile organism (if not simultaneously, then in alternative physiological states). Second, the same reactions may be shared by organisms that are quite distantly related and possibly quite different in other aspects of their physiology. Third, if bacteria are to be detected and isolated based on a particular functional role, it may be necessary to use molecular probes directed against functional components (such as nitrogenase and ammonia mono-oxygenase) rather than those based on phylogenetic position (16S rRNA), since closely related organisms may have different physiologies.

CONCLUSION

I have argued that it is both possible and useful to consider the abundance of individual bacterial species within the soil microbial community. It is possible because bacteria probably do have discrete, biologically defined species, and because DNA sequencing tells us how these are related to each other and can provide us with tools to identify and quantify them. It is useful because an appreciation of the dynamics and evolutionary potential of individual species is necessary in order to predict and understand the response of the biomass to environmental changes.

References

Amann, R., Ludwig, W. and Schleifer, K.H. 1988. Beta-subunit of ATP-synthase: A useful marker for studying the phylogenetic relationship of eubacteria. *J. General Microbiology* 134: 2815-21.

Baron, L.S., Gemski, P., Johnson, E.M. and Wohlhieter, J.A. 1986. Intergeneric bacterial matings. *Bacteriological Reviews* 32: 362-69.

DuBose, R.F., Dykhuizen, D.E. and Hartl, D.L. 1988. Genetic exchange among natural isolates of bacteria: Recombination within the *phoA* gene of *Escherichia coli*. *Proc. National Academy of Sciences of the USA* 85: 7036-40.

Head, I .M., Hiorns, W.D., Embley, T.M., McCarthy, A.J. and Saunders, J.R. 1993. The phylogeny of autotrophic ammonia-oxidising bacteria as determined by 16S ribosomal DNA sequence analysis. *J. General Microbiology* 139: 1147-53.

Hedges, R.W. 1972. The pattern of evolutionary change in bacteria. *Heredity* 28: 39-48.

Kondorosi, A., Vincze, E., Johnston, A.W.B. and Beringer, J.E. 1980. A comparison of three *Rhizobium* gene maps. *Molecular and General Genetics* 178: 403-08.

Ludwig, W., Weizenegger, M., Betzl, D., Leidel, E., Lenz, T., Ludvigsen, A., Mollenhoff, D., Wenzig P. and Schleifer, K.H. 1990. Complete nucleotide sequences of seven eubacterial genes coding for the elongation factor Tu: Functional, structural and phylogenetic evaluations. *Archives of Microbiology* 153: 241-47.

Maynard Smith, J. 1990. The evolution of prokaryotes: Does sex matter? *Annual Review of Systematics and Ecology* 21: 1-12.

Rayssiguier, C., Thaler, D.S. and Radman, M. 1989. The barrier to recombination between *Escherichia coli* and *Salmonella typhimurium* is disrupted in mismatch-repair mutants. *Nature* 324: 396-401.

Selander, R.K., Caugant, D.A., Ochman, H., Musser, J.M., Gilmour, M.N. and Whittam, T.S. 1986. Methods of multilocus enzyme electrophoresis for bacterial population genetics and systematics. *Applied and Environmental Microbiology* 51: 873-84.

Sharp, P.M. 1991. Determinants of DNA sequence divergence between *Escherichia coli* and *Salmonella typhimurium:* Codon usage, map position, and concerted evolution. *J. Molecular Evolution* 33: 23-33

Sonea, S. and Panisset, M. 1983. *A New Bacteriology*. Boston, Massachusetts, USA: Jones and Bartlett.

Tsuji, K., Tsien, H.C., Hanson, R.S., DePalma, S.R., Scholtz, R. and LaRoche, S. 1990. 16S ribosomal RNA sequence analysis for determination of phylogenetic relationship among methylotrophs. *J. General Microbiology* 136: 1-10.

Wheatcroft, R. and Watson, R.J. 1988. Distribution of insertion sequence IS*Rm1* in *Rhizobium meliloti* and other Gram-negative bacteria. *J. General Microbiology* 134: 113-21.

Willems, A. and Collins, M.D. 1992. Evidence for a close genealogical relationship between *Afipia* (the causal agent of cat scratch disease), *Bradyrhizobium japonicum* and *Blastobacter denitrificans*. *FEMS Microbiology Reviews* 75: 241-46.

Woese, C.R. 1987. Bacterial evolution. *Microbiological Reviews* 51: 221-71.

Young, J.P.W. 1989. The population genetics of bacteria. In Hopwood, D.A. and Chater, K.F. (eds) *Genetics of Bacterial Diversity*. London, UK: Academic Press.

Young, J.P.W. 1992. Phylogenetic classification of nitrogen-fixing organisms. In Stacey, G., Burris, R.H. and Evans, H.J. (eds) *Biological Nitrogen Fixation*. New York, USA: Chapman and Hall.

Young, J.P.W., Downer, H.L. and Eardly, B.D. 1991. Phylogeny of the phototrophic rhizobium strain BTAi1 by polymerase chain reaction-based sequencing of a 16S rRNA gene segment. *J. Bacteriology* 173: 2271-77.

Young, J.P.W. and Wexler, M. 1988. Sym plasmid and chromosomal genotypes are correlated in field populations of *Rhizobium leguminosarum*. *J. General Microbiology* 134: 2731-39.

PART III

Tracking specific components of microbial communities

Beyond the Biomass
Edited by K. Ritz, J. Dighton and K.E. Giller
© 1994 British Society of Soil Science (BSSS)
A Wiley-Sayce Publication

CHAPTER 12

Analyses of DNA extracted from microbial communities

D. HARRIS

The recent development and application of techniques for the extraction of relatively intact DNA from complex microbial communities such as those in soil offers new ways to resolving previously intractable problems in microbial ecology. Many of these problems arise from our failure to culture, and thus examine in isolation, almost all the individual organisms seen in soil by microscopy. We do not know the extent to which this uncultured majority comprises moribund examples of commonly isolated species or contains novel and culturally fastidious organisms. However, evidence from community DNA re-annealing experiments (Torsvik et al., 1990; Torsvik, *Chapter 4, this volume*) suggests that the number of species which may occur in a single soil is very large (in excess of 10 000?). If this is true there must be many soil bacterial species which are never isolated in culture. Unless there is *a priori* knowledge of a selective medium it is also very difficult to detect organisms which occur at low frequency in the microbial community using culture-based methods. The complex mixture of sequences obtained by extraction of DNA from soil has been termed the 'community genome' (Dockendorf et al; 1992) and, ideally, contains a gene pool that is qualitatively and quantitatively characteristic of the microbial community. Analysis of this community DNA can provide information at a variety of levels, ranging from broad descriptions of community composition, which may be as general as the ratio of fungal to bacterial DNA in the soil, to the highly specific detection of particular sequences.

The relative abundance of DNA of different base composition (% G + C) within the sample is also being used to provide a 'fingerprint' of overall community composition (Holben et al., 1993). Highly specific identification and quantification of specific sequences by hybridisation with probes which have phylogenetic significance can be used to identify DNA sequences which are characteristic of various levels of phylogeny, from kingdom to subspecies. Similarly, probes for sequences of functional significance can be used to detect and quantify the genes coding for particular enzyme systems at several levels of specificity. This alone does not tell us enough about the expression of the gene or the activity of the enzyme and must be combined with other measurements before we can evaluate its functional significance.

DNA CONTENT OF SOILS

Most mineral soils from surface horizons contain about 10^9 bacteria/g, which at 4 fg DNA per bacteria (1 fg = 10^{-15} g) (Ellenbroek and Cappenberg, 1991) corresponds to 4 µg DNA/g soil, or 1 µg DNA/ 2.5×10^8 bacteria. Soil fungal populations are much more variable and the ratio of hyphal biomass to DNA is indeterminate, so it is difficult to predict average fungal hyphal DNA contents for soils. In a soil containing 200 m/g of hyphae, we found that the DNA content of the isolated hyphae was about 1 µg/g (*see* Table 12.1); extraction of this DNA from the hyphae required exhaustive grinding in liquid nitrogen before applying a hot detergent lysis procedure, a much more vigorous extraction than that normally applied to soils. This suggests that fungal hyphal DNA is probably not extracted by the procedures normally used, where emphasis is more often placed on bacterial DNA. It is possible that a large proportion of fungal DNA in soil is present in spores; to date, there appear to be no measurements of this contribution to total soil DNA content. The biomass of other soil eukaryotes is too low to contribute significantly to the total DNA content of most bulk soils. Some soils may contain significant amounts of 'dead' extracellular DNA preserved in the soil organic matter-clay colloid complex. In my experience, contributions to total DNA content of soils from this source have been slight (< 0.1 µg/g soil).

Table 12.1   Bacterial and fungal biomass and DNA content of soils

	Bacteria	Fungi
Number/length (per g)	2.5×10^9	205 m
Biovolume (µm^3/g)[a]	4.9×10^8	1.4×10^9
Biomass C (µg/g)[b]	69	218
DNA (µg/g)		
theoretical[c]	10	?
measured[d]	?	1.3
extracted[e]	10	1

Note: a   Calculated using a mean bacterial volume of 0.196 µm^3; this value was the mean volume of 480 soil
            bacteria measured individually
      b   Calculated assuming specific gravity = 1.07, dry matter = 30% and C content = 44% (Bakken and
            Olsen, 1983)
      c   Theoretical bacterial DNA content assumes 4×10^{-15} g DNA/cell (Ellenbroek and Cappenberg, 1991)
      d   DNA content of extracted fungal hyphae measured using the diphenylamine method
      e   Bacterial DNA extracted after lysis in whole soil; fungal DNA extracted from isolated hyphae

EXTRACTION OF DNA FROM SOILS

Almost every laboratory engaged in the extraction of DNA from soils has developed its own extraction methods. The procedures are, however, broadly similar variants of two basic approaches:

- *Direct lysis.* In this approach, the microbes are lysed in the soil and the free DNA is extracted and purified. Most direct lysis procedures are based on the work of Ogram et al. (1987). Developments and specialisations have led to the reduction of soil sample sizes by 10-500 times from 100 g in the original procedure (for example, Picard et al., 1992; Holben, 1993). The direct lysis procedures appear to recover DNA quantitatively from favourable soils (*see* Table 12.1) but at much lower efficiency from high clay soils (Holben, 1993).

- *Microbial fractionation.* In this approach, the microorganisms are separated by repeated differential centrifugation from most of the soil before lysis (Torsvik, 1980; Holben et al., 1988; Holben, 1993). Because only 25-30% of the soil bacteria are obtained using the fractionation procedure, larger soil samples (usually 50 g) must be used. This approach has also been used to obtain fungal hyphal DNA from soil (D. Harris, unpubl.) after separation by a wet sieving and differential sedimentation procedure (Bingle and Paul, 1986).

Each method has certain advantages and the choice has to be made according to the nature of the soils and the questions being addressed. Important considerations are:

- Direct lysis recovers more DNA than microbial fractionation, particularly from low clay soils, and is therefore preferred, especially when biomass is low or sample size limited.

- Direct lysis probably extracts a wider sample of the total community DNA than microbial fractionation.

- Direct lysis is faster and can be performed on more samples simultaneously.

- Humic contamination of direct lysis-DNA is usually greater than in microbial fractionation-DNA. This may reduce its suitability for studies using polymerase chain reactions (PCR) or restriction endonuclease digestion and may obviate its use in organic soils.

- Microbial fractionation-DNA is usually of higher molecular weight than direct lysis-DNA.

It is also important to note that all DNA extraction methods which are sufficiently gentle to allow the isolation of high-molecular-weight DNA from soil will introduce some form of bias. We know that many groups of bacteria (for example, methanogens) are extremely resistant to lysis and will be under-represented in the final DNA sample obtained by most extraction procedures. If the purpose is to detect specific organisms in the soil community for which the DNA extraction can be tested and shown to be efficient, then this bias may be unimportant or advantageous and negative findings informative. However, if the purpose is to examine the community for the presence of sequences which may occur in many species, there is a high probability of low extraction efficiency from some organisms, and thus false negatives or biased ratios. This difficulty does not invalidate the approach but requires experiments designed to accommodate these restrictions.

ANALYSIS OF COMMUNITY DNA

Fractionation by base composition

The DNA binding dye bisbenzimide (Hoechst 33258) has the property of binding to double-stranded DNA in a complex variety of modes. At high salt concentration and suitable stoichiometry, regions of DNA rich in A-T pairs are preferred binding sites (Loontiens et al., 1990). The buoyant density of the resulting DNA-bisbenzimide complex is reduced in proportion to the amount of dye bound. This enables the separation of DNA according to its % G + C by centrifugation to near isopycnic equilibrium in CsCl gradients (Holben and Harris, 1991). The resulting CsCl gradient can be fractionated and a profile of the relative abundance of DNA in each % G + C fraction generated after standardisation against DNA mixtures of known composition (*see* Figure 12.1). The resulting profile is a low resolution fingerprint which is characteristic of the DNA extracted from the soil. Figure 12.2 shows a typical profile juxtaposed to a box plot of the distribution of % G + C in some common soil genera (Krieg and Holt, 1984). Most aerobic soils that we have studied have shown broadly similar patterns, with a single dominant peak at about 68% G + C. The profiles can be used to follow the effects of perturbations such as waterlogging or substrate amendment which cause large changes in the relative abundance of components of the soil microbial community.

Figure 12.1   **Separation of standard mixture of DNAs from *Clostridium perfringens* (31% G + C),**
                    ***Escherichia coli* (51% G + C) and *Micrococcus luteus* (71% G + C), using isopycnic**
                    density gradient centrifugation of DNA-bisbenzimide in CsCl gradients

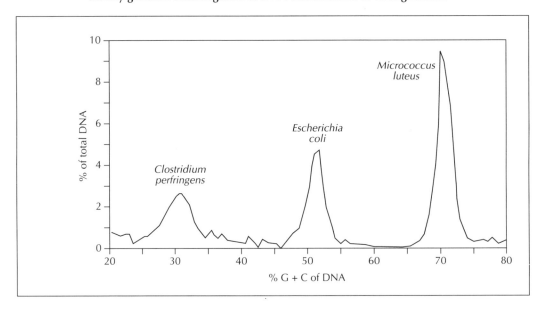

This kind of low-resolution profile can offer relatively simple methods of characterising whole communities and provides overall measurements of the relative abundance of members of the microbial community which are represented in the DNA extract. The ability to sort DNA according to some general differential property such as % G + C is potentially a very powerful tool for the microbial ecologist because it allows the very complex total soil DNA mixture to be partitioned into smaller groups which can then be separately analysed in greater detail. To date, no other techniques which are comparable to bisbenzimide gradient fractionation have been developed but such techniques are sure to be developed in the future. The most important restriction is that the separation should be insensitive to the molecular weight of the DNA because the DNA is randomly sheared during extraction.

Figure 12.2   Relative abundance of DNA of differing % G + C in microbial community DNA extracted from an agricultural soil (2.3% organic matter, 12% clay) compared with the distribution of % G + C in known species of some common soil genera

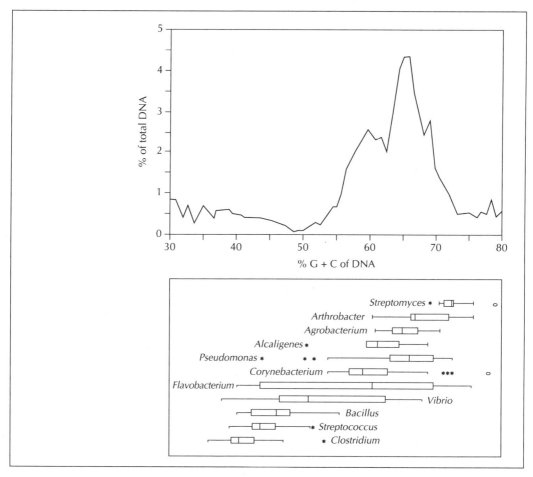

Community DNA fractionation can be made much more powerful in combination with other measurements or assays which provide additional dimensions to the data. One example of such a combination in our studies is the fractionation of community DNA from a soil which had been labelled for 2 hours with ^{3}H thymidine (*see* Figure 12.3 *overleaf*). The profiles of total DNA and radioactivity are remarkably similar. This means that the average specific replication rates of DNA from organisms in all fractions were uniform. This indicates that microbial growth in soil is not confined to a very small active group of species but is widespread amongst the genera represented in the profile. The uniformity between fractions is probably a reflection of the extreme complexity of the soil microbial community. Such uniformity is expected if each % G + C fraction contains many, perhaps hundreds, of species and their replication rates are uncorrelated with % G + C. Each fraction would then contain a large random sample of the whole community with respect to replication rate, and fraction means would tend towards the mean of the whole DNA population.

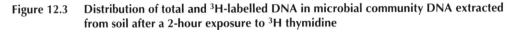

Figure 12.3   Distribution of total and ^{3}H-labelled DNA in microbial community DNA extracted from soil after a 2-hour exposure to ^{3}H thymidine

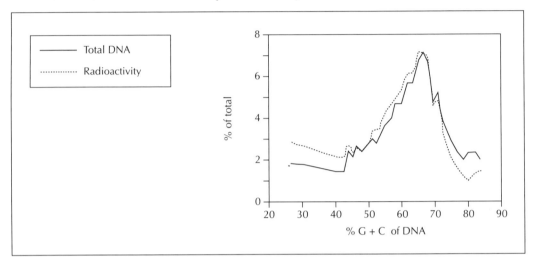

Hybridrisation techniques

Nucleic acid probe techniques which were originally developed in clinical microbiology are now widely used in microbial ecology for the detection and enumeration of specific genes and genotypes in complex communities such as soil. Radioactive or fluorescent labelled probes are used against target DNA either in extracts or against fixed cells *in situ*. The subject of probe design and preparation and the conduct of hybridisation experiments has been well covered by Matthews and Kricka (1988), Dockendorf et al. (1992) and Knight et al. (1992). Here, only some general factors which affect the application of probe techniques to DNA extracted from soil will be considered, along with some illustrative examples.

Two aspects of the application of probe techniques to DNA extracted from soil are especially pertinent to the general problem of detecting and enumerating particular sequences in the community genome. The first of these are practical detection limits for target sequences. It has been shown that 0.02 pg of target sequence can be detected on slot blots containing 1 µg of soil community DNA using a single-stranded DNA probe of very high specific activity (Holben et al., 1988). The hydridisation signal can be measured quantitatively to give an estimate of the number of target sequences in the DNA sample. The number of organisms that this amount of target represents depends upon size of the target sequence, number of copies in the genome and size of the genome. If the target were, for example, a single-copy 1 kb segment from a bacterial genome of 5000 kb total size, 0.02 pg of target would correspond to 0.1 ng of genomic DNA or one part in 10^4 of the total in the blot. With DNA derived from a bacterial population of 10^9/g soil, this detection limit would correspond to 10^5 organisms/g soil. Detection limits for probes prepared by conventional techniques such as nick translation, which result in probes of much lower specific activity, are at least an order of magnitude greater. Thus, direct searches for specific organisms in whole community DNA are restricted to those that are present in rather high numbers unless either very high molecular weight probes are used or multicopy target sequences can be found. Target sequences can be amplified by PCR to increase the potential sensitivity considerably (Herrick et al., 1993). However, great care must be taken to avoid artefacts arising in the PCR, or the possibility for direct quantitative measurement is lost.

The second important aspect is the extreme complexity of the DNA mixture obtained from soil. This rapidly becomes evident where probes with broad specificity are used. Examples include attempts to assess the structure of soil communities through analysis of restriction fragment length polymorphism (RFLP) and related techniques where gel electrophoresis and Southern hybridisation are used to separate restriction fragments of characteristic molecular weight which include a common target sequence. The resolution of these techniques can easily be overwhelmed by the complexity of the pattern of restriction fragments obtained, such that the Southern blot becomes an undifferentiated smear.

A strategy which helps both to increase the sensitivity of detection of specific sequences and to reduce the complexity of the DNA mixture is to fractionate the sample before attempting hybridisation studies. Currently, the fractionation according to % G + C, obtained using bisbenzimide, is the only suitable technique available. There are two sets of advantages to its use in combination with gene probes. First, any particular fraction contains a small subset of the original DNA sample, and thus the concentration of a desired sequence will increase in one or more fractions and decrease in others, relative to the original sample. The extent of this effect will vary in different regions of the profile. In the peak region typically found at 65-70% G + C (*see* Figure 12.1), the concentration increase may be only about tenfold. In regions of the profile containing small proportions of the total DNA, the relative increase in concentration of a particular genome may be very large (1000 times?). This reduces the detection limits for all target organisms or genes and greatly reduces the limit for those with genomes at a % G + C of low abundance. Similarly, the complexity of the mixture in any particular fraction will probably be reduced in comparison with the original sample. The magnitude of this effect on complexity will have to be tested empirically, since it need have little relationship to the amount of DNA in the fraction.

The localisation of a hybridisation signal within a community profile is illustrated by the work of Holben et al. (1993) who used a probe for a gene encoding an enzyme of the 2,4-D degradation pathway and showed that the hybridisation signal occurred only at a characteristic % G + C in a community DNA sample. The fractionation of the DNA could be also be used to separate the DNA of different organisms which hybridised to the same probe. For example it should be possible to detect the occurrence of particular enzyme systems in different fractions, and thus different organisms, within a bulk DNA sample using functional probes.

CONCLUSION

DNA-based methods for the detection of phylogenetic groups, specific organisms and functional genes in the soil community show great potential. However, their use to date has been confined mainly to methodological demonstration experiments. This is not suprising in view of the enormous complexity of the problem. This complexity poses the central technical problem in the detection and analysis of the community components. Techniques such as % G + C fractionation of community DNA are useful at both broad and specific levels of detail. At the broad level, the fractionation provides a profile of the relative abundance of DNA from different components of the community that are represented in the extract and is characteristic of the soil community from which it was extracted. At more specific levels, each fraction represents a particular subset of the whole community and is thus more easily analysed by the whole battery of hybridisation and reassociation techniques that are available.

Acknowledgements

The work described in this chapter was funded by a grant (No. BIR 9120006) from the National Science Foundation.

References

Bakken, L.R. and Olsen, R.A. 1983. Buoyant density and dry matter contents of microorganisms. The conversion of measured biovolume into biomass. *Applied and Environmental Microbiology* 45: 1188-95.

Bingle, W.H. and Paul, E.A. 1986. A method for separating fungal hyphae from soil. *Canadian J. Microbiology 32*: 62-66.

Dockendorf, T.C., Breen, A., Ogunseitan, O.A., Packard, J.G. and Sayler, G.S. 1992. Practical considerations of nucleic acid hybridisation and reassociation techniques in environmental analysis. In Levin, M.A., Seidler, R.J. and Rogul, M. (eds) *Microbial Ecology Principles, Methods and Applications.* New York, USA: McGraw Hill.

Ellenbroek, F.M. and Cappenberg, T.E. 1991. DNA synthesis and tritiated thymidine incorporation by hetero-trophic freshwater bacteria in continuous culture. *Applied and Environmental Microbiology* 57: 1675-82.

Herrick, J.B., Madsen, E.L., Batt, C.A. and Ghiorse, W.C. 1993. Polymerase chain reaction amplification of napthalene-catabolic and 16SrRNA gene sequences from indigenous sediment bacteria. *Applied and Environmental Microbiology* 59: 687-94.

Holben, W.E. and Harris, D. 1991. Monitoring changes in microbial community composition by fractionation of total community DNA based on % G+C content. In *Proc. 91st Annual Meeting of the American Society of Microbiology.* New Orleans, USA: American Society of Microbiology.

Holben, W.E. 1993. Isolation and purification of bacterial DNA from soil. In Keeney, D.R. (ed) *Methods of Soil Analysis.* Madison, Wisconsin: American Society of Agronomy.

Holben, W.E., Calabrese, V.G.M., Harris, D., Ka, J.O. and Tiedje, J.M. 1993. Analysis of structure and selec-tion in microbial communities by molecular methods. In *Proc. 6th International Symposium on Micro-bial Ecology., Barcelona, Spain.* Washington DC, USA: American Society for Microbiology. (in press).

Holben, W.E., Jansonn, J.K., Chelm, B.K. and Tiedje, J.M. 1988. DNA probe method for the detection of spec-ific microorganisms in the soil bacterial community. *Applied and Environmental Microbiology* 54: 703-11.

Knight, I.T., Holben, W.E., Tiedje, J.M. and Colwell, R.R. 1992. Nucleic acid techniques for detection, identification, and enumeration of microorganisms in the environment. In Levin, M.A., Seidler, R.J. and Rogul, M. (eds) *Microbial Ecology Principles, Methods and Applications.* New York, USA: McGraw Hill.

Krieg, N.R. and Holt, J.G. (eds) 1984. *Bergey's Manual of Systematic Bacteriology.* Baltimore, Maryland, USA: Williams and Wilkins.

Loontiens, F.G., Regenfuss, P., Zechel, A., Dumortier, L. and Clegg, R.M. 1990. Binding characteristics of Hoechst 33258 with calf thymus DNA, poly[d(A-T)], and d(CCGGAATTCCGG): Multiple stoichiometries and determination of tight binding with a wide spectrum of site affinities. *Biochemistry* 29: 9029-39.

Matthews, J. and Kricka, L.J. 1988. Analytical strategies for the use of DNA probes. *Analytical Biochemistry* 169: 1-25.

Ogram, A., Sayler, G.S. and Barkay, T. 1987. The extraction and purification of microbial DNA from sediments. *J. Microbiological Methods* 7: 57-66.

Picard, C., Ponsonnet, C., Paget, E., Nesme, X. and Simonet, P. 1992. Detection and enumeration of bacteria in soil by direct DNA extraction and polymerase chain reaction. *Applied and Environmental Microbiology* 58: 2717-22.

Torsvik, V. 1980. Isolation of bacterial DNA from soil. *Soil Biology and Biochemistry* 12: 15-21.

Torsvik, V., Goksøyr, J. and Daae, F.L. 1990. High diversity in DNA of soil bacteria. *Applied and Environmental Microbiology* 56: 782-87

Beyond the Biomass
Edited by K. Ritz, J. Dighton and K.E. Giller
© 1994 British Society of Soil Science (BSSS)
A Wiley-Sayce Publication

CHAPTER 13

Plasmid transfer in agricultural soils as influenced by sucrose and ground sugar beet

W. KLINGMULLER

Gene transfer between bacteria in the environment is an important biological phenomenon (Levy and Miller, 1989). However, it is difficult to study, particularly in natural soil. There are a large number of biotic and abiotic factors that can affect the process (Fry and Day, 1990; Gauthier, 1992; Wellington and van Elsas, 1992). The problems involved and the methods currently available for such studies have been reviewed by Stotzky et al. (1990, 1991) and Stotzky (1992).

A particularly fascinating aspect of bacterial gene transfer is conjugative plasmid transfer. We have focused on this for agricultural soil, and in earlier studies measured the influence of nutrients and plant remains on homologous and heterologous conjugative plasmid transfer (Klingmüller et al., 1990; Klingmüller, 1991a, b, 1992, 1993). The bacteria used were nitrogen-fixing *Enterobacter* strains, isolated from the rhizosphere of wheat and barley (Kleeberger et al., 1983; Singh et al., 1983; Singh and Klingmüller, 1986; Singh et al., 1988), *Pseudomonas* (Temann et al., 1992) and *Escherichia coli*. The plasmids studied are pEA9, a large, self-transmissible *Enterobacter* plasmid of narrow host range with the gene set for nitrogen fixation (Klingmüller et al., 1989; Steibl et al., 1993), and the resistance plasmid RP4 with a wide host range and high transfer rates (Datta et al., 1971; Thomas, 1981; Guiney and Lanka, 1989). As complex substrates gave complex results, we decided to study pure substrates (for example, sucrose, glucose, fructose and citrate) as additives to the soil. This chapter presents the results of this work.

MATERIALS AND METHODS

Strains

Enterobacter agglomerans 339 was available in our collection (Kleeberger et al., 1983). It was cured of its own plasmid pEA9 (Singh et al., 1983); if used as a donor, it contained plasmid RP4 instead. *Pseudomonas aeruginosa* (prototrophic) was obtained from Dr U. Römling, Germany; again, if used as a donor, it contained plasmid RP4. All strains, if used as donors, were nal[r]; if used as recipients they were sm^r. In addition, the *Enterobacter* strains were rif^r. Plasmid RP4 is a plasmid of 60 kb, coding resistance for ap, tc and km (Guiney and Lanka, 1989). It is identical to RK2 (Inc P-1 group).

Mating and screening for exconjugants

In standard matings, donor and recipient cells were grown as overnight cultures in liquid luria broth (LB) medium with appropriate antibiotics (the concentrations are given below). They were then washed, resuspended in saline solution and mixed 1/1 (volume and cell titres). For the mating, 4 ml of cell mixture with (usually) 10^9 cells of the donor and 10^9 cells of the recipient, in saline, was dripped with a pipette onto the surface of 50 g soil in a 250 ml Erlenmeyer flask. The flasks were sealed with cotton and incubated at 22°C for 24 hours. In the experiments shown in Table 13.1, the cell mixture was administered after replacing 2 ml of the saline by 1.7 ml minimal solution M56 (8x) and 0.3 ml of the respective sugars (or citrate) as 20% stock solution. Thus, the final amount of sugars (or citrate) per flask at zero time was 0.06 g. The soil was later suspended with 50 ml saline solution, stirred using a rotary shaker (Infors TR-125) at 180 rpm for 30 minutes and allowed to settle for 20 minutes. Aliquots were taken from the supernatant for dilution series, and platings made on LB plates, supplemented with antibiotics as required to identify donors, recipients and exconjugants. Presumptive exconjugant colonies were further screened for rifampicin resistance, and in critical cases for nalidixic acid sensitivity, citrate utilisation and the physical presence of RP4. Transfer rates were expressed as number of (true) exconjugants per donor cells surviving at the time of sampling.

Soil

The soil used was agricultural sandy loam (1.5% organic carbon; 13.2 C/N; 7.2 meq CEC/100 g; 78% base saturation; pH 5.2; 13.2 mg P_2O5, 52.1 mg K_2O and 11.9 mg MgO/100 g). It was air-dried to 7.6% remaining water content, and passed through a sieve (2 mm).

Growth media and solutions

The LB contained (g/l demineralised water): tryptone 10; yeast extract 5; and NaCl 5. For plates, 1.5% Difco agar was added. The saline solution was 0.85% NaCl in water. M56 was used at 8 times normal strength. As such, it had (per litre demineralised water): 43.2g Na_2HPO_4 x 2 H_2O; 28.4g K_2HPO_4 x 3 H_2O; 8.0g $(NH_4)_2SO_4$; 8.0 ml $MgSO_4$ x 7H_2O (10% w/v); 2.8 ml Ca $(NO_3)_2$ x 4H_2O (1% w/v); 4 ml $FeSO_4$ x 7 H_2O (0.05% w/v) (Nguyen et al., 1983).

Antibiotics

In the selective plates, kanamycin was applied at 20 µg/ml, nalidixic acid, cycloheximide and nystatin at 50 µg/ml, neomycin and rifampicin at 100 µg/ml and streptomycin at 200 µg/ml. For example, plates to select for *Enterobacter* exconjugants against a non-sterilised soil suspension background contained km, neo, sm, cycloheximide and nystatin. The spontaneous rate of mutation of the *Enterobacter* donor to sm^r was < 10^{-10}.

In situ exconjugants

In the experiments with RP4, at high plating densities of the conjugation mixture on the plates to select for exconjugants, a number of the developing colonies could, in spite of the antibiotics added, be attributable to (undesired) *in situ* matings on the plates (Smit and van Elsas, 1990; Walter et al., 1991; W. Stelzer and E. Schneider, pers. comm.). Their rate was enumerated in parallel platings of suspensions, from bacteria separately incubated in soil but mixed on the selective plates, and deducted when calculating the rates in the table.

Sugar beet

Sugar beets were the Victoria variety (Kleinwanzlebener Saatzucht, Germany). They were ground using a blender and kept frozen at -20°C. Immediately before use, the samples were ground to a powder in liquid nitrogen, using a pestle and mortar. The powder was then dusted on top of the soil samples, before adding the inoculation mixture.

Statistics

All the experiments were set up at least twice, independently. Where two independent experiments were conducted, the averages were calculated; where more than two were conducted, the means and the standard error of the mean were calculated. In Tables 13.1 and 13.3 and in Figure 13.1 the single standard error of the mean is given.

RESULTS

During the first 24 hours of incubation of the bacterial inoculation mixtures at 22°C, in non-sterile soil, no cell propagation was observed, either of the donors or of the recipients, nor did exconjugants appear (Klingmüller et al., 1990; Klingmüller, 1991a, b; data not shown). However, in the presence of sucrose and ground sugar beet, we observed a rapid increase of bacterial numbers after 24 hours, the titre of donors and recipients exceeding the inoculation titre by a factor of 10 for *Enterobacter* + sucrose, by a factor of 10 for *Pseudomonas* + sucrose and by a factor of about 30 for *Pseudomonas* + ground sugar beet. This increase was linked to an increase in plasmid transfer, from zero to 10^{-3} per donors/g soil and more after the first day for *Enterobacter* matings, and from zero to between 10^{-2} and 10^{-1} for *Pseudomonas* matings (*see* Table 13.1 *overleaf*).

In contrast to the *Enterobacter* strain used, the *Pseudomonas* strain was unable to utilise sucrose, and thus its performance in soil with sucrose could not be explained initially. Studies with glucose, fructose and citrate, instead of sucrose, gave additional information (*see* Table 13.1). With these substrates, growth and plasmid transfer was registered as well, not only for matings of *Enterobacter* with the homologous partner (line a in Table 13.1), but also for matings of *Pseudomonas* (line b). These substrates, in contrast to sucrose, were utilised by both organisms (*see* Table 13.2 *overleaf*).

The problem with *Pseudomonas* (that is, its growing and transferring plasmid RP4 with sucrose as substrate) was resolved by following sucrose degradation in flasks with natural soil, without inoculation. Here, there was a rapid degradation of sucrose (*see* Figure 13.1 *overleaf*), together with

Table 13.1   Rates of plasmid RP4 transfer[a] obtained in soil, with different sugars or related substrates, for two pairs of bacterial donor and recipient strains: (a) sucrose utilising and (b) sucrose non-utilising

		M56				
		1 + sucrose	2 + glucose	3 + fructose	4 + citrate	5 + sugar beet[b]
(a)	Enterobacter agglomerans (E.a.) (RP4) x E.a.	7.1 ± 2.0 x 10^{-3}	7.7 ± 1.3 x10^{-3}	7.4 ± 3.0 x 10^{-3}	1.7 ± 0.3 x 10^{-2}	1.4 ± 1.0 x 10^{-3}
(b)	Pseudomonas aeruginosa (prototrophic) (PAO) (RP4) x PAO	1.4 ± 0.3 x 10^{-1}	1.1 ± 0.4 x 10^{-1}	1.9 ± 0.6 x 10^{-3}	5.1 ± 2.1 x 10^{-2}	3.8 ± 0.7 x 10^{-2}

Note: a   Calculated per donor; 0.06 g sugar (citrate) given to 50 g soil at zero time
        b   Beet contained 21.9% sucrose, 0.19% glucose and 0.27% fructose (fresh weight)

Table 13.2   Growth of bacterial inoculants of the indicated strains in liquid culture + M56 ions, and indicated energy source at 1% each, or in luria broth (LB) at normal strength, after 2 days rolling at 30°C[a]

		M56			
	+ LB	+ sucrose	+ glucose	+ fructose	+ citrate
Enterobacter agglomerans (E.a.)	+	+	+	+	+
E.a. (RP4)	+	+	+	+	+
Pseudomonas aeruginosa (prototrophic) (PAO)	+	-	+	+	+
PAO (RP4)	+	-	+	+	+
Escherichia coli (E.c.)	+	-	+	+	+
E.c. (RP4)	+	-	+	+	+

Note: a   The inoculum at day zero delivered 10^{-3} of culture which had been washed overnight
            (about 10^{-6} cells/ml final concentration)

the occurrence of glucose and fructose. In autoclaved soil, the content of added sucrose remained constant (*see* Table 13.3) and only traces of glucose and fructose were found. If the kinetics of sucrose degradation, cell growth and plasmid transfer is followed with *Pseudomonas* in non-sterile soil + sucrose, in one experiment over 24 hours, data such as those in Figure 13.2 (*overleaf*) can be obtained. They demonstrate the link, albeit after a lag phase, between bacterial growth and plasmid transfer, on the one hand, and the degradation of sucrose and the occurrence and subsequent disappearance of glucose and fructose, on the other.

Figure 13.1   Decomposition of sucrose and occurrence of glucose and fructose over time in
a non-sterile soil incubated at 22°C for 18 hours

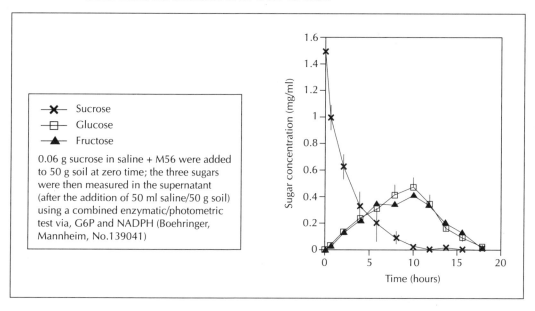

Table 13.3   Stability of sucrose in sterilised soil[a]

| | Incubation (hours) | | |
	0	6	24
Sucrose (g/l)	1.08 ± 0.16	1.02 ± 0.16	1.20 ± 0.06
Glucose	0.01	0.03	0.03
Fructose	0.01	0.00	0.03

Note: a   Addition of sucrose to the soil and determination of the sucrose, glucose and fructose as described in
Figure 13.1; means of four samples each (two independent experiments) and SEM

DISCUSSION

Since the *Pseudomonas* strains used were unable to utilise sucrose, as shown in Table 13.2, the stimulation of bacterial growth and plasmid transfer after the addition of this sugar (or of ground sugar beet material) to the soil samples cannot be the result of sucrose utilisation by the inoculum. However, in non-sterile soil samples there are indigenous microflora, the action of which may cause invertase to split the sucrose. Also, there are soil enzymes in free or particle-bound form, including soil invertase (Beck, 1973; Kiss et al., 1975; Burns, 1978, 1986; Tabatabai, 1981; Nannipieri et al., 1990). Such invertase can easily and immediately provide glucose and fructose from the degradation of sucrose. We have initiated studies to examine the two possibilities (microflora and soil enzyme contribution) and to

Figure 13.2   Decomposition of sucrose and occurrence of glucose and fructose over time, correlated with growth dynamics and plasmid transfer of *Pseudomonas* (non-sterile soil was incubated at 22°C for 24 hours)

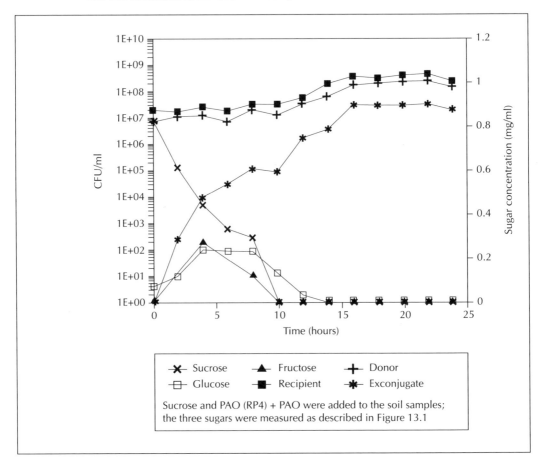

check several soils, particularly those from sugar beet fields, for their content of invertase-producing microbes and free soil invertase activity.

Since sucrose will also stimulate the growth of the indigenous microbes in general, which then will compete with the inoculum for sucrose and its decomposition products, it seems premature to present stoichiometric calculations on the concentration of the several sugars, in addition to the kinetics of growth of the inoculants and plasmid transfer in the soil investigated here. However, it does seem clear that the performance of *Pseudomonas* in soil containing sucrose is a secondary phenomenon, depending upon the degradation of sucrose by the indigenous microflora and/or by free soil invertase present in the non-sterile soil used. In the course of sucrose degradation, glucose appears in the soil, together with fructose, so that even bacteria such as *Pseudomonas* that are unable to utilise sucrose will be able to grow and exchange plasmids, using glucose and fructose.

Soil is a highly complex system, and non-sterile soil even more so. It would be desirable to be able to predict and direct what happens to bacteria released into the soil for agricultural purposes. In an earlier publication (Klingmüller, 1992) it was suggested that genetically determined properties of soil bacteria (for example, their ability or inability to utilise particular organic substrates or energy sources) might offer a means of predicting or directing their fate upon release. The data presented here invalidate that assumption for an agriculturally important example. Since even bacteria which are unable to utilise sucrose can, in non-sterile soil, grow on this sugar and exchange plasmids, albeit indirectly, the fate and risk provided by such bacteria when released into a field cannot be predicted at all with the present level of knowledge.

Acknowledgements

The author wishes to thank the BMFT, Bonn, for financially supporting the work described here, M. John for invaluable technical assistance, Professor H. Goldbach for the soil analyses, M. Daniel for assistance in translation, and A. Lechler and S. Winkler for their assistance with the figures.

References

Beck, T. 1973. The measurement of biological activity in soils by using soil model systems. *Bayerisches Landwirtschaftliches Jahrbuch* 50: 270-88.

Burns, R.G. 1978. *Soil Enzymes*. New York, USA: Academic Press.

Burns, R.G. 1986. Interaction of enzymes with soil minerals and organic colloids. In Huang P.M. and Schnitzer, M. (eds) *Interactions of Soil Minerals with Natural Organics and Microbes*. Madison, Wisconsin, USA: Soil Science Society of America.

Datta, N., Hedges, R.W., Shaw, E.J., Sykes, R.B. and Richmond, M.H. 1971. Properties of an R factor from *Pseudomonas aeruginosa*. *J. Bacteriology* 108: 1244-49.

Fry, J.C. and Day, M.J. (eds) 1990. *Bacterial Genetics in Natural Environments*. London, UK: Chapman and Hall.

Gauthier, M.J. (ed) 1992. *Gene Transfers and Environment*. New York, USA/Berlin, Germany: Springer-Verlag.

Guiney, D.G. and Lanka, E. 1989. Conjugative transfer of IncP-plasmids. In Thomas, C. (ed) *Promiscous Plasmids of Gram-Negative Bacteria*. London, UK: Academic Press.

Kiss, S., Dragan-Bularda, M. and Radulescu, D. 1975. Biological significance of enzymes accumulated in soil. In *Advances in Agronomy* (Vol. 27). New York, USA: Academic Press.

Kleeberger, A., Castorph, H. and Klingmüller, W. 1983. The rhizosphere microflora of wheat and barley with special reference to Gram-negative bacteria. *Archives of Microbiology* 136: 306-11.

Klingmüller, W. 1991a. Plasmid transfer in natural soil: A case-by-case study with nitrogen fixing *Enterobacter*. *FEMS Microbiology Ecology* 85: 107-16.

Klingmüller, W. 1991b. Contained experiments and risk assessment for releases of N_2-fixing *Enterobacter*, with special focus on possible plasmid spread. In MacKenzie, D.R. and Henry, S.C. (eds) *Biological Monitoring of Genetically Engineered Plants and Microbes*. Bethesda, Maryland, USA: Agricultural Research Institute.

Klingmüller, W. 1992. Risk assessment in releases of nitrogen fixing *Enterobacter* in soil: Survival and gene transfer as influenced by agricultural substrates. In Stewart-Tull, D.E. and Sussman, M. (eds) *The Release of Genetically Modified Microorganisms (REGEM II)*. New York, USA: Plenum Press.

Klingmüller, W. 1993. Plasmid transfer in natural soil, as stimulated by sucrose, wheat rhizosphere and ground sugar beets: A study with nitrogen-fixing *Enterobacter*. *Microbial Releases* 1: 229-35.

Klingmüller, W., Herterich S. and Min, B.W. 1989. Self transmissible *nif*-plasmids in *Enterobacter*. In Skinner, F.A., Boddey, R.M. and Fendrik, I. (eds) *Nitrogen Fixation with Non-Legumes*. Dordrecht, The Netherlands: Kluwer.

Klingmüller, W., Dally, A., Fentner, C. and Steinlein, M. 1990. Plasmid transfer between soil bacteria. In Fry, J.C. and Day M.J. (eds) *Bacterial Genetics in Natural Environments*. London, UK: Chapman and Hall.

Levy, S.B. and Miller, R.V. (eds) 1989. *Gene Transfer in the Environment*. New York, USA: McGraw-Hill.

Nannipieri, P., Grego, S. and Ceccanti, B. 1990. Ecological significance of the biological activity in soil. In Bollag, J.M. and Stotzky, G. (eds). *Soil Biochemistry* (Vol. 6). New York, USA: Marcel Dekker.

Nguyen, N.D., Göttfert, M., Singh, M. and Klingmüller, W. 1983. *Nif*-hybrids of *Enterobacter cloacae*: Selection for *nif*-gene integration with *nif*-plasmids containing the Mu transposon. *Molecular and General Genetics* 192: 439-43.

Singh, M. and Klingmüller, W. 1986. Cloning of pEA3, a large plasmid of *Enterobacter agglomerans* containing nitrogenase structural genes. *Plant and Soil* 90: 235-42.

Singh, M., Kleeberger, A. and Klingmüller, W. 1983. Location of nitrogen fixation (*nif*) genes on indigenous plasmids of *Enterobacter agglomerans*. *Molecular and General Genetics* 190: 373-78.

Singh, M., Kreutzer, R., Acker, G. and Klingmüller, W. 1988. Localisation and physical mapping of a plasmid-borne 23 kb *nif*-gene cluster from *Enterobacter agglomerans* showing homology to the entire *nif*-gene cluster of *Klebsiella pneumoniae* M5a1. *Plasmid* 19: 1-12.

Smit, E. and van Elsas, J.D. 1990. Determination of plasmid transfer frequency in soil: Consequences of bacterial mating on selective plates. *Current Microbiology* 21: 151-57.

Steibl, H.-D., Siddavattam, D. and Klingmüller, W. 1993. Similar *nif*-clusters on dissimilar plasmids of *Enterobacter agglomerans* strains isolated from the rhizosphere of wheat. *Plasmid* (in press).

Stotzky, G. 1992. Gene transfer among and ecological effects of genetically modified bacteria in soil. In Casper, R. and Landsmann, J. (eds) *The Biosafety Results of Field Tests of Genetically Modified Plants and Microorganisms*. Braunschweig, Germany: Biologische Bundesanstalt für Land und Forstwirtschaft.

Stotzky, G., Devanas, M.A. and Zeph, L.R. 1990. Methods for studying bacterial gene transfer in soil by conjugation and transduction. *Advances in Applied Microbiology* 35: 157-69.

Stotzky, G., Zeph, L.R. and Devanas, M.A. 1991. Physicochemical and biological factors affect the transfer of genetic information among microorganisms in soil. In Ginzburg, L. (ed) *Assessing Ecological Risks of Biotechnology*. London, UK: Butterworth-Heinemann.

Tabatabai, M.A. 1981. Soil enzymes. In Page, A.L., Miller, R.M. and Keeney, D.R. (eds). *Methods of Soil Analysis: Chemical and Microbiological Properties. Part 2. Agronomy* (2nd edn). Madison, Wisconsin, USA: American Society of Agronomy.

Temann, U.A., Hösl, R. and Klingmüller, W. 1992. Plasmid transfer between *Pseudomonas aeruginosa* strains in natural soil. In Stewart-Tull, D.E. and Sussman, M. (eds) *The Release of Genetically Modified Microorganisms (REGEM II)*. New York, USA: Plenum Press.

Thomas, C.M. 1981. Molecular genetics of broad host range plasmid RK2. *Plasmid* 5: 10-19.

Walter, M.V., Porteous, L.A., Vieland, V.P., Seidler, R.J. and Armstrong, J.L. 1991. Formation of transconjugants on plating media following *in situ* conjugation experiments. *Canadian J. Microbiolology* 37: 703-07.

Wellington, E. and van Elsas, J.D. (eds) 1992. *Genetic Interactions among Microorganisms in the Natural Environment*. Oxford, UK: Pergamon Press.

Beyond the Biomass
Edited by K. Ritz, J. Dighton and K.E. Giller
© 1994 British Society of Soil Science (BSSS)
A Wiley-Sayce Publication

CHAPTER 14

Use of molecular methods to enumerate *Frankia* in soil

D.D. Myrold, A.B. Hilger, K. Huss-Danell and K.J. Martin

Our understanding of soil microbiology has often expanded as a result of technological advances. This is true of the study of microbial biomass as a whole, where methods have changed from plate (viable) counts to direct (total) counts to chloroform fumigation-incubation to chloroform fumigation-extraction. It is also true of studies on selected segments of the microbial community, where the use of selective media or indicators gave way to immunological techniques and, most recently, to nucleic acid methods. Our work with *Frankia* is an example of these trends.

In this chapter we describe some of the characteristics of *Frankia*, discuss the influence of these characteristics on the selection of appropriate detection methods, give examples of the use of standard and molecular methods to quantify *Frankia* populations in soil, and discuss likely advances in applying other, newer techniques in the study of *Frankia* populations.

ATTRIBUTES OF *FRANKIA*

Perhaps the most distinctive feature of *Frankia* species is their ability to form nitrogen-fixing root nodules on a wide range (over 20 genera) of plants. These actinorhizal symbioses can fix substantial amounts of nitrogen, but *Frankia* themselves have unique characteristics that influence their study.

Frankia species are actinomycetes. They have a filamentous growth form and, like most actinomycetes, the hyphae can differentiate to form sporangia which contain numerous ovoid spores. These spores are dormant resting stages that, under appropriate environmental conditions, germinate to form hyphae. Unique to *Frankia* is that hyphae also differentiate to form vesicles, the spherical, thick-walled structures that are the site of nitrogen fixation. Like spores, isolated *Frankia* vesicles can develop into hyphae. In addition to their pleiomorphic growth, *Frankia* species grow slowly *in vitro*, have no identified selectable traits (except nitrogen fixation and their ability to nodulate certain plants), and are generally present in low numbers in soil. All these features have limited the types of assays that can be used to enumerate *Frankia* in soil.

127

DETECTION METHODS

Many traditional microbiological methods, such as plate counts and fluorescent antibody stains, do not work with *Frankia* (Myrold, 1994). Plate counts are unsuccessful because *Frankia* grow slowly and there is no selective medium. Attempts involving the use of immunological techniques have failed to produce antibodies of sufficient specificity and the relatively low *Frankia* populations in soil would probably limit the utility of this approach. Only two methods have had sufficient specificity and sensitivity to measure *Frankia* populations in soil: plant bioassay and polymerase chain reaction (PCR) amplification of *Frankia* rRNA genes (Myrold, 1994). To date, these methods have been used only to monitor populations of *Frankia* that infect alders (*Alnus* species).

Plant bioassay

Quispel (1954) used a plant bioassay for *Frankia* based on the assumption that the number of nodules formed on test seedlings was proportional to the number of *Frankia* propagules capable of infecting that host plant. This plant bioassay was further refined by van Dijk (1984), who used the term 'nodulation capacity' to describe the infective *Frankia* population. We have employed a somewhat different approach in using plant bioassays to enumerate *Frankia* (Hilger et al., 1991b; Huss-Danell and Myrold, 1994) based on the concept of most probable numbers (MPN). As in the nodulation capacity bioassay, we inoculate replicate seedlings with serial soil dilutions, but each replicate seedling is simply scored for presence of nodules. Nodule counts are not made. The population estimate for *Frankia* obtained by the MPN plant bioassay is referred to as 'nodulation units' (NUs). The MPN approach is commonly used in microbiology and, unlike the nodulation capacity bioassay, the mathematics and statistics for calculating population sizes and their variances are well established (for example, Koch, 1981; Alexander, 1982). Nevertheless, the estimated populations of *Frankia* given using the two approaches are usually in agreement (Myrold and Huss-Danell, 1994).

PCR amplification assay

Initial work with *Frankia* DNA demonstrated that *Frankia*-specific probes could be developed (Simonet et al., 1988; Hahn et al., 1989a). Direct applications of such probes to detect *Frankia* DNA or RNA extracted from soil was hampered, however, by a relatively low sensitivity of about 10^4 cells (2 µg protein)/g soil (Hahn et al., 1990). This led to the use of the PCR involving oligonucleotide primers specific to *Frankia* to enhance the sensitivity of detection.

Our first attempts at using the PCR to quantify *Frankia* in soil showed that one PCR amplification (30-40 cycles) was insufficient to increase sensitivity much beyond 10^4 genomic units (GUs)/g soil (Myrold et al., 1990), which was similar to the detection limit with rRNA (Hahn et al., 1990). A GU is defined as the amount of *Frankia* containing a single genome, which in *Frankia* appears to contain two rDNA operons (Normand et al., 1992). Sensitivity was greatly enhanced to about 10 GUs/g soil when the PCR was optimised for Mg^{2+} concentration and the booster PCR protocol, with more amplification cycles (Ruano et al., 1989), was used (Myrold et al., 1990). The use of booster PCR or nested PCR (Mullis and Faloona, 1987) seems necessary to guarantee sufficient sensitivity to detect native *Frankia* populations in soil.

Having obtained adequate sensitivity, the next step was to develop a means of making the PCR quantitative. Hilger et al. (1991a) evaluated the use of an internal standard, or competitive PCR

(Gilliland et al., 1990), and found it to be unreliable when copy numbers of the template DNA were less than about 100 per sample. Adaptation of the MPN concept to the PCR allows for the quantification of low copy numbers of target genes (Hilger and Myrold, 1992; Picard et al., 1992). To date, two modifications of this PCR MPN methodology have been used for enumerating *Frankia* in soil.

Nested PCR MPN

We have found that the direct extraction method described by Hilger and Myrold (1991), which relies on chemical cell lysis followed by agarose gel electrophoresis to separate DNA from humic materials, yields DNA of sufficient quality for PCR amplification from many soils. Soils high in organic matter require the slight modification of casting the soil extract in a higher-concentration agarose block prior to electrophoresis. This is a relatively gentle extraction procedure that results largely in intact DNA. The purified DNA extract is then serially diluted (we often use threefold dilutions of an initial 1/100 dilution), and replicates of these dilutions (we usually use three) are used in a nested PCR.

We have used several variations of the nested PCR procedure, but the following example is illustrative (Myrold and Huss-Danell, 1994). The first 50 µL reaction was done in 1x PCR amplification buffer (10x buffer contains 100 mM Tris HCl [pH 8.3], 500 mM KCl, 20 mM MgCl$_2$ and 0.1% [w/v] gelatin), 100 µM of each dNTP, 5 pmol of each primer (FR183 and FR1401'; *see* Table 14.1), 1 U of *Taq* DNA polymerase and 10 µL of template. A hot-start protocol (Erlich et al., 1991) was used in which all reagents, except the *Taq* DNA polymerase, were added and covered with mineral oil prior to the first extension cycle. Thirty cycles of denaturing (94°C, 30 seconds), annealing (55°C, 1 minute)

Table 14.1 Oligonucleotide primer sequences used to amplify *Frankia* rDNA (*rrn*) quantitatively using the polymerase chain reaction (PCR) method

Label[a]	Specificity[b]	Numbering[c]	Sequence[d]
FGPS59	*Frankia*	59 to 76	AAG TCG AGC GGG GAG CTT
FR183	*Frankia*	183 to 204	CTG GTG GTG TGG AAA GAT TTA T
FGPS305'	*Frankia*	281 to 305	CCA GTG TGG CCG GTC GCC CTC TCA G
FR485	Eubacteria	485 to 502	CAG CAG CCG CGG TAA TAC
FR1009'	*Frankia*	990 to 1009	TGC AGG ACC CTT ACG GA(C/t) CC
FR1401'	Gram-positive	1377 to 1401	TTC GGG TGT TAC CGA CTT TCG TGA C

Note: a No suffix designates a forward primer, homologous to the coding strand; ' designates a reverse primer, complementary to the coding strand; all primers with an FGPS prefix are from Picard et al. (1992); FR183 and FR1009' are based on Hahn et al. (1989b), who found them to be specific for *Frankia* (FR183) and for *Frankia* that infect alders (FR1009'). FR183 is based on an eubacterial consensus sequence and FR1401' is based on a Gram-positive consensus sequence (Woese, 1987)

b Primer specificity determined from empirical tests with other bacterial species and strains, or by comparison with *rrn* sequence information

c Numbering based on sequence for DNA of *Frankia* CeD isolated from *Casuarina equisetifolia* (Normand et al., 1992); this sequence is accessible from the GenBank databank, accession number M55343

d Lower case bases designate disagreement with the *rrn* sequence of *Frankia* CeD (Normand et al., 1992); bases in parentheses designate an equal mixture of the two bases at this position and extension (72°C, 2 minutes) were used, with a final extension step (72°C, 10 minutes). This first reaction is specific for a 1218 base pairs (bp) product. The second step of the PCR used a 50 μL reaction in 1 x PCR amplification buffer, with 100 μ*M* of each dNTP, 10 pmol of each primer (FR485 and FR1009'; *see* Table 14.1), 1 U of *Taq* DNA polymerase and 1 μL of the first amplification reaction. The same thermocycling protocol was used for the second step of the nested PCR. A 524 bp product resulted from a positive second reaction. This procedure is summarised in Figure 14.1 and typical results from the second reaction are shown in Figure 14.2.

Figure 14.1   **Flow chart of the nested polymerase chain reaction (PCR) used by Myrold and Huss-Danell (1994)**

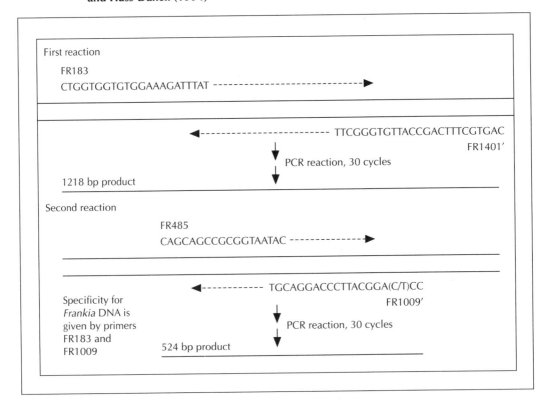

Booster PCR MPN

Picard et al. (1992) extracted soil DNA using successive physical disruption methods of ultrasonication, thermal shocks and microwaving. This gave high yields of low-weight DNA (100-500 bp), which could be effectively cleaned of humic materials using resin column purification. Their PCR MPN was also based on triplicate, threefold dilutions of purified soil DNA.

The booster PCR procedure described by Picard et al. (1992) used a single set of oligonucleotide primers (*see* Figure 14.3 *overleaf*). The first reaction was done in a total volume of 50 μL in 1x PCR amplification buffer (10x buffer contained 100 m*M* Tris HCl [pH 8.3], 500 m*M* KCl, 30 m*M* MgCl$_2$,

Figure 14.2 Ethidium bromide stained 2% agarose gel of a nested most probable number (MPN) polymerase chain reaction (PCR) of DNA extracted from a Swedish soil under alder species

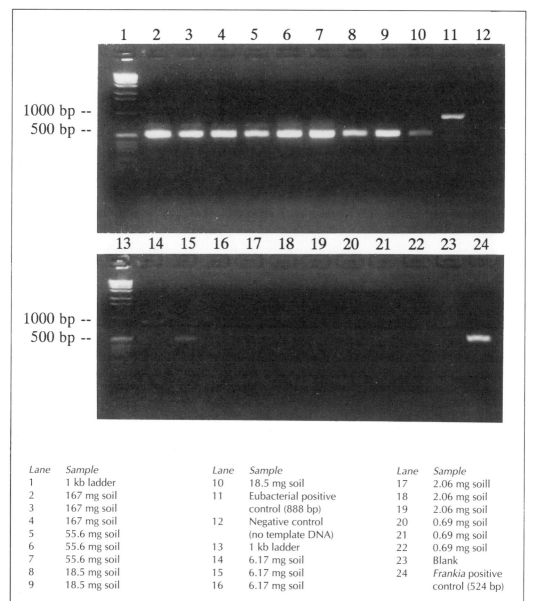

Lane	Sample	Lane	Sample	Lane	Sample
1	1 kb ladder	10	18.5 mg soil	17	2.06 mg soill
2	167 mg soil	11	Eubacterial positive	18	2.06 mg soil
3	167 mg soil		control (888 bp)	19	2.06 mg soil
4	167 mg soil	12	Negative control	20	0.69 mg soil
5	55.6 mg soil		(no template DNA)	21	0.69 mg soil
6	55.6 mg soil	13	1 kb ladder	22	0.69 mg soil
7	55.6 mg soil	14	6.17 mg soil	23	Blank
8	18.5 mg soil	15	6.17 mg soil	24	*Frankia* positive
9	18.5 mg soil	16	6.17 mg soil		control (524 bp)

The MPN score of 3-3-3-1-0-0 positive reactions results in 4700 GUs/g soil. The larger of the two additional bands (916 bp) seen in lanes 2-11 and 14-15 is probably the product of primers FR485 and FR1401', which is specific to Gram-positive, including *Frankia*. The smaller additional band seen in lanes 2-11 may be the product of primers FR183 and FR1009' (826 bp), which is specific for *Frankia*. These products appear because some of the primers FR183 and FR1401' are carried over from the first reaction.

Figure 14.3   Flow chart of the booster polymerase chain reaction (PCR) procedure used by Picard et al. (1992)

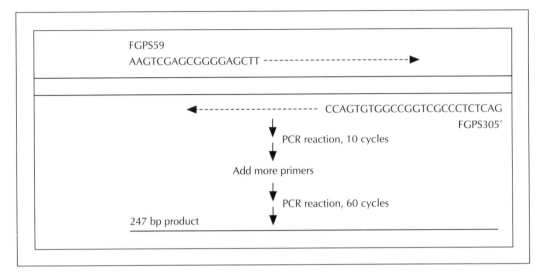

0.1% [w/v] gelatin and 10 % formamide), 200 μM of each dNTP, 50 pmol of each primer (FGPS59 and FGPS305'; *see* Table 14.1), 2.5 U of *Taq* DNA polymerase and 1 µL of template DNA. An initial denaturation (95°C, 3 minutes) was followed by 10 PCR cycles of denaturing (95°C, 1 minute), annealing (55°C, 1 minute) and extension (72°C, 1 minute). Primer concentrations were then increased to 5 pmol each and amplification resumed for an additional 60 cycles using the following protocol: denaturation (90°C, 1 minute), annealing (55°C, 1 minute) and extension (72°C, 1 minute). A 247 bp product was produced in a positive reaction.

POPULATIONS OF *FRANKIA* IN SOIL

Most published data on *Frankia* populations in soil have been obtained using some variation of the plant bioassay. Surveys of *Frankia* NUs in forest soils of Finland, Sweden and north-western USA range from 0 to 4600 NUs/g soil. Although it is uncertain what factors control the population size of these infective *Frankia*, two interesting observations have been made. First, *Frankia* NUs have sometimes been observed to be higher in the rhizospheres of non-host plants, such as birch (Smolander and Sundman, 1987). Second, *Frankia* NUs increase with increasing pH, at least in the range of pH 3.4 to 4.4 (Smolander and Sundman, 1987).

We have applied both the plant bioassay and the PCR MPN methods to a variety of forest soils to examine the relationship between NUs (which reflect only those *Frankia* physiologically capable of nodulation) and GUs (which should be proportional to the total population of *Frankia*). Table 14.2 presents a selection of this work.

Frankia NUs varied enormously in the studies examined; however, tens to a few hundreds per gram of soil was most typical. Both laboratory (for example, Smolander and Sarsa, 1990) and field studies (for example, Myrold and Huss-Danell, 1994) suggest that *Frankia* NUs are relatively dynamic and responsive to environmental factors.

Table 14.2 *Frankia* populations in selected soils of the world as measured by plant bioassay and/or polymerase chain reaction (PCR) most probable number (MPN) techniques

Location	Plant bioassay (NUs/g soil)	PCR MPN (GUs/g soil)	DNA extracted (μg/g soil)
Finland[a]	0.2 - 2940	—	—
Washington, USA[b]	62	10 000	5.1
Sweden[c]	46 - 74	2450 - 3020	8.5 - 11.4
Oregon, USA[d]	593	92 000	9.5
France[e]	—	20 000	50

Note: a   From Smolander and Sundman (1987); higher numbers from soils under birch and/or soils with high pH; data assume a bulk density of 1.0; this study used only the plant bioassay method
  b   From Hilger and Myrold (1992) and unpublished data; the soil was from a Douglas fir forest on a glacially derived sandy loam
  c   From Myrold and Huss-Danell (1994); the range represents seasonal differences for June and September sampling; the glacially derived fine sandy soil was collected under grey alders
  d   Unpublished data
  e   From Picard et al. (1992); the silt loam soil was collected from an alder stand; no plant bioassay was done in this study

The fraction of *Frankia* capable of nodulation was just a small portion of the total population of *Frankia* as measured by GUs, which ranged from 2000 to 92 000 g/soil. The relatively small range in GUs among these very different sites and the small changes in GUs over time (Myrold and Huss-Danell, 1994) suggest that the total *Frankia* population is fairly constant in soils. The ratio of NUs to GUs in soil may be a useful indicator of *Frankia* activity in soil.

FUTURE TRENDS IN *FRANKIA* RESEARCH

Molecular methods hold the key to much of the future research on *Frankia* ecology, although plant infectivity assays will continue to provide important information. Enumeration studies will focus on comparisons of methods.

Plant bioassays are dependent upon reproducible growing plants that are capable of infection and of maintaining culture conditions conducive to nodulation. These caveats have been discussed by Huss-Danell and Myrold (1994), who suggest that further refinement may be necessary to standardise this method. Another difficulty in quantifying *Frankia* with plant bioassays is determining what represents an NU. A single spore that germinates, and which presumably contains 1 GU, could conceivably form a nodule. Similarly, it is uncertain what length of hyphae, and consequently what number of GUs, is required for infection. Thus, the degree of hyphal fragmentation occurring during the preparation of serial dilutions may inflate (if only a small hyphal fragment is needed for nodulation) or deflate (if hyphae are broken into units too small to initiate nodulation) the numbers of NUs from a given assay. Beyond these fundamental questions, there is a need to adapt the bioassay method for species other than alders.

The PCR MPN approach is new and needs further evaluation. Of particular importance here is determining soil DNA extraction efficiency and improving DNA purification. The physical disruption methodology described by Picard et al. (1992) is more rigorous than our chemical lysis approach

(Hilger and Myrold, 1991) and may extract DNA more completely from soils; however, the small fragments generated by physical disruption may carry potential risks. As the size of the isolated soil DNA approaches that of the target sequence, the probability that DNA fragments contain only partial target sequences increases markedly. These partial sequences may not be efficiently amplified, as shown by Picard et al. (1992), who detected only about 15% of the 247 bp target sequences when genomic DNA was fragmented to 100-500 bp.

An implicit assumption in all work to date is that all *Frankia* DNA extracted from soil comes from living cells. Dead *Frankia* cells still containing DNA, or even free DNA in the soil which is capable of PCR amplification, could distort the results of any attempts to quantify *Frankia* with the PCR MPN or other molecular approaches.

The continuing cataloguing of more *Frankia* rDNA sequences (for example, Nazaret et al., 1991; Bosco et al., 1992; Rouvier et al., 1992) should allow for the selection of host-group, genospecies or strain-specific oligonucleotide primers that can be used for PCR MPN. It must be remembered that the specificity of primers currently in use, or those yet to be developed, is only as certain as the strains or sequences that they are tested against. One advantage of the nested PCR over the booster PCR approach is that a second set of primers is required, which provides a second screen for specificity.

Both nested and booster PCR are very sensitive, with the potential to detect a single GU/g soil, but they are both limited in practice by the purity of the extracted soil DNA and the presence of inhibiting substances. Without extensive purification, we were able to amplify *Frankia* DNA from a soil DNA extract equivalent to 50 mg soil but not one equivalent to 500 mg soil. This is equivalent to a detection limit of 20 GUs/g soil, assuming that 1 GU would be amplified as successfully as the 570 GUs measured in this volume of soil DNA extract using the nested PCR MPN protocol. More difficult soils or other purification procedures may result in lower sensitivity (for example, Picard et al., 1992).

Methods will be developed that more directly monitor the activity of *Frankia* in soil. Initially, this will probably involve the application of *in situ* 16S ribosomal RNA (rRNA) hybridisation techniques, although relatively low populations of *Frankia* and the requirement for actively growing cells that contain multiple copies of rRNA may limit the usefulness of this approach (Hahn et al., 1992). A newer and perhaps more promising technique is to monitor mRNA as an indicator of gene expression (for example, Tsai et al., 1991).

Significant information has been produced on the diversity of *Frankia*, largely with isolates from root nodules (for example, Nazaret et al., 1991; Rouvier et al., 1992). It should be relatively straightforward to use *Frankia* genus-specific oligonucleotide primers to amplify *Frankia* DNA selectively directly from nodules or even soil (for example, Nick et al., 1992; Liesack and Stackebrandt, 1992). These DNA products could then be sequenced and potentially applied to studies of *Frankia* autecology in soil.

Lastly, there is the long-standing goal of inserting genetic markers into *Frankia* strains. This has not yet been accomplished, but recent progress (for example, Cournoyer and Normand, 1992) suggests that this goal may be achieved.

CONCLUSION

This review of the study of *Frankia* ecology in soil highlights the interplay between the development of new methods and their application in studies of the ecology of microorganisms in soil. The traditional method of plant bioassays has generated tantalising questions about the ecology of *Frankia* in soil. The currently used molecular methods, and newer methods soon to be used, will probably answer some of these questions (for example, whether *Frankia* are metabolically active in soil and how such activity is influenced by environmental factors) but some of the answers will, of course, give rise to even more intriguing questions about the ecology of *Frankia* in soil.

Acknowledgements

The research reported in this chapter (Technical Paper Number 10 265 of the Oregon Agricultural Experiment Station) was supported by grants from the US National Science Foundation (BSR-8657269, BSR-8906527, INT-9025112 and DEB-9119809) and the Swedish Forestry and Agriculture Research Council.

References

Alexander, M. 1982. Most probable number method for microbial populations. In Page, A.L., Miller, R.H. and Keeney, D.R. (eds) *Methods of Soil Analysis. Part 2. Chemical and Microbiological Properties.* (2nd edn) Agronomy Monograph No. 9. Madison, Wisconsin, USA: American Society of Agronomy.

Bosco, M., Fernandez, M.P., Simonet, P., Materassi, R. and Normand, P. 1992. Evidence that some *Frankia* sp. strains are able to cross boundaries between *Alnus* and *Elaeagnus* host specificity groups. *Applied and Environmental Microbiology* 58:1569-76

Cournoyer, B. and Normand, P. 1992. Electroporation of *Frankia* intact cells to plasmid DNA. *Acta Oecologica* 13: 369-78.

Erlich, H.A., Gelfand, D. and Sninsky, J.J. 1991. Recent advances in the polymerase chain reaction. *Science* 252: 1643-51.

Gilliland, G., Perrin, S., Blanchard, K. and Bunn, H.F. 1990. Analysis of cytokine mRNA and DNA: Detection and quantification by competitive polymerase chain reaction. *Proc. National Academy of Science, USA* 87: 2725-29.

Hahn, D., Dorsch, M., Stackebrandt, E. and Akkermans, A.D.L. 1989a. Synthetic nucleotide probes for identification of *Frankia* strains. *Plant and Soil* 118: 211-19.

Hahn, D., Lechevalier, M.P., Gischer, A. and Stackebrandt, E. 1989b. Evidence for a close phylogenetic relationship between members of the genera *Frankia*, *Geodermatophilus* and *Blastococcus* and emendation of the family *Frankiaceae*. *Systematics and Applied Microbiology* 11: 236-42.

Hahn, D., Kester, R., Starrenburg, M.J.C. and Akkermans, A.D.L. 1990. Extraction of ribosomal RNA from soil for detection of *Frankia* with oligonucleotide probes. *Archives of Microbiology* 154: 239-335.

Hahn, D., Amann, R.I., Ludwig, W., Akkermans, A.D.L. and Schleifer, K.-H. 1992. Detection of micro–organisms in soil after *in situ* hybridisation with rRNA-targeted, fluorescently labelled oligonucleotides. *J. General Microbiology* 138: 879-87.

Hilger, A.B. and Myrold, D.D. 1991. Method for extraction of *Frankia* DNA from soil. *Agriculture, Ecosystems and Environment* 34: 107-13.

Hilger, A.B. and Myrold, D.D. 1992. Quantification of soil *Frankia* by bioassay and gene probe methods: Response to host and non-host rhizospheres and liming. *Acta Oecologica* 13: 505-06.

Hilger, A.B., Martin, K.J., Strauss, S.H. and Myrold, D.D. 1991a. PCR for quantification of low copy number *Frankia* DNA. In *Abstracts of the 91st Annual Meeting of the American Society for Microbiology.* Washington, DC, USA: American Society for Microbiology.

Hilger, A.B., Tanaka, Y. and Myrold, D.D. 1991b. Inoculation of fumigated nursery soil increases nodulation and yield of bare-root red alder (*Alnus rubra* Bong.). *New Forests* 5: 35-42.

Huss-Danell, K. and Myrold, D.D. 1994. Intrageneric variation in nodulation of *Alnus*: Consequences for quantifying *Frankia* nodulation units in soil. *Soil Biology and Biochemistry* (in press).

Koch, A.L. 1981. Growth measurement. In Gerhardt, P. (ed) *Manual of Methods for General Bacteriology*. Washington, DC, USA: American Society for Microbiology.

Liesack, W. and Stackebrandt, E. 1992. Occurrence of novel groups of the domain bacteria as revealed by analysis of genetic material isolated from an Australian terrestrial environment. *J. Bacteriology* 174: 5072-78.

Mullis, K.B. and Faloona, F.A. 1987. Specific synthesis of DNA *in vitro* via a polymerase-catalysed chain reaction. *Methods in Enzymology* 155: 335-51.

Myrold, D.D. 1994. *Frankia* and the actinorhizal symbiosis. In Weaver, R.W., Bottomley, P.J. and Angle, J.S. (eds) *Methods of Soil Analysis. Part 2. Microbiological and Biochemical Properties*. Madison, Wisconsin, USA: Soil Science Society of America (in press).

Myrold, D.D. and Huss-Danell, K. 1993. Population dynamics of *Alnus*-infective *Frankia* in a forest soil with and without host trees. *Soil Biology and Biochemistry* (in press).

Myrold, D.D., Hilger, A.B. and Strauss, S.H. 1990. Detecting *Frankia* in soil using PCR. In Gresshoff, P.M., Roth, L.R., Stacey, G. and Newton, W.E. (eds) *Nitrogen Fixation: Achievements and Objectives*. New York, USA: Chapman and Hall.

Nazaret, S., Cournoyer, B., Normand, P. and Simonet, P. 1991. Phylogenetic relationships among *Frankia* genomic species determined by use of amplified 16S rDNA sequences. *J. Bacteriology* 173: 4072-78.

Nick, G., Paget, E., Simonet, P., Moiroud, A. and Normand, P. 1992. The nodular endophytes of *Coriaria* spp. form a distinct lineage within the genus *Frankia*. *Molecular Ecology* 1: 175-81.

Normand, P., Cournoyer, B., Simonet, P. and Nazaret, S. 1992. Analysis of a ribosomal RNA operon in the actinomycete *Frankia*. *Gene* 111: 119-24.

Picard, C., Ponsonnet, C., Paget, E., Nesme, X. and Simonet, P. 1992. Detection and enumeration of bacteria in soil by direct DNA extraction and polymerase chain reaction. *Applied and Environmental Microbiology* 58: 2717-22.

Quispel, A. 1954. Symbiotic nitrogen fixation in non-leguminous plants. II. The influence of the inoculation density and external factors on the nodulation of *Alnus glutinosa* and its importance to our understanding of the mechanism of infection. *Acta Botanica Neerlandica* 3: 512-32.

Rouvier, C., Nazaret, S., Fernandez, M.P., Picard, B., Simonet, P. and Normand, P. 1992. *rrn* and *nif* intergenic spacers and isoenzyme patterns as tools to characterise *Casuarina*-infective *Frankia* strains. *Acta Oecologica* 13: 487-95.

Ruano, G., Fenton, W. and Kidd, K.K. 1989. Biphasic amplification of very dilute DNA samples via 'booster' PCR. *Nucleic Acids Research* 17: 5407.

Simonet, P., Le, N.T., du Cros, E.T. and Bardin, R. 1988. Identification of *Frankia* strains by direct DNA hybridisation of crushed nodules. *Applied and Environmental Microbiology* 54: 2500-03.

Smolander, A. and Sundman, B. 1987. *Frankia* in acid soils of forests devoid of actinorhizal plants. *Physiologia Plantarum* 70: 297-303.

Smolander, A. and Sarsa, M.-L. 1990. *Frankia* strains of soil under *Betula pendula*: Behaviour in soil and in pure culture. *Plant and Soil* 122: 129-36.

Tsai, Y., Park, M.J. and Olson, B.H. 1991. Rapid method for direct extraction of mRNA from seeded soils. *Applied and Environmental Microbiology* 57: 765-68.

van Dijk, D. 1984. *Ecological Aspects of Spore Formation in the* Frankia-Alnus *Symbiosis*. PhD thesis. Leiden University, The Netherlands.

Woese, C.R. 1987. Bacterial evolution. *Microbiological Reviews* 51: 221-71.

Beyond the Biomass
Edited by K. Ritz, J. Dighton and K.E. Giller
© 1994 British Society of Soil Science (BSSS)
A Wiley-Sayce Publication

CHAPTER 15

In situ detection of bacteria in the environment

D. Hahn and J. Zeyer

Specific detection and enumeration of bacteria has usually been carried out using techniques which depend upon the isolation of the target organisms: selective plating and immunofluorescence (Hill and Gray, 1967; Bohlool and Schmidt, 1968; Bohlool and Schmidt, 1980; Bohlool, 1987; Postma et al., 1988; Thompson et al., 1990). However, although selective isolation procedures exist for many microorganisms (Grant and Holt, 1977; Wellington et al., 1987), most members of the natural bacterial communities are non-culturable and their identity remains unknown (Brock, 1987; Torsvik et al., 1990).

Over the past decade, molecular methods based on DNA or RNA sequence analysis have reached a high level of acceptance in microbial ecology as new techniques for the identification of bacteria unbiased by the limitations of culturability (Holben et al., 1988; Fuhrman and Lee, 1989; Ward, 1989; Sayler et al., 1989; and reviews by Pickup, 1991, and Knight et al., 1992). Molecular protocols for the detection or identification of bacteria focus mainly on the detection of DNA or RNA by labelled complementary probes. Nucleic acids of sufficient quality to serve as targets in hybridisation experiments have been extracted from aquatic environments (Fuhrman et al., 1988; Sommerville et al., 1989; Burnison and Nuttley, 1990), sediments and soils (Ogram et al., 1987; Holben et al., 1988; Steffan et al., 1988; Hahn et al., 1990b; Tsai and Olsen, 1991; Tsai et al., 1991), as well as from blood, tissue and manure (Boom et al., 1990; Tavangar et al., 1990; Xu et al., 1990). The limited sensitivity of the detection of nucleic acids isolated from natural environments has recently been improved by the use of the polymerase chain reaction (PCR) that amplifies the target sequences (Steffan and Atlas, 1988; Chaudhry et al., 1989; Steffan et al., 1989; Simonet et al., 1990; Bej et al., 1991). This technique also allows sequence information to be retrieved from non-culturable members of complex communities (Weller and Ward, 1989; Giovannoni et al., 1990; Relman et al., 1990; Ward et al., 1990a, b; Amann et al., 1991), and therefore specific detection of these organisms has become possible.

Ribosomal RNA (rRNA) molecules have exceptional advantages as targets because of their ubiquitous occurrence, their natural amplification up to several thousands of copies per cell, their composition of conserved and variable regions and their single-stranded nature (Olsen et al., 1986). Elaborated databases have been constructed for the rRNAs (5S, 16S and 23S) by taxonomists, using the sequences of these molecules to reveal the phylogeny of microorganisms (Woese, 1987). Based on comparative sequence analysis, rRNA-directed oligonucleotide probes of differing specificity have been designed (Kohne et al., 1986; Göbel et al., 1987; Hahn et al., 1989; Witt et al., 1989) and have been used successfully to detect or identify organisms with or without prior isolation (Stahl et al., 1988; Giovannoni et al., 1990; Hahn et al., 1990a, b) and to monitor population dynamics (Stahl et al., 1989).

IN SITU DETECTION

The approach to the analysis of microbial community structure based on nucleic acid detection can be combined with conventional microscopy. *In situ* detection experiments focus on the microscopic detection of labelled complementary probes hybridised to target sequences in fixed organisms. Because cells are exposed to elevated temperatures, detergents and osmotic gradients during hybridisation, fixation is essential to stabilise the morphological structure of the bacterial cells. Different fixation protocols are recommended, such as fixation in 4% paraformaldehyde in phosphate-buffered saline (PBS, composed of 0.13 M NaCl, 7 mM Na_2HPO_4, 3 mM Na_2HPO_4, pH 7.2) (Amann et al., 1990a), in 50% ethanol/PBS (Spring et al., 1992) or in 90% ethanol/3.7% formaldehyde (Braun-Howland et al., 1992). Fixed cells are washed in PBS and then stored in 50% ethanol/PBS at -20°C until further use.

A standard hybridisation protocol can be used for the detection of bacteria in pure culture or in heterogeneous systems. Fixed samples are applied to gelatin-coated slides (0.1% gelatin, 0.01% $KCr(SO_4)_2$), and the preparations are air-dried and then hybridised in 9 µl of hybridisation buffer (0.9 M NaCl, 0.1% SDS, 20 mM Tris, pH 7.2) and 1 µl of labelled oligonucleotide probe (25-50 ng) at 46°C for at least 1 hour. After hybridisation, the slides are washed in hybridisation buffer at 46°C for 20 minutes, rinsed with distilled water and air-dried (Amann et al., 1990a, b; Hahn et al., 1992).

Four different labelling strategies are currently used to detect specific hybridisation. In addition to the laborious method using probes labelled with radioisotopes in *in situ* hybridisation experiments with subsequent autoradiography (Giovannoni et al., 1988; Wiel et al., 1990; Ultsch et al., 1991; Yang et al., 1991), non-radioactive detection methods have been developed. Ribosomal RNA-targeted oligonucleotide probes have been labelled with fluorescent dyes (DeLong et al., 1989 a, b; Amann et al., 1990 a, b; Amann et al., 1992a), enzymes (Amann et al., 1992b) and non-radioactive reporter molecules such as biotin or digoxigenin (Zarda et al., 1991, Hahn et al., 1993), and used for the identification of individual cells. The choice of the label and the detection procedure, however, can cause methodological problems which are concerned mainly with non-specific signals resulting from autofluorescence or non-specific binding of probes, or with restricted cell permeability because of the size of probe and label. These constraints are usually more pronounced when the *in situ* detection strategies are applied in heterogeneous environmental systems such as soil or plant material (for example, root nodules).

IN SITU DETECTION WITH FLUORESCENTLY LABELLED OLIGONUCLEOTIDE PROBES

The *in situ* detection of bacteria with fluorescently labelled probes is based on the microscopic detection of the fluorescent dye, such as tetramethylrhodamine isothiocyanate (Tritc), which is covalently coupled to a primary amino group at the 5' end of the synthetic oligonucleotide (Amann et al., 1990a; Stahl and Amann, 1991). After hybridisation, the slides are usually mounted in Citifluor® solution (Citifluor, UK) and probe-conferred fluorescence is examined under a microscope (for example, Zeiss® Axiophot microscope; Zeiss, Oberkochen, Germany) fitted for epifluorescence with a high-pressure mercury bulb and filter sets for the appropriate dyes.

In situ hybridisation with fluorescently labelled probes allows the specific detection of bacteria in soils by an unaltered *in situ* hybridisation protocol for pure cultures (*see* Figure 15.1) (Hahn et al., 1992; Schleifer et al., 1992). Soil minerals do not disturb specific detection, and non-specific binding of labelled probes to soil particles does not occur. Detection of low numbers of bacteria, however, is influenced by the intensive autofluorescence of organic material in the preparation, especially in soils such as peat. Low amounts of organic material (up to 8%) do not significantly influence the detection of hybridisation signals (Hahn et al., 1992). It is difficult to view cells attached to organic material, however, because the autofluorescence of the organic material will be brighter than probe-conferred fluorescence. Therefore, quantitative detection of specific bacteria will be achieved only with reliable methods that release cells attached to soil particles or entrapped in soil aggregates (Kingsley and Bohlool, 1981; Fry, 1990). The dispersal of soil aggregates and dissociation of microorganisms from soil particles can be affected by the addition of surfactants (Bohlool and Schmidt, 1973; Reyes and Schmidt, 1979), hydrolysed gelatine (Kingsley and Bohlool, 1981) or chelating resins to the dilution

Figure 15.1 Detection of *Pseudomonas aeruginosa* introduced into a clay soil after *in situ* hybridisation with fluorescently labelled (a, right) or digoxigenin-labelled (b, right) 16S rRNA-targeted oligonucleotide probe Eub338 (Stahl et al., 1988) with the corresponding phase contrast photographs (left)

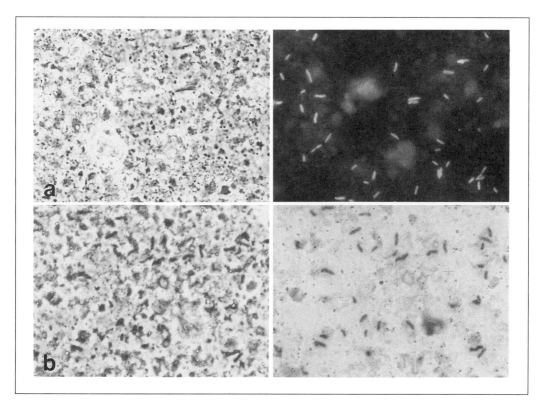

buffer (Herron and Wellington, 1990; Hopkins et al., 1991) or by the choice of the dilution buffer (Bakken, 1985). Separating bacterial cells from soil material, however, is not necessary.

The quality of the *in situ* hybridisation is influenced by the choice of the fixation procedure (for example, glutaraldehyde-fixed cells normally exhibit much larger autofluorescence than paraformaldehyde-fixed cells; Stahl and Amann, 1991; Braun-Howland et al., 1992). Fixatives cause physical changes to cellular and extracellular constituents and increase the porosity of the cell membranes and the cytoplasm. Fixation in coagulant fixatives such as mercuric chloride-based fixatives generally results in larger pore sizes than fixation in non-coagulant fixatives such as paraformaldehyde-based fixatives (Boenisch, 1989) which fix by cross-linking basic amino acids. The intensity of cross-linking depends upon the fixation time and the fixation conditions which can therefore directly influence the permeability for macromolecules (that is, probes). Because of the relatively small size of fluorescently labelled oligonucleotides (with a molecular weight of approximately 6500), *in situ* hybridisation on metabolically active, paraformaldehyde-fixed cells is not inhibited by restricted permeability. Pure cultures of Gram-negative bacteria (for example, pseudomonads) are usually permeable, whereas pure cultures of Gram-positive bacteria (*Streptomyces, Frankia*) need enzymatic pre-treatments to overcome a permeability barrier (Hahn et al., 1992; Hahn

Figure 15.2   Hyphae of *Streptomyces scabies* growing in sterile amended loamy sand visualised after *in situ* hybridisation with fluorescently labelled (a, left) or digoxigenin-labelled (b, left) probe Eub338 without enzymatic pre-treatments to increase cell permeability, showing (right) spores not detected by *in situ* hybridisation but well stained with DAPI

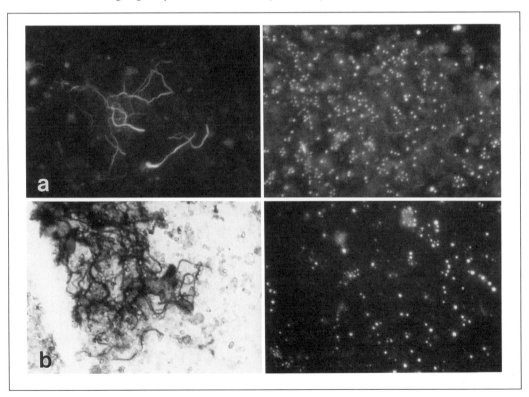

et al., 1993). The same bacteria growing in soil amended with nutrients do not show any need for pre-treatments. Metabolically active Gram-negative (*Pseudomonas aeruginosa*) or Gram-positive (*Streptomyces scabies*) cells are permeable for fluorescently labelled oligonucleotide probes (*see* Figure 15.2) (Hahn et al., 1992). Spores of *Streptomyces*, however, which still contain quite large amounts of rRNA, cannot be detected by *in situ* hybridisation, probably because of a restricted permeability of the spores. Spores can be viewed only after staining with the DNA specific dye 4'6-diamidino-2-phenylindole (DAPI) (*see* Figure 15.2).

In plant material such as nodules of *Alnus glutinosa* harbouring the nitrogen-fixing actinomycete *Frankia*, autofluorescence is even more pronounced than in soil (Hahn et al., 1993). The autofluorescent signal of even low amounts of contaminating organic material is brighter than potential hybridisation signals. Detection of the nitrogen-fixing actinomycete *Frankia* in nodule homogenates by *in situ* hybridisation with fluorescently labelled oligonucleotide probes is therefore extremely difficult. These difficulties cannot be overcome by the choice of different fluorescent dyes or different filter systems because suboptimal detection conditions will reduce not only the autofluorescence but also the intensity of probe-conferred signals. A non-fluorescent assay, as described below, is therefore more appropriate for the detection of *Frankia*.

IN SITU DETECTION WITH DIGOXIGENIN-LABELLED OLIGONUCLEOTIDE PROBES

An alternative labelling strategy to fluorescent dyes, which offers the opportunity of circumventing problems of autofluorescence of cells and matrix, is based on the detection of non-fluorescent reporter molecules such as biotin or digoxigenin that are attached to the probes. The formation of stable hybrids in the hybridisation is shown by labelled binding proteins (for example, avidin, streptavidin or antibodies) that specifically bind to these reporters. Detection of the digoxigenin reporter molecule is often done using an antibody conjugated to an enzyme, such as alkaline phosphatase. This enzyme hydrolyses a substrate (for example, napthol phosphate esters) to phenolic compounds and phosphates. The phenols combine with colourless diazonium salts (chromogen) to produce insoluble, coloured azo dyes, which can be examined by brightfield microscopy. This procedure has been used successfully with rRNA-targeted oligonucleotide probes, PCR amplification products or *in vitro* transcripts in the detection and identification of single bacterial cells (Zarda et al., 1991), the detection of genes on the chromosome (Celeda et al., 1992) or in the localisation of RNA in thin sections (Bochenek and Hirsch, 1990; Holtke and Kessler, 1990; Wiel et al., 1990), respectively.

The standard *in situ* hybridisation protocol used for pure cultures and fluorescently labelled oligonucleotides is also applicable for digoxigenin-labelled probes in heterogeneous environmental systems. Digoxigenin-labelled probes visualised via antibody/alkaline phosphatase conjugates and substrate precipitation can detect bacterial cells even though they are entrapped in aggregates of dense soil particles (*see* Figures 15.1 and 15.2). This does not reduce the accessibility of the target sites for probes and antibodies but can disturb a quantitative determination of cell numbers. Non-specific binding of probes and antibody to soil minerals does not occur. Flocs of organic material, however, often show some background staining that is caused mainly by binding of the antibody/enzyme conjugate. Non-specific binding is sufficiently reduced by minor changes to the standard procedure: first, the samples are pre-blocked in a maleic acid-based buffer (150 mM NaCl, 100 mM maleic acid/NaOH, pH 7.5) containing 1% blocking reagent (Boehringer) prior to incubation with the antibody/enzyme conjugate; second, the incubation with the antibody/enzyme conjugate is performed in the same buffer instead of the standard Tris-based buffer (Zarda et al., 1991).

In pure cultures of many Gram-negative bacteria, short pre-treatments with lysozyme enable reliable detection of the digoxigenin reporter molecule by the antibody/enzyme conjugate (Zarda et al., 1991). For Gram-positive bacteria, this pre-treatment is not sufficient for penetration of the antibody/enzyme conjugate. Here, additional pre-treatments with detergents and solvents (toluene, Nonidet P-40) can enhance the reliability of the permeabilisation but at the same time enhance non-specific binding of probe and antibody and increase the risk of total lysis (Hahn et al., 1993). However, when bacterial cells growing in amended soil are used as target organisms, cell permeabilisation is no longer required. Gram-negative (*P. aeruginosa*) as well as Gram-positive (*S. scabies*) (*see* Figure 15.2) organisms can be detected without pre-treatments. Spores are not visualised.

Detection of Gram-negative *P. aeruginosa* cells, however, is not quantitatively achieved. Only a small proportion of cells which are present after several days of incubation can be detected by the hybridisation-based assay, whereas the major population can be visualised only by DAPI staining. Enzymatic pre-treatments cannot increase the number of cells visualised because even very short treatments with lysozyme reduce the amount of cells significantly. This picture does not change when

Figure 15.3   Bright micrograph of *Frankia* vesicles in nodule homogenates after H_2O_2 treatment, subsequent lysozyme pre-treatment, additional Nonidet P-40 and toluene treatments, and *in situ* hybridisation with digoxigenin-labelled oligonucleotides EFP (Hahn et al., 1989) showing (right) subsequent anitbody/alkaline phosphatase detection and (left) the corresponding phase contrast photograph

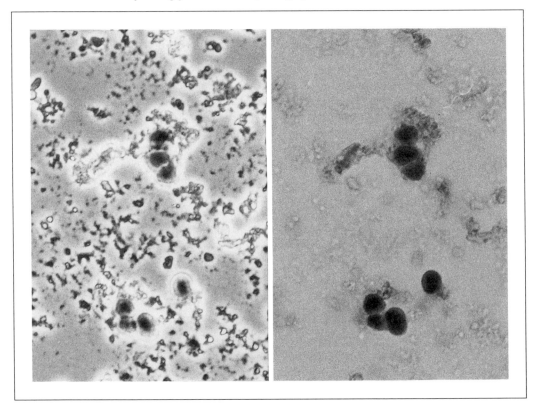

cells are activated by the addition of nutrients. The numbers which are detected by *in situ* hybridisation with digoxigenin-labelled probes remain extremely low compared with the numbers of cells obtained using *in situ* detection with fluorescently labelled probes or DAPI staining. Because the amount of rRNA is correlated with the activity of cells (DeLong et al., 1989) and cell walls of many inactive cells are altered, both the amount of rRNA per cell and the permeability for probes become limiting factors for the sensitivity of detection. The low sensitivity of the digoxigenin-based detection protocol compared with detection using fluorescently labelled probes is obviously based on the restricted permeability of the cells for the antibody/alkaline phosphatase conjugate. Because the permeability problem cannot be solved by standard pre-treatment protocols, the *in situ* detection of even less physiologically active bacteria is therefore more likely to be achieved by using *in situ* hybridisation with fluorescently labelled oligonucleotides than by using digoxigenin-labelled probes. This obvious drawback also reduces the possibility of quantifying soil bacteria by the digoxigenin-based detection protocol.

In plant material, background has not caused any problems after *in situ* hybridisation with digoxigenin-labelled probes and subsequent enzymatic detection on tissue sections (Bochenek and Hirsch, 1990; Wiel et al., 1990). *In situ* detection of *Frankia* in nodule homogenates is not influenced by background signals either. However, hybridisation is significantly hampered by restricted cell permeability caused by additional capsular material surrounding the vesicles in nodule material. Identification of *Frankia* vesicles in nodule homogenates is possible only after the initial removal of the polysaccharide capsule surrounding the vesicles. Because of the fairly large size of the antibody/enzyme conjugate (molecular weight about 100 000), the *in situ* hybridisation strategy with digoxigenin-labelled oligonucleotides still has poor efficiency, since cell walls and cell membranes form penetration barriers (Zarda et al., 1991; Hahn et al., 1993). Incubation with H_2O_2 prior to lysozyme and detergent treatments is found to facilitate specific hybridisation (*see* Figure 15.3). Hyphae or spores cannot be detected in nodule homogenates by *in situ* hybridisation. However, since digoxigenin-labelled probes visualised via antibody/alkaline phosphatase conjugates overcome the problems arising from the autofluorescence of cells and plant material, they are more suitable for *in situ* detection of *Frankia* than fluorescently labelled probes.

CONCLUSION

The application of rRNA-targeted oligonucleotides for *in situ* detection of natural bacterial communities in heterogeneous systems offers an alternative to serological techniques. The combination of DNA/rRNA extraction procedures from soil, the selective amplification of rRNA sequences and the subsequent sequence analysis of cloned amplification products shows a high potential for obtaining specific probes against uncultured and unculturable microorganisms (Giovannoni et al., 1990; Ward et al., 1990a, b; Amann et al., 1991). However, the application of non-radioactive rRNA probing in heterogeneous environments is seriously hampered by problems arising from the heterogeneity of the environment.

Reliable detection depends upon the design of specific probes. Specificity of oligonucleotide probes can generally be freely adjusted (Amann et al., 1990a). However, specificity tests conducted on a limited number of laboratory strains do not necessarily exclude non-specific binding to cells and soil particles and cross-hybridisation to bacteria other than the target organism (Stahl et al., 1988), or lower specificity to additional target organisms, such as different strains of one species (Hahn et al., 1990).

The detection of specific target organisms is also influenced by the physiological condition of the target organism. In soil, bacterial cells are often metabolically inactive and embedded in a matrix of polysaccharides (Roszak and Colwell, 1987). Both the amount of rRNA per cell and the permeability of the cells for probes therefore become limiting factors for the sensitivity of detection. *In situ* hybridisation with rRNA-targeted probes on natural populations of bacteria grown in soil does not result in significant hybridisation signals (Hahn et al., 1992). After activating these microorganisms with the addition of nutrients, however, a significant increase in the amount of detectable cells is obtained (Hahn et al., 1992). The *in situ* hybridisation technique with fluorescently labelled or digoxigenin-labelled oligonucleotide probes is therefore currently restricted to physiologically active bacteria; that is, those found in environments rich in nutrients such as syntrophic organisms or bacterial endosymbionts (Amann et al., 1991; Distel et al., 1991; Amann et al., 1992; Spring et al., 1992).

Acknowledgements

The work reported in this chapter was supported by grants from the ETH Zurich and the Swiss Federal Office of Environment, Forests and Landscape (BUWAL).

References

Amann, R.I., Krumholz, L. and Stahl, D.A. 1990a. Fluorescent-oligonucleotide probing of whole cells for determinative, phylogenetic, and environmental studies in microbiology. *J. Bacteriology* 172: 762-70.

Amann, R.I., Binder, B.J., Olsen, R.J., Chisholm, S.W., Devereux, R. and Stahl, D.A. 1990b. Combination of 16S rRNA-targeted probes with flow cytometry for analysing mixed microbial populations. *Applied and Environmental Microbiology* 56: 1919-25.

Amann, R.I., Springer, N., Ludwig, W., Görtz, H.-D. and Schleifer, K.-H. 1991. Identification *in situ* and phylogeny of uncultured bacterial endosymbionts. *Nature* 351: 161-64.

Amann, R.I., Stromley, J., Devereux, R., Key, R. and Stahl, D.A. 1992a. Molecular and microscopic identification of sulfate-reducing bacteria in multispecies biofilms. *Applied and Environmental Microbiology* 58: 614-23.

Amann, R.I., Zarda, B., Stahl, D.A. and Schleifer, K.-H. 1992b. Identification of individual prokaryotic cells by using enzyme-labeled, rRNA-targeted oligonucleotide probes. *Applied and Environmental Microbiology* 58: 3007-11.

Bakken, L.R. 1985. Separation and purification of bacteria from soil. *Applied and Environmental Microbiology* 49: 1482-87.

Bej, A.K., DiCesare, J.L., Haff, L. and Atlas, R.M. 1991. Detection of *Escherichia coli* and *Shigella* spp. in water by using the polymerase chain reaction and gene probes for *uid*. *Applied and Environmental Microbiology* 57: 1013-17.

Bochenek, B. and Hirsch, A. 1990. *In situ* hybridisation of nodulin mRNAs in root nodules using non-radioactive probes. *Plant Molecular Biology Report* 8: 237-48.

Boenisch, T. 1989. Background. In Naish, S.J. (ed) *Handbook Immunochemical Staining Methods*. Carpinteria, Califonia, USA: DAKO Corporation.

Bohlool, B.B. 1987. Fluorescence methods for study of *Rhizobium* in culture and *in situ*. In Elkan, G.H. (ed). *Symbiotic Nitrogen Fixation Technology*. New York, USA: Marcel Dekker.

Bohlool, B.B. and Schmidt, E.L. 1968. Nonspecific staining: Its control in immunofluorescence examination of soil. *Science* 162: 1012-14.

Bohlool, B.B. and Schmidt, E.L. 1973. A fluorescent antibody technique for determination of growth rates of bacteria in soil. *Bulletins from the Ecological Research Committee (Stockholm)* 17: 336-38.

Bohlool, B.B. and Schmidt, E.L. 1980. The immunofluorescence approach in microbial ecology. In Alexander, M. (ed) *Advances in Microbial Ecology.* New York, USA: John Wiley.

Boom, R., Sol, C.J.A., Salimans, M.M.M., Jansen, C.L., Wertheim-van Dillen, P.M.E. and van der Noordaa, J. 1990. Rapid and simple method for purification of nucleic acids. *J. Clinical Microbiology* 28: 495-503.

Braun-Howland, E.B., Danielsen, S.A. and Nierzwicki-Bauer, S.A. 1992. Development of a rapid method for detecting bacterial cells *in situ* using 16S rRNA-targeted probes. *BioTechniques* 13: 928-33.

Brock, T.D. 1987. The study of microorganisms *in situ*: Progress and problems. *Symposia of the Society for General Microbiology* 41: 1-17.

Burnison, B.K. and Nuttley, D.J. 1990. Purification of DNA for bacterial productivity estimates. *Applied and Environmental Microbiology* 56: 362-65.

Celeda, D., Bettag, U. and Cremer, C. 1992. PCR amplification and simultaneous digoxigenin incorporation of long DNA probes for fluorescence in *in situ* hybridisation. *BioTechniques* 12: 98-102.

Chaudhry, G.R., Toranzos, G.A. and Bhatti, A.R. 1989. Novel method for monitoring genetically engineered microorgannisms in the environment. *Applied and Environmental Microbiology* 55: 1301-04.

DeLong, E.F., Wickham, G.S. and Pace, N.R. 1989a. Phylogenetic stains: Ribosomal RNA-based probes for the identification of single cells. *Science* 243: 1360-63.

DeLong, E.F., Schmidt, T.M. and Pace, N.R. 1989b. Analysis of single cells and oligotrophic picoplankton populations using 16S rRNA sequences. In Hattori, T., Ishida, Y., Maruyama, Y., Morita, R. and Uchida, A. (eds) *Recent Advances in Microbial Ecology.* Tokyo, Japan: Japan Scientific Societies Press.

Distel, D.L., DeLong, E.F. and Waterbury, J.B. 1991. Phylogenetic characterisation and *in situ* localisation of the bacterial symbiont of shipworms (Teredinidae: Bivalvia) by using 16S rRNA sequence analysis and oligonucleotide probe hybridisation. *Applied and Environmental Microbiology* 57: 2376-82.

Fry, J.C. 1990. Direct methods and biomass estimation. *Methods in Microbiology* 22: 41-86.

Fuhrman, J.A., Comeau, D.E., Hagström, A. and Chan, A.M. 1988. Extraction from natural planktonic microorganisms of DNA suitable for molecular biological studies. *Applied and Environmental Microbiology* 54: 1426-29.

Fuhrman, J.A. and Lee, S.H. 1989 Natural microbial species variations studies at the DNA level. In Hattori, T., Ishida, Y., Maruyama, Y., Morita, R. and Uchida, A. (eds) *Recent Advances in Microbial Ecology.* Tokyo, Japan: Japan Scientific Societies Press.

Giovannoni, S.J., DeLong, E.F., Olsen, G.J. and Pace, N.R. 1988. Phylogenetic group-specific oligonucleotide probes for identification of single microbial cells. *J. Bacteriology* 170: 720-26.

Giovannoni, S.J., Britschgi, T.B., Moyer, C.L. and Field, K.G. 1990. Genetic diversity in Sargasso Sea bacterioplankton. *Nature* 345: 60-63.

Göbel, U.B., Geiser, A. and Stanbridge, E.J. 1987. Oligonucleotide probes complementary to variable regions of ribosomal RNA discriminate between *Mycoplasma* species. *J. General Microbiology* 133: 1969-74.

Grant, M.A. and Holt, J.G. 1977. Medium for the selective isolation of members of the genus *Pseudomonas* from natural habitats. *Applied and Environmental Microbiology* 33: 1222-24.

Hahn, D., Dorsch, M., Stackebrandt, E. and Akkermans, A.D.L. 1989. Synthetic oligonucleotide probes in identification of *Frankia* strains. *Plant and Soil* 118: 211-19.

Hahn, D., Starrenburg, M.J.C. and Akkermans, A.D.L. 1990a. Oligonucleotide probes that hybridise with rRNA as a tool to study *Frankia* strains in root nodules. *Applied and Environmental Microbiology* 56: 1342-46.

Hahn, D., Kester, R., Starrenburg, M.J.C. and Akkermans, A.D.L. 1990b. Extraction of ribosomal RNA from soil for detection of *Frankia* with oligonucleotide probes. *Archives of Microbiology* 154: 329-35.

Hahn, D., Amann, R.I., Ludwig,W., Akkermans, A.D.L. and Schleifer, K.-H. 1992. Detection of microorganisms in soil after *in situ* hybridisation with rRNA targeted fluorescently labelled oligonucleotides. *J. General Microbiology* 138: 879-87.

Hahn, D., Amann, R.I. and Zeyer, J. 1993. Whole cell hybridisation of *Frankia* strains with fluorescence- or digoxigenin-labeled 16S rRNA targeted oligonucleotide probes. *Applied and Environmental Microbiology* 59: 1709-16.

Herron, P.R. and Wellington, E.M.H. 1990. New method for extraction of streptomycetes spores from soil and application to the study of lysogeny in sterile amended and nonsterile soil. *Applied and Environmental Microbiology* 56: 1406-12.

Hopkins, D.W., MacNaughton, S.J. and O'Donell, A.G. 1991. A dispersion and differential centrifugation technique for representatively sampling microorganisms from soil. *Soil Biology and Biochemistry* 23: 217-25.

Hill, I.R. and Gray, T.R.G. 1967. Application of the fluorescent-antibody technique to an ecological study of bacteria in soil. *J. Bacteriology* 93: 1888-96.

Holben, W.E., Jansson, J.K., Chelm, B.K. and Tiedje, J.M. 1988. DNA probe method for the detection of specific microorganisms in the soil bacterial community. *Applied and Environmental Microbiology* 54: 703-11.

Holtke, H.-J., and Kessler, C. 1990. Non-radioactive labeling of RNA transcripts *in vitro* with the hapten digoxigenin (DIG); hybridisation and ELISA-based detection. *Nucleic Acids Research* 18: 5843-51.

Kingsley, M.T. and Bohlool, B.B. 1981. Release of *Rhizobium* spp. from tropical soils and recovery for immunofluorescence enumeration. *Applied and Environmental Microbiology* 42: 241-48.

Knight, I.T., Holben, W.E., Tiedje, J.M. and Colwell, R.R. 1992. Nucleic acid hybridisation techniques for detection, identification, and enumeration of microorganisms in the environment. In Levin, M.A., Seidler, R.J. and Rogul, M. (eds) *Microbial Ecology: Principles, Methods, and Applications*. New York, USA: McGraw-Hill.

Kohne, D., Hogan, J., Jonas, V., Dean, E. and Adams, T.H. 1986. Novel approach for rapid and sensitive detection of microorganisms: DNA probes to rRNA. In Leive, L. (ed) *Microbiology-1986*. Washington, DC, USA: American Society for Microbiology.

Ogram, A., Sayler, G.S. and Barkay, T. 1987. The extraction and purification of microbial DNA from sediments. *J. Microbiological Methods* 7: 57-66.

Olsen, G.J., Lane, D.J., Giovannoni, S.J., Pace, N.R. and Stahl, D.A. 1986. Microbial ecology and evolution: A ribosomal RNA approach. *Annual Reviews of Microbiology* 40: 337-65.

Pickup, R.W. 1991. Development of molecular methods for the detection of specific bacteria in the environment. *J. General Microbiology* 137: 1009-19.

Postma, J., van Elzas, J.D., Govaert, J.M. and van Veen, J.A. 1988. The dynamics of *Rhizobium leguminosarum* biovar *trifolii* introduced into soil as determined by immunofluorescence and selective plating techniques. *FEMS Microbiology Ecology* 53: 251-60.

Relman, D.A., Loutit, J.S., Schmidt, T.M., Falkow, S. and Tompkins, L.S. 1990. The agent of bacillary angiomatosis. *New England J. Medicine* 323: 1573-80.

Reyes, V.G. and Schmidt, E.L. 1979. Population densities of *Rhizobium japonicum* strain 123 estimated directly in soil and rhizospheres. *Applied and Environmental Microbiology* 37: 854-58.

Roszak, D.B. and Colwell, R.R. 1987. Survival strategies of bacteria in the natural environment. *Microbiological Reviews* 51: 365-79.

Sayler, G.S., Fleming, J., Applegate, B., Werner, C. and Nikbakht, K. 1989 Microbial community analysis using environmental nucleic acid extracts. In Hattori, T., Ishida, Y., Maruyama, Y., Morita, R. and Uchida, A. (eds) *Recent Advances in Microbial Ecology*. Tokyo, Japan: Japan Scientific Societies Press.

Schleifer, K.- H., Amann, R.I., Ludwig, W., Rothemund, C., Springer, N. and Dorn, S. 1992. Nucleic acid probes for the identification and *in situ* detection of Pseudomonads. In Galli, E., Silver, S. and Witholt, B. (eds) Pseudomonas: *Molecular Biology and Biotechnology*. Washington DC, USA: ASM.

Simonet, P., Normand, P., Moiroud, A. and Bardin, R. 1990. Identification of *Frankia* strains in nodules by hybridisation of polymerase chain reaction products with strain-specific oligonucleotide probes. *Archives of Microbiology* 153: 235-40.

Sommerville, C.C., Knight, I.T., Straube, W.L. and Colwell, R.R. 1989. Simple, rapid method for direct isolation of nucleic acids from aquatic environments. *Applied and Environmental Microbiology* 55: 5548-54.

Spring, S., Amann, R.I., Ludwig, W., Schleifer, K.-H. and Petersen, N. 1992. Phylogenetic diversity and identification of nonculturable magnetotactic bacteria. *Systematic and Applied Microbiology* 15: 116-22.

Stahl, D.A., Flesher, B., Mansfield, H.R. and Montgomery, L. 1988. Use of phylogenetically based hybridisation probes for studies of ruminal micobial ecology. *Applied and Environmental Microbiology* 54: 1079-84.

Stahl, D.A., Devereux, R., Amann, R.I., Flesher, B., Lin, C. and Stromley, J. 1989. Ribosomal RNA based studies of natural microbial diversity and ecology. In Hattori, T., Ishida, Y., Maruyama, Y., Morita, R. and Uchida, A. (eds) *Recent Advances in Microbial Ecology*. Tokyo, Japan: Japan Scientific Societies Press.

Stahl, D.A. and Amann, R.I. 1991. Development and application of nucleic acid probes. In Stackebrandt, E. and Goodfellow, M (eds), *Nucleic Acid Techniques in Bacterial Systematics*. New York, USA: John Wiley.

Steffan, R.J. and Atlas, R.M. 1988. DNA amplification to enhance detection of genetically engineered bacteria in environmental samples. *Applied and Environmental Microbiology* 54: 2185-91.

Steffan, R.J., Goksøyr, J., Bej, A.K. and Atlas, R.M. 1988. Recovery of DNA from soils and sediments. *Applied and Environmental Microbiology* 54: 2908-15.

Steffan, R.J., Breen, A., Atlas, R.M. and Sayler, G.S. 1989. Application of gene probe methods for monitoring specific microbial populations in freshwater ecosystems. *Canadian J. Microbiology* 35: 681-85.

Tavangar, K., Hoffman, A.R. and Kraemer, F.B. 1990. A micromethod for the isolation of total RNA from adipose tissue. *Analytical Biochemistry* 186: 60-63.

Thompson, I.P., Cook, K.A., Lethbridge, G. and Burns, R.G. 1990. Survival of two ecologically distinct bacteria (*Flavobacterium* and *Arthrobacter*) in unplanted and rhizosphere soil: Laboratory studies. *Soil Biology and Biochemistry* 22: 1029-37.

Torsvik, V., Goksøyr, J. and Daae, F.L. 1990. High diversity in DNA of soil bacteria. *Applied and Environmental Microbiology* 56: 782-87.

Tsai, Y.L. and Olsen, B.H. 1991. Rapid method for direct extraction of DNA from soils and sediments. *Applied and Environmental Microbiology* 57: 1070-74.

Tsai, Y.L., Park, M.J. and Olsen, B.H. 1991. Rapid method for direct extraction of mRNA from seeded soils. *Applied and Environmental Microbiology* 57: 765-68.

Ultsch, A., Schuster, C.M., Betz, H. and Wisden, W. 1991. *In situ* hybridisation with oligonucleotides: A simplified method to detect *Drosophila* transcripts. *Nucleic Acids Research* 19: 3746.

Ward, D.M. 1989 Molecular probes for analysis of microbial communities. In Characklis, W.G. and Wilderer, P.A. (eds) *Structure and Function of Biofilms*. New York, USA: John Wiley.

Ward, D.M., Weller, R. and Bateson, M.M. 1990. 16S rRNA sequences reveal uncultured inhabitants of a well-studied thermal community. *FEMS Microbiology Reviews* 75: 105-16.

Ward, D.M., Weller, R. and Bateson, M.M. 1990. 16S rRNA sequences reveal numerous uncultured microorganisms in a natural community. *Nature* 345: 63-65.

Weller, R. and Ward, D.M. 1989. Selective recovery of 16S rRNA sequences from natural microbial communities in the form of cDNA. *Applied and Environmental Microbiology* 55: 1818-22.

Wellington, E.M.H., Al-Jawadi, M. and Bandoni, R. 1987. Selective isolation of *Streptomyces* species-groups from soil. *Developments in Industrial Microbiology* 28: 99-104.

Wiel, C.van de, Norris, J.H., Bochenek, B., Dickstein, R., Bisseling, T. and Hirsch, A. 1990. Nodulin gene expression and ENOD2 localisation in effective, nitrogen-fixing and ineffective, bacteria-free nodules of alfalfa. *Plant Cell* 2: 1009-17.

Witt, D., Liesack, W. and Stackebrandt, E. 1989. Identification of streptomycetes by 16S rRNA sequences and oligonucleotide probes. In Hattori, T., Ishida, Y., Maruyama, Y., Morita, R. and Uchida, A. (eds) *Recent Advances in Microbial Ecology*. Tokyo, Japan: Japan Scientific Societies Press.

Woese, C.R. 1987. Bacterial evolution. *Microbiological Reviews* 51: 221-71.

Xu, L., Harbour, D. and McCrae, M.A. 1990. The application of polymerase chain reaction to the detection of rotaviruses in faeces. *J. Virological Methods* 27: 29-38.

Yang, W.-C., Horvath, B., Hontelez, J., van Kammen, A. and Bisseling, T. 1991. *In situ* localisation of *Rhizobium* mRNAs in pea root nodules: *nif* A and *nif* H localisation. *Molecular Plant-Microbe Interactions* 4: 464-68.

Zarda, B., Amann, R.I., Wallner, G. and Schleifer, K.-H. 1991. Identification of single bacterial cells using digoxigenin-labelled, rRNA-targeted oligonucleotides. *J. General Microbiology* 137: 2823-30.

Beyond the Biomass
Edited by K. Ritz, J. Dighton and K.E. Giller
© 1994 British Society of Soil Science (BSSS)
A Wiley-Sayce Publication

CHAPTER 16

Molecular markers as tools to study the ecology of microorganisms

K.J. WILSON, A. SESSITSCH and A. AKKERMANS

One of the greatest barriers to the study of microbial ecology is the difficulty in identifying the organisms under study. This is generally achieved through the use of molecular markers; that is, molecules that can be used to distinguish one type of (micro)organism from another. These could be endogenous molecules, such as characteristic metabolic enzymes, a spontaneous antibiotic resistance marker or antigens expressed on the cell surface that can be recognised by specific antisera. More recently, techniques have been developed whereby a specific marker gene can be introduced into the organism being studied. Such a gene will code for a product that can be readily assayed, most commonly an enzyme that can act on different substrates to give readily detectable coloured or fluorescent products. The development of such marker genes facilitates monitoring of the fate of specific bacterial strains. Additionally, marker genes can be used as reporters linked to gene promoters that respond to a variety of environmental signals.

We are interested in studying bacteria that interact with plants. A prerequisite for choosing an appropriate marker gene for this is that there should be no detectable background activity in the bacteria or the plants. We have worked for some time on using the *gusA* gene from *Escherichia coli* (formerly *uidA*, Jefferson et al., 1986), which codes for the hydrolytic enzyme beta-glucuronidase (GUS), as a marker for the study of competitive ability of different rhizobial strains. The strategy is to incorporate *gusA* into a derivative of transposon Tn5 using *E. coli*, the standard laboratory workhorse, as a host. On mating the donor *E. coli* strain and a recipient *Rhizobium* (or other bacterial) strain of interest, the transposon with *gusA* becomes permanently inserted into the recipient strain's chromosome. Root nodules occupied by the marked strain can then be identified by developing a blue-coloured product when supplied with the enzyme substrate X-GlcA (5-bromo-4-chloro-3-indolyl glucuronide).

An integral component of the competitive ability of rhizobial and other soil bacterial strains is their ability to compete and survive in the free-living state in the soil or rhizosphere. In order to study different aspects of competition, therefore, we have made a number of genetic constructs in which expression of *gusA* is favoured either in the free-living or the symbiotic state. In this chapter, we discuss initial experiments demonstrating the utility of the GUS marker in studies on the ecology of free-living soil bacteria.

MATERIALS AND METHODS

Bacterial strains and plasmids

Plasmid pKW107 is a mobilisable suicide plasmid carrying *gusA* under the control of the constitutive *aph* promoter, inserted in the transposon Tn5 making transposon Tn5*gusA*KW107; therefore, the transposon also confers kanamycin (kan) and neomycin (neo) resistance (Wilson et al., 1991) (kanamycin and neomycin are similar antibiotics which are sensitive to the same inactivating enzyme encoded by Tn5). *E. coli* strain S17-1 (Simon et al., 1983) was used for mobilising pKW107 into other bacteria. *Pseudomonas putida* strain JH1 was provided by J. van Vuurde.

GUS assays

Histochemical substrates for GUS, 5-bromo-4-chloro-3-indolyl glucuronide (X-GlcA) and 6-bromo-4-chloro-3-indolyl glucuronide (Magenta-GlcA), were obtained from Biosynth AG in Switzerland. X-GlcA was used at 50 µg/ml and Magenta-GlcA at 150 µg/ml in plates.

Isolation of bacteria from soil

Ten grams of soil were mixed with 10 g gravel and 90 ml 0.1% (w/v) sodium pyrophosphate in a conical flask (500 ml), and shaken at 250 rpm at room temperature for 10 minutes (Smit and van Elsas, 1990). A series of tenfold dilutions was then plated on 1/10 Tryptic Soy Agar (TSA) containing 100 µg/ml cycloheximide, with these additions where appropriate: X-GlcA, 50 µg/ml; kanamycin, 100 µg/ml; gentamicin, 50 µg/ml; neomycin, 50 µg/ml; and nalidixic acid (nal), 10 µg/ml. The plates were incubated at 28-30°C.

Transfer of pKW107 to *Pseudomonas putida* strain JH1

The endogenous drug resistances of strain JH1 were first determined. The strain proved resistant to nalidixic acid (10 µg/ml) and to rifampicin (rif) (150 µg/ml), but sensitive to neomycin (50 µg/ml). Overnight cultures of JH1 and of *E. coli* S17-1 (pKW107) were grown in LB (10 g tryptone, 5 g yeast extract, 5 g NaCl per litre) and LB kanamycin (50 µg/ml), respectively. One millilitre of each culture was pelleted, washed once in LB without antibiotics, and resuspended in 250 µl LB; 100 µl of each was then mixed and plated together on King's B medium (20 g pentone, 1.5 g K_2HPO_4, 1.5 g $MgSO_4$, 10 g glycerol per litre) without antibiotics; 100 µl of each culture was also plated separately on King's B. All plates were incubated overnight at 30°C. The mating mixture was then resuspended by scraping the plate with 2 ml sterile 0.1% NaPPi, and 100 µl of a tenfold dilution of this initial mating mixture was plated on YM (0.5 g K_2HPO_4, 0.2 g $MgSO_4$, 0.1 g NaCl, 1 g yeast extract, 10 g mannitol per litre) $nal_{10}neo_{50}$X-GlcA$_{50}$ and on YM $rif_{150}neo_{50}$XGlcA$_{50}$ plates, respectively. The two parental cultures were also streaked on these plates. After 3 days of incubation at 30°C, about 20 colonies were visible on the selective plates, about 30% of which were blue, indicating that they had received the full Tn5*gusA*KW107 transposon. No growth was observed on the selective plates from either parental culture on its own. One of these GUS-marked JH1 colonies was purified for use in population studies.

Monitoring changes in *Pseudomonas* populations in soil

An exponentially growing culture of Tn5gusAKW107-marked strain JH1 was used as the inoculum, and 1 ml of the original culture and of three tenfold dilutions was added to separate 10 g soil samples. The soil used was a loamy soil from Ede, the Netherlands (*see* van Elsas et al., 1986) and the experiment was repeated in duplicate in untreated soil and in soil that had been sterilised by gamma-irradiation. Labels on samples containing different initial population densities of the marked strain were randomised, to create a 'single-blind' experiment and to determine whether population densities measured after 24 hours of incubation in the soil could be related to initial inoculum sizes. After 24 hours of incubation at room temperature, bacteria were isolated from soil as described above and serial tenfold dilutions were plated on 0.1 x TSA plates with the following additions: X-GlcA$_{50}$; X-GlcA$_{50}$, kan$_{50}$; and kan$_{50}$, nal$_{10}$. The plates were incubated at 30°C for 2 days prior to analysis of the colonies.

RESULTS

Measurement of background activity in soil

One of the key factors underlying the sensitivity of any method for detecting microorganisms in soil is the level of background activity of the marker being used. To examine this, samples of a loamy soil were collected from Ede and appropriate dilutions were plated on 0.1 x TSA with the following additions: none, to determine the total population of colony-forming units (CFUs); kanamycin, to determine the frequency of kanamycin-resistant CFUs; kanamycin + gentamycin, to determine the frequency of CFUs with resistance to both antibiotics; X-GlcA, for the population of GUS$^+$ CFUs; and X-GlcA + kanamycin, for the population of kanamycin-resistant GUS$^+$ CFUs. The results are given in Table 16.1.

Table 16.1   Background population of colony-forming units (CFUs) exhibiting various antibiotic and non-antibiotic markers, detected in soil from Ede, the Netherlands[a]

Phenotype	CFUs/g soil
Total	10^6
Kanamycin resistant	2×10^4
Kanamycin and gentamycin resistant	10^3
GUS positive	2×10^5
GUS positive and kanamycin resistant	0

Note: a   Data are the means of two independent platings after 48 hours of growth at 28°C

It can be seen from these data that, while the frequency of CFUs exhibiting GUS activity is about 20%, there are no CFUs showing resistance to kanamycin and exhibiting GUS activity. This contrasts with the observation that, even with the addition of a second antibiotic, gentamicin, the background population of CFUs expressing a dual antibiotic resistance remains high relative to desired detection levels for a microorganism.

This observation that there is no detectable background in dilution plating experiments when the combination of the GUS marker and a single antibiotic is used has been repeated in Delhi, India (no blue colonies on kanamycin + X-GlcA) plates) and Canberra, Australia (no magenta colonies on spectinomycin + Magenta-GlcA plates).

Use of the GUS marker to determine populations of a genetically engineered microorganism

To test the effectiveness of GUS as a marker for detecting a genetically modified organism in soil, the *P. putida* strain JH1 was marked with transposon Tn5*gusA*KW107, which confers kanamycin and neomycin resistance and GUS activity. Soil microcosms were set up with four different initial inoculum densities in both gamma-irradiated and untreated soil; after 24 hours of incubation at room temperature, bacteria were isolated from the soil and different dilutions were plated on media with the following additions: $XGlcA_{50}$, to determine the total CFUs with and without GUS activity; $neo_{50}nal_{10}$, to detect the modified *P. putida* strain using two of its antibiotic resistance markers (one endogenous and one introduced); and $neo_{50}XGlcA_{50}$, to detect the modified *P. putida* strain using both introduced markers. The experiment was performed as a single-blind experiment so that the people determining the populations of the marked JH1 in the different soil samples did not know the initial inoculum density in each microcosm.

The total population of CFUs in the untreated soil was found to be greater than 10^8/g soil, with a high proportion of GUS$^+$ CFUs (the exact proportion was not measurable because the plates were overgrown even at the greatest dilution). In the gamma-irradiated soil, there was in fact a background population of approximately 10^5 CFUs/g soil, of which less than 1% were GUS$^+$. The bacteria forming these colonies were presumably either resistant to the gamma-irradiation or had colonised the soil following the irradiation treatment, which had occurred about 6 months prior to the experiment.

The results of plating on other selective media are shown in Table 16.2. Again, these results indicate clearly that the combined use of GUS and an antibiotic resistance marker is more effective than the use of two antibiotic resistance markers in eliminating background activity. Growth from the uninoculated, untreated soil on the neomycin/nalidixic acid plates indicated that there was a background population of 8×10^4 CFUs/g soil that was resistant to neomycin and nalidixic acid in the untreated soil. By contrast, no blue colonies were obtained on X-GlcA/neomycin plates from the uninoculated, untreated soil, indicating that there were no background CFUs that were GUS$^+$ and resistant to neomycin (*see* Table 16.2). The population sizes of the marked *P. putida* strain were then estimated in each sample from the numbers of blue colonies forming on the X-GlcA/neomycin plates. It can be seen that the ranking of population sizes in each sample corresponded to the ranking of the initial inoculum densities, and that the population of marked bacteria apparently increased in the untreated soil and decreased in the sterile soil during the 24-hour incubation period.

DISCUSSION

The results presented above demonstrate that the combined use of GUS and kanamycin/neomycin resistance as markers is very efficient for the detection of a genetically modified microorganism through the use of dilution plating. In experiments on three continents, no endogenous bacteria were found exhibiting the combination of GUS and an antibiotic resistance marker (*see* Tables 16.1 and

Table 16.2   Changes in the population of a genetically marked *Pseudomonas putida* strain in gamma-irradiated soil and in untreated soil, detected on media using different selection methods

| | | 0.1 x Tryptic Soy Agar (TSA) plates containing | | |
Soil type	Initial inoculum density[a]	neo$_{50}$nal$_{10}$[b] (CFUs/g)	neo$_{50}$X-GlcA$_{50}$ blue (CFUs/g)	neo$_{50}$X-GlcA$_{50}$ non-blue (CFUs/g)
gamma-irradiated	5×10^6	7.7×10^4	6.4×10^4	0
gamma-irradiated	5×10^5	1.0×10^4	1.1×10^4	0
gamma-irradiated	5×10^4	7.0×10^2	3.0×10^2	0
gamma-irradiated	Uninoculated	0	0	0
Untreated	5×10^6	$> 10^8$	$> 10^8$	$> 10^8$
Untreated	5×10^5	6.0×10^7	6.0×10^7	5.0×10^5
Untreated	5×10^4	1.0×10^7	1.2×10^7	3.3×10^5
Untreated	5×10^3	1.2×10^6	1.3×10^6	4.5×10^5
Untreated	Uninoculated	8.0×10^4	0	6.5×10^5

Note: a   All data are given as CFUs/g soil fresh weight; the data are from one single-blind experiment, with single platings at different dilutions; very similar results were obtained in a second experiment, where the microcosms were incubated for 2 days prior to plate counting

       b   Neo = neomycin; nal = nalidixic acid

16.2, and unpublished data referred to above), enabling detection rates of 10^2 cells/g soil or less. This contrasts with the use of two antibiotic resistance markers, where background populations can still be 10^3-10^4 CFUs/g soil (*see* Tables 16.1 and 16.2; Drahos, 1991). Presumably, the reason for these differing background frequencies is that spontaneous antibiotic resistance is acquired at relatively high frequencies, whereas *de novo* acquisition of a complete enzyme activity is extremely rare, and likely to occur only through gene transfer.

The practical utility of the GUS + kanamycin/neomycin marker was demonstrated in a simple experiment monitoring population changes of a marked *P. putida* strain in soil microcosms. In two independent single-blind experiments, similar results were obtained by two groups of scientists who neither knew the expected outcome of the experiment nor had used the technique previously. Interestingly, population levels of the marked *P. putida* strain were seen to decline rapidly in soil that had been sterilised through gamma-irradiation, whereas they increased substantially in soil from the same source that had not been sterilised. This implies either that the population of organisms that grew after gamma-irradiation was more antagonistic to strain JH1 than the original flora or that other changes occurred in the soil upon irradiation which made it less supportive of the survival and growth of this strain of *P. putida*.

Thus, the *gusA* gene can be added to the collection of marker genes that are available for monitoring genetically modified microbes in the environment. Other available marker systems include *lacZY* encoding beta-galactosidase and the galactoside permease, *xylE*, encoding catechol-2,3,-dioxygenase, the *lux* genes which catalyse the production of light, and *ina*, which encodes an ice-nucleation protein (reviewed in Drahos, 1991; Pickup, 1991). A striking advantage of GUS over these other systems is that it is absent from higher plants, and therefore can be used very effectively to study bacteria in the rhizosphere and in symbiotic association with plants. Use of the chromogenic substrate X-GlcA, which produces a blue precipitate on cleavage by GUS, allows the position of major concentrations of bacteria on the root to be visualised, as well as the early stages of infection (Sharma and Signer, 1990; Wilson et al. 1991; Wilson, 1994). Although in theory the *lux* system also makes such analysis possible, in practice it has proved difficult to obtain high enough levels of expression for easy visualisation without resorting to elaborate and expensive equipment for the detection of the light signal (O'Kane et al. 1988; de Weger et al., 1991). Thus, at present the GUS marker is the most affordable, versatile and reliable system for studying the ecology of bacteria and other microorganisms that interact with plants.

One of the questions frequently raised is whether the introduced marker will compromise the fitness of the organism being studied. Some examples have shown that fitness is compromised (for example, van Elsas et al., 1991), whereas in other cases it may be impaired little, if at all (for example, Orvos et al., 1990). It is clear that this factor must be taken into account when designing experiments, as must any genotypic or environmental change that is introduced into an experiment designed to study microbial ecology. For example, one can also argue that the use of artificial microcosms renders data of uncertain relevance to the real, undisturbed field situation. It is probably only the use of nucleic acid based techniques to analyse population structure in the environment that can truly be said to leave both the organisms being studied and the environment undisturbed (for example, Giovannoni et al., 1990; Ward et al., 1990; and other chapters in this volume).

It is certainly the case that one of the traditional types of marker systems used (that is, the use of spontaneous antibiotic-resistant mutants) can have a substantial impact on the fitness of the organism because many such mutations affect fundamental processes of the cell's metabolism (Compeau et al., 1988). The introduction of exogenous genes is likely to have a lessor effect on the fitness of the organism. Lam et al. (1990) analysed over 1200 mutants of *P. putida* containing a transposon Tn5 derivative for their ability to colonise roots, and found isolates with both increased colonisation ability and severely decreased colonisation ability. The majority of isolates, however, showed a colonisation ability that was very close to the wild type strain. In a study of several independent isolates of a rhizobial strain marked with a GUS transposon, we have found isolates with reduced fitness, the same fitness and even enhanced fitness, as assessed by competition for nodule occupancy in soil-grown plants (A. Sessitsch, upubl.).

The most important parameter in designing any experiment is consideration of the question being asked, to ensure that the best tool for the problem is selected, rather than vice versa. The advantages of the use of marker genes are that they can be used to generate a large amount of data relatively easily, inexpensively and by people with microbiological rather than molecular biological expertise. Another advantage is that marker genes can be used creatively to ask specific questions by engineering the marker gene to be responsive to a particular environmental signal. For example, Lam et al. (1990) used *lacZ* fusions to isolate promoters that responded to rhizosphere exudates, and such markers could then be used to analyse the role of exudate response in successful root colonisation. Similarly, Heitzer et al. (1992) have developed a simple bioassay for the bioavailability of naphthalene in soil using the *lux* genes. Thus, rather than using a marker gene simply as a molecular tag, the fact that its expression can be controlled to respond to environmental signals can and should be exploited in the design of ecological experiments utilising marker gene systems.

To this end, we have constructed a number of new *gusA*-containing transposons in which the *gusA* gene is under the control of different DNA promoter sequences. These facilitate optimal detection of marked rhizobia either in the free-living state or, by using a different promoter sequence for *gusA*, in the symbiotic state (Wilson, 1994) and this will be of great utility in studying the competitive ability of marked rhizobial strains. In future work we plan to extend this approach by developing additional marker systems that can also be used to analyse the symbiotic interaction with plants. This will enable competition between two or more marked strains to be studied at once, using substrates for the different reporter genes that yield different coloured products.

Acknowledgements

This work was supported by a grant from the British Royal Society to the first author. Some of the experiments were carried out during a course on the 'Introduction of genetically modified organisms into the environment: Biosafety aspects', organised by the European Environmental Research Organisation in Wageningen, the Netherlands, in 1991.

References

Compeau, G., Al-Achi, B.J., Platsouka, E. and Levy, S.B. 1988. Survival of rifampin-resistant mutants of *Pseudomonas fluorescens* and *Pseudomonas putida* in soil systems. *Applied and Environmental Microbiology* 54: 2432-38.

de Weger, L.A., Dunbar, P., Mahafee, W.F., Lugtenberg, B.J.J. and Sayler, G.S. 1991. Use of bioluminescence markers to detect *Pseudomonas* spp. in the rhizosphere. *Applied and Environmental Microbiology* 57: 3641-44.

Drahos, D. 1991. Current practices for monitoring genetically engineered microbes in the environment. *AgBiotech News and Information* 3: 39-48.

Giovannoni, S.J., Britschgi, T.B., Moyer, C.L. and Field, K.G. 1990. Genetic diversity in Sargasso Sea bacterioplankton. *Nature* 345: 60-63.

Heitzer, A., Webb, O.F., Thonnard, J.E. and Sayler, G.S. 1992. Specific and quantitative assessment of naphthalene and salicylate bioavailability by using a bioluminescent catabolic reporter bacterium. *Applied and Environmental Microbiology* 58: 1839-46.

Jefferson, R.A., Burgess, S.M. and Hirsh, D. 1986. β-glucuronidase from *E. coli* as a gene-fusion marker. *Proc. National Academy of Sciences, USA* 83: 8447-51.

Lam, A.T., Ellis, D.M. and Ligon, J.M. 1990. Genetic approaches for studying rhizosphere colonisation. *Plant and Soil* 129: 11-18.

O'Kane, D.J., Lingle, W.L., Wampler, J.E., Legozki, M., Legozki, R.P. and Szalay, A.A. 1988. Visualisation of bioluminescence as a marker of gene expression in rhizobium-infected soybean root nodules. *Plant Molecular Biology* 10: 387-99.

Orvos, D.R., Lacy, G.H. and Cairns. J. Jr. 1990. Genetically engineered *Erwinia carotovora*: Survival, intraspecific competition, and effects upon selected bacterial genera. *Applied and Environmental Microbiology* 56: 1689-94.

Pickup, R.W. 1991. Development of molecular methods for the detection of specific bacteria in the environment. *J. General Microbiology* 137: 1009-19.

Sharma, S.B. and Signer, E.R. 1990. Temporal and spatial regulation of the symbiotic genes of *Rhizobium meliloti* in planta revealed by transposon Tn5-*gusA*. *Genes and Development* 4: 344-56.

Simon, R., Priefer, U. and Pühler, A. 1983. A broad host-range mobilisation system for *in vivo* genetic engineering: Transposon mutagenesis in Gram-negative bacteria. *Bio/Technology* 1: 784-91

Smit, E. and van Elsas, J.D. 1990. Determination of plasmid transfer frequency in soil: Consequences of bacterial mating on selective agar media. *Current Microbiology* 21: 151-57.

van Elsas, J.D., Dijkstra, A.F., Govaert, J.M. and van Veen, J.A. 1986. Survival of *Pseudomonas fluorescens* and *Bacillus subtilis* introduced into two soils of different texture in field microplots. *FEMS Microbiology Ecology* 38: 151-69.

van Elsas, J.D., van Overbeek, L.S., Feldmann, A.M., Dullemans, A.M. and de Leeuw, O. 1991. Survival of genetically engineered *Pseudomonas fluorescens* in soil in competition with the parent strain. *FEMS Microbiology Ecology* 85:53-64.

Ward, D.M., Weller, R. and Bateson, M.M. 1990. 16SrRNA sequences reveal numerous uncultured microorganisms in a natural community. *Nature* 345: 63-65.

Wilson, K.J. 1994. Molecular techniques for the study of rhizobial ecology in the field. *Soil Biology and Biochemistry* (in press).

Wilson, K.J., Giller, K.E. and Jefferson, R.A. 1991. β-glucuronidase (GUS) operon fusions as a tool for studying plant-microbe interactions. In Hennecke, H. and Verma, D.P.S. (eds) *Advances in Molecular Genetics of Plant-Microbe Interactions* (Vol. 1). Dordrecht, the Netherlands: Kluwer.

PART IV

Functional interactions of soil communities

Beyond the Biomass
Edited by K. Ritz, J. Dighton and K.E. Giller
© 1994 British Society of Soil Science (BSSS)
A Wiley-Sayce Publication

CHAPTER 17

Biological interactions between fauna and the microbial community in soils

M.-M. Couteaux and P. Bottner

The role of soil fauna in soil processes is of great importance and operates mainly through interactions with the microflora. Although behaviour in both communities varies widely, as does the response to availabilities or constraints, the basic interaction consists of a two-way relationship between a specific animal and a specific microbe controlled by the biotic and abiotic environment. Because of the diversity, synchrony and interdependence of these basic functions, they are hidden in the complexity of interactions. Experiments based on a simplified system of interactions therefore show a biological potential which is not apparent in complex situations. However, some unexpected biological miens can emerge from the synergy of complex relationships.

In order to understand and model the biological mechanisms, two approaches can be used to describe the carbon fluxes in the soil:

- all biological interactions are integrated in a general rate and the compartments are defined by their specific decomposition rate

- all organic matter passes through the different levels of the food web and each biological group is viewed as a compartment

The goal of this review is to analyse the role of soil fauna. These animals can alter the composition and function of the microbial community by their basic processes. We also broaden the analysis to examine the possible responses of these mechanisms to predicted climate changes.

BASIC PROCESSES OF SOIL FAUNA

Several soil faunal processes lead to changes in the population structure and in interactions with the microbial communities. These include predation and grazing, comminution and propagule dissemination, and microbial dynamics in animal guts. These processes are simultaneous and their relative importance depends upon the general characteristics of the ecosystem. To identify these mechanisms, it is necessary to conduct laboratory experiments using simplified biological models.

Predation and grazing

Although many soil animals are considered to have a variable diet of microorganisms, two main groups can be distinguished — bacterivorous and fungivorous — resulting in different food webs (Hendrix et al., 1986). The protozoa/bacteria relationship has long been used as a basis for predator-prey modelling (Volterra, 1927; Jost et al., 1973). For soil, Steinberg et al. (1987) proposed a model which includes the logistic growth of the bacterial population on a limiting substrate, the protozoan growth on the bacteria, the exponential mortality of the predator and the substrate excreted by the protozoa or produced by cell lysis. This is an example of feedback interaction in a two-level food chain where the product of the predator stimulates the growth of the prey.

Selective grazing pressure

In a more complex system, the structure of the microbial community is controlled by selective animal grazing and depends upon the availability of the preferred prey. Most feeders can survive on alternative food but their growth and fecundity are reduced. The main factor of selection is the prey species (Parkinson et al. 1979; Moore et al. 1988) but other factors include age and biochemical composition of the prey (Bengtsson and Rundgren, 1983; Leonard, 1984). Moore et al. (1987) showed that the collembolan *Folsomia candida* preferred younger and actively metabolising hyphae to less active hyphae. Its growth and fecundity also increased when it was fed on fungi grown on a nitrogen-rich medium (Booth and Anderson, 1979).

In some cases, the preference does not stem from better palatability. Some fungi appear to produce toxins which can be lethal for animals. Coûteaux and Dévaux (1983) interpreted the depressed growth of testate amoebae in soil humus as being partly due to inhibitory exudates from some tested fungi. Darbyshire et al. (1992) suggested that within the soil pore network, clusters of fungal spores may release sufficient toxins to discourage predation of the fungi by nearby *Cercomonas* species. A similar toxic effect of an unidentified basidiomycete on a collembolan was observed by Parkinson et al. (1979). Presumably, a volatile toxin was involved since no hyphae of the fungus were observed in the guts of dead animals.

The impact of grazers differs according to the size of their population, either stimulating or inhibiting prey activity. Hanlon and Anderson (1979) observed a stimulating effect on the microbial activity in experimental units with five collembola (*F. candida*); when there were more than 10, the grazing pressure resulted in decreased microbial biomass and activity. Trofymow and Coleman (1982) showed that inoculation of high numbers of fungal-feeding nematodes into soil microcosms containing a single species of fungus reduced respiration rates, while soil inoculated with low numbers of nematodes had a higher rate of respiration. Bacterivorous nematodes increased both the rate of respiration and bacterial populations, compared with bacteria alone. It seems that there is an optimal size of the grazer population, as shown for *Oniscus asellus* (isopods) and *Glomeris marginata* (millipedes). Up to this size, the reduced microbial biomass resulting from grazing is likely to be compensated for by stimulation of microbial activity (Hanlon and Anderson, 1980).

The animal consumption rate also controls the grazing pressure. During the growth phase of soil microcosms, microorganisms constitute a high-quality food, and under these conditions the generation time of the protozoa is only a few hours, with bacterial consumption rates of 10^3-10^5 cells/division (Clarholm, 1985). Nematode consumption rates for bacteria range from 2 to 72 x 10^5 cells/day per nematode (Ingham et al., 1985), with a generation time of 10-70 days (Scheimer, 1983). The combined effect of the size of the grazer population and its consumption rate reduces the microbial biomass, which compensates for this reduction by its growth.

Effect on size of microbial biomass

The dynamics of the microbial community are the result of the balance between growth and grazing pressure. This balance can lead to increased, stable or reduced microbial density. If the microbial density increases or remains stable, this means that microbial growth is stimulated by grazing. To some extent, even a reduction can involve a stimulation of microbial growth. A system in which predators are present may produce more microbial biomass than a system without predators. Depending upon the nutrient availability for the prey, a cryptic growth of the prey population may occur if the animal consumption is higher than the difference between systems with and without predators. It is possible to identify this cryptic growth by measuring the grazing pressure in culture in saline solution, where bacterial growth is more inhibited than in culture with nutrients (see Table 17.1). As shown in the table, when addition of chloramphenicol (control) prevented the bacteria from growing and from using the waste products of predation, 99.8 % of the bacteria were consumed by the ciliates. Without chloramphenicol, the use of the waste products induced a slight cryptic growth, resulting in the grazed

Table 17.1   Density of *Pseudomonas fluorescens* (MPN counts) in the presence and absence of *Colpoda aspersa*, percentage of grazed to ungrazed populations and percentage of grazed population attributable to cryptic growth in a saline solution + chloramphenicol or nutrients[a]

Medium	Inoculum	Initial density (no./ml)	Density after 14 days (no./ml)	% of ungrazed population	% due to cryptic growth
Saline solution + chloramphenicol (50 mg/l)	Bacteria alone	$2.47 \pm 0.85 \times 10^8$	$2.47 \pm 0.97 \times 10^8$		
	Bacteria + ciliates	$2.47 \pm 0.85 \times 10^8$	$4.53 \pm 4.05 \times 10^5$	0.18	0.00
Saline solution	Bacteria alone	$1.04 \pm 0.20 \times 10^7$	$3.20 \pm 0.20 \times 10^7$		
	Bacteria + ciliates	$1.04 \pm 0.20 \times 10^7$	$5.10 \pm 1.17 \times 10^5$	1.59	1.41
Saline solution + asparagine (0.27 mg/l) + glucose (1 mg/l)	Bacteria alone	$2.60 \pm 0.12 \times 10^7$	$1.02 \pm 0.07 \times 10^9$		
	Bacteria + ciliates	$2.60 \pm 0.12 \times 10^7$	$1.07 \pm 0.58 \times 10^8$	10.49	10.31

Note: a   ± confidence interval (p = 0.05, n = 3)
Source: Calculated from Palka (1988)

population being 1.4% higher proportionally than the control with chloramphenicol; when some substrate was added (glucose and aspargine), even in limiting conditions the cryptic growth on the waste products and the substrate led to a 10.3% increase in the population compared with the control. If there is no nutrient limitation (*see* Table 17.2), the cryptic growth will counteract the grazing pressure and there will be no difference in size between grazed and ungrazed bacterial biomass. This stimulation and cryptic growth are due to both the active metabolism of young strains which are preferentiallly chosen and the recycling of nutrients through the predator excretion. The same effect on the size of the fungal standing crop was shown with collembolan grazing but mycelial growth was modified morphologically (Faber et al., 1992). The amount of visible, superficially growing, mycelial weft was smaller in grazed microcosms than in the controls. Reductions in microbial biomass will occur when cryptic growth is constrained by environmental stress (Scholle et al., 1992).

Table 17.2   Density of *Pseudomonas fluorescens* (x 10^9/g dry weight) and *Colpoda aspersa* (x 10^4/g dw) (MPN counts) after 14 days of incubation in an organic soil with an initial inoculum of 2.6 x 10^5/g dw and 9.5 x 10^5/g dw in the ungrazed and grazed units, respectively[a]

Amendment	Inoculum	Density	
		Bacteria	Ciliates
C-glucose (1 mg/g dw)	Bacteria alone	9.6 ± 3.8	
	Bacteria + ciliates	10.5 ± 7.7	22.4 ± 1.3
N-NH$_4$SO$_4$ (0.1 mg/g dw)	Bacteria alone	6.4 ± 1.4	
	Bacteria + ciliates	8.3 ± 2.0	20.6 ± 3.9
C-glucose (1 mg/g dw)	Bacteria alone	10.3 ± 3.4	
N-NH$_4$SO$_4$ (0.1 mg/g dw)	Bacteria + ciliates	10.2 ± 4.0	21.9 ± 2.8
Control	Bacteria alone	5.7 ± 0.7	
	Bacteria + ciliates	7.4 ± 3.5	19.1 ± 4.5

Note: a   ± confidence interval (p = 0.05, n = 3)
Source: Henkinet et al. (1990)

The change in the density of some species according to selective grazing induces competitive or mutualistic responses which can modify soil functioning. Newell (1984) showed that litter was decomposed by the basiodiomycete *Mycena galopus* at one-third to one-half the rate of *Marasmius androsaceus*. The selective grazing of *Onychiurus latus* (collembolan) on *M. androsaceus* altered the outcome of competition between the fungi such that the decomposition was carried out primarily by *M. galopus* and was reduced by the presence of the animal. Switches in existing competitive relationships by selective grazing were reported by Parkinson et al. (1979) in an experimental competitive situation in which selective grazing by collembolans on sterile dark fungi permitted better colonisation by basidiomycetes. Switches in dominance were also shown between dark and white-rot fungi in response to the nitrogen content of the substrate. The incubation for 10 months of two kinds of litter, one produced in a doubled CO_2 atmosphere and with a low nitrogen content (N = 0.5 %) and a control produced in normal atmosphere (N = 1 %) with animal food webs of different complexities (Coûteaux et al., 1991a), showed no effect of grazing on the respiration of the litter with high nitrogen content (*see* Figure 17.1). In the litter with low nitrogen content, after a first stage of low respiration which involved a low competitive pressure, white-rot fungi were able to grow; this invasion occurred earlier in the microcosms where the animal groups were more diverse, indicating a shift in microbial communities favoured by animal diversity.

Figure 17.1 Cumulative respiration over time from litter as influenced by the activity of food webs of different complexity

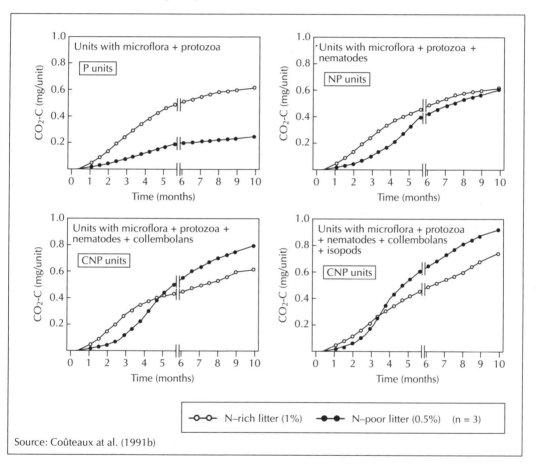

In some cases, animal grazing can counteract the mutualistic relationship between mycorrhizae and plants. Chakraborty et al. (1985) showed that inoculation of pine-seedling cultures with amoebae reduced the colonisation of roots by an ectomycorrhizal fungus. Mycophagous nematodes have been shown to reduce ectomycorrhizal biomass (Fogel, 1988), as they choose hyphal tips, penetrate the fungal mantle on root tissues and, by wounding, may facilitate the entry of root pathogens.

Grazing on vesicular-arbuscular mycorrhizae (VAM) can affect the phosphorus uptake process. Warnock et al. (1982) added *Folsomia candida* to *Allium porum* culture associated with the VAM *Glomus fasiculatus*. After 12 weeks, for the collembolan the production of the mycorrhizal leeks was 50% lower and the phosphorus concentration of the plant was 55% lower than in the ungrazed mycorrhizal controls. This effect is particularly apparent in phosphorus-limiting conditions, in natural habitats and in low-input agriculture where soil phosphorus content is low. VAM consumption by earthworms has also been observed (Rabatin and Stinner, 1988; Reddel and Spain, 1991). Microarthropods, even non-specialised feeders, can feed on ubiquitous VAM and it has been suggested that they may adversely affect plant production in some cases by reducing mycorrhizal infections or hyphal connections (Moore et al., 1985).

Effect on metabolic activity

Animal grazing controls the size of the microbial biomass but its role in metabolic activity is more obvious. It is now well established that predation enhances carbon turnover and nitrogen mineralisation and this effect is dependent upon grazing pressure and the age and the biochemical composition of the microbial biomass. Specific activity has been shown to be enhanced by grazing; Faber et al. (1992) demonstrated this for cellullase activity, for example. Different species can have different effects, depending upon the kind of activity. Teuben and Roelofsma (1990) reared *Tomocerus minor* and an isopod, *Philoscia muscorum*, in microcosms containing pine litter. In both cases, CO_2 output and exchangeable phosphate concentrations were increased; however, only with *T. minor* was there increased dehydrogenase, cellulase activity and nitrate concentration. Hyphal morphology and extracellular enzyme production were studied in the same species by Hedlund et al. (1991). Grazing induced switching from the 'normal' hyphal mode, with appraised growth and sporulating hyphae, to sectors of fast-growing and non-sporulating mycelium which developed an extensive area of mycelium. Specific protease and alpha-amylase activities were several times higher in grazed cultures where switching occurred than in plates without switching. Switching to a fast-growing hyphal mode could explain the compensatory growth of grazed fungi.

The mechanism of stimulation of the microbial activity by grazing also relates to the production of chemical mediators. The antagonism between *Pseudomonas putida* and *Fusarium oxysporum* is known to arise from the production of sidephores by the bacteria. Levrat et al. (1992) compared this pigment production by *P. putida* in King's B medium, with the production in King's medium diluted 10 times and amended with the filtrate of an axenic culture of *Acanthamoeba castellanii* (*see* Figure 17.2). The quantity of pigment (pyoverdin) produced was higher when the culture medium was amended with filtrate at 10% than at 1%. This stimulation also occurred in metabolic activities such as respiration and ammonium production. Amoebae produce stimulatory compounds responsible for the enhancement of bacterial activity.

Comminution and propagule dissemination

Animals move through the soil. Mesofauna and, in particular, macrofauna have a considerable effect by grinding plant residues and increasing the surface area of the detritus, by moving soil organic matter through mineral horizons, by burrowing in the soil and improving its structure, and by disseminating propagules.

Figure 17.2   Pyoverdin production calculated for 1.10⁹/ml of cells of *Pseudomonas putida* in King's B medium diluted 10 times with various amendments

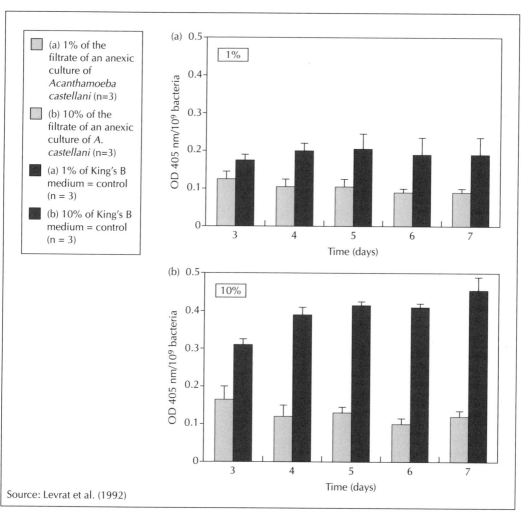

Source: Levrat et al. (1992)

From litter bag experiments, litter comminution by mesofauna was estimated to increase the decomposition rate by 4-69% (Seastedt, 1984). The ingestion by millipedes in a temperate oak forest was estimated by David (1987) to range between 9 and 16% of the annual litterfall. The mechanical fragmentation of the litter offers microorganisms new substrates and new surfaces to be invaded. Kheirallah (1990) measured the rate of fragmentation of sycamore leaves by the leaf litter-feeding millipede, *Julus scandinavius*. The increase in the fragmentation rate (consumed surface/produced surface) was 72×10^6. Particle size was found to be an important factor (Hanlon, 1981) in determining bacterial and fungal colonisation of a litter resource. Bacterial growth flourished on finely ground litter of a particle size similar to that found in isopod faeces (< 0.2 mm diameter), and fungal growth (*Coriolus versicolor*) was more prevalent on larger litter fragments.

Burrowing changes the spatial distribution of soil organic matter and improves soil structure by increasing pore volume, field water-holding capacity and stable aggregates. Dietz and Bottner (1979) studied the decomposition of a uniformly ^{14}C-labelled litter in the presence of the anecic earthworm, *Nicodrilus meridionalis*, and its incorporation into the profile. After 2 years, the labelled material was distributed in the 0-10 cm layer in casts and in superficial burrows. In the longer term, the labelled carbon may reach deeper layers (Stout, 1983).

Microorganisms decompose substrate and deplete the soil of essential nutrients locally, and then enter senescence. Through their ability to move, animals can cover greater distances than microorganisms and carry propagules either actively by feeding or passively on their bodies. For saprophytic fungi growing in a patchy environment, this dissemination allows them to reach more nutrient-rich sites and renew their metabolism. Rabatin and Stinner (1988), in an examination of the guts of 335 invertebrates, found that 28% contained VAM fungal spores and sporocarps. Higher densities and a greater diversity of VAM fungi were found in earthworms than in any other macroinvertebrate group examined. In sites with predominantly herbaceous vegetation, 97% of the earthworms samples contained VAM fungal spores. Reddel and Spain (1991) showed that the diversity of VAM spore types found in casts was similar to that in adjacent, uningested soil but the spores were more concentrated in the casts of *Pontoscolex corethrurus* and *Diplotrema heteropora*, which are able to initiate infection on *Sorghum bicolor*. According to these authors, the transport of VAM propagules by endogeic earthworms must be considered as a relatively short-distance phenomenon, perhaps occurring at a rate of a few meters per year.

Microbial dynamics in animal guts and faeces

In the soil, the animal guts and faeces constitute a particular environment for bacterial growth, which contributes to the patchy distribution of the microorganisms. The question is whether they host specific symbiotic communities which have specific functions. Stefaniak and Seniczak (1981) pointed to the microbiological origin of the digestive enzymes in moss mites, arguing that bacteria able to synthesise proteolytic and amynolytic enzymes were a regular component of the gut microflora of mites. Anderson and Bignell (1980) showed that there was extensive bacterial growth in the gut of *G. marginata* compared with food leaves. The stimuli provided within the gut may include the secretion or production of readily metabolisable materials (such as excretory products, enzymes and autolytic products), peristalsis buffering to a more favourable pH and redox potential. By comparison with bacterial counts on food litter, it was suggested that the increased bacterial populations resulted from the proliferation of the indigenous litter bacteria rather than a population of specific gut symbionts. However, in a study by Ineson and Anderson (1985) on *G. marginata* and *O. asellus*, in which the bacteria found in the gut were compared with those on the ingested litter, the data for *O. asellus* indicated that the gut favours those bacteria able to grow anaerobically and to utilise chitin. In the gut of *F. candida*, Borkott and Insam (1990) counted five times as many chitinolytic bacteria than that usually found in soil. Teuben and Verhoef (1992) suggested that *T. minor* may have a role in nitrification; the reduced NH_4^+ and increased NO_3^- after gut passage suggested that nitrifying gut bacteria may have been present and that collembolan faeces directly increased NO_3^- in the forest floor by a factor of 2.4.

This enhancement of microbial activity in the guts is not necessarily common to all animals and is probably related to their diet. In a study of the gut content of *Fridericia striata* (Enchytraeidae) using transmission electron microscopy, Toutain et al. (1982) observed that the bacterial and fungal microflora disappeared in the gut, probably digested by the animal. Are propagules part of their food? In their studies, Rouelle and Pussard (1985) demonstrated that active protozoa were an alimentary requirement of earthworms and were lysed in the gut. There was no evidence of microfloral activity in the gut but the passage of the soil through the gut led to an increase in soil respiration of about 90% over a period of 4-weeks (Scheu, 1987, 1992). On the other hand, Kristufek et al. (1992) identified a number of bacteria, actinomycetes and micromycetes in the gut of two earthworm species, *Lumbricus rubellus* and *Aporrectodea caliginosa*. The differences they observed between the changes in microbial densities in these two species may have been the result of the differences in chemical composition of the food materials. The consumed food material of *L. rubellus* was rich in readily available substrate and thus the digestive canal could be considered as a fermentor causing rapid bacterial growth; the soil organic matter which was ingested by *A. caliginosa*, however, contained more resistant humic substances, resulting in slower growth of gut microbial communities. Thus, the host animal, during its digestive processes, also utilises the consumed soil bacteria as a nutrient source. Barois (1992) showed the importance of water content and mucus production in the microbial activity in the guts of earthworms.

Lavelle and Martin (1992) and Lavelle et al. (1992) found that microbial biomass increased only in freshly deposited casts (by six times) and decreased in aging faeces by about 10% compared with the surrounding soil. These authors showed that the inorganic nitrogen concentration was five times higher in the fresh casts of *P. corethrurus* than in non-ingested soil. Nitrogen also accumulated in microbial biomass.

The greatest effect of the passage of plant residues through the gut is a shift in the balance of fungal and bacterial biomass in faecal material compared with uncomminuted litter. This effect is attributable to the sensitivity of the fungal thallus to disruption and a more favourable environment for bacteria. Thus, even in highly acid, fungi-dominated soils, most litter and soil organic matter passes through a phase of intense bacterial activity with generation times of only a few hours (Anderson, 1988).

SPATIAL DISTRIBUTION

Biological interactions in soil are confined to particular zones where the nutrient conditions are favourable. The rhizosphere and the litter are such zones of intensive biological activity. A conceptual model of root and bulk soil was proposed by Coleman et al. (1983) where different zones of microbial activity were distinguished along the root, from the root cap, allowing for root movement forward through the soil followed by the microorganisms. The carbon exudated from the cap stimulates the microbial biomass which, in turn, immobilises the inorganic nitrogen. Near the root-hair zone, carbon availability decreases, while predation by the microfauna enhances mineralisation and uptake by the plant. Nevertheless, there is no evidence that the microfauna is more localised in the root-hair zone. Direct observation has indicated that amoebae are most numerous near the root cap (Coûteaux et al., 1988).

The heterogeneity of nutrient distribution results in the patchiness of the activity. Cromack et al. (1988) showed that ectomycorrhizal fungal mats (*Hysterangium setchelli*) are sites of greater microbial biomass and increased concentrations of soil carbon and nitrogen. Soil fauna such as mites, collembola, nematodes and protozoa are more abundant in mat-colonised soil. With this patchiness of fungal mycelium, grazing has a positive effect on the fungal standing crop. Bengtsson and Rundgren (1983) showed that *Mortierella isabellina* periodically grazed by *Onychiurus armatus* responded with significantly higher respiration than when it was continuously grazed. In arid climates or in annually burnt savanna, plants are distributed in tussocks surrounded by bare soil. Therefore, the biological interactions are patchily distributed, as shown by testate amoebae communities in *Loudetia* and *Hyparrhenia* savanna (Coûteaux, 1976, 1978).

The vertical distribution of the biological interactions is linked to both the nutrient and water regime. Unlike microorganisms, which are located in a particular zone, animals are able to move through the profile in response to stress or favourable conditions. In semi-arid grassland, Leetham and Milchunas (1985) observed two peaks in microarthropods in the soil profile, one associated with the surface concentration of root biomass and the other with stability of soil water. In more mesic mixed-grass and tall grass habitats, the greater concentration of microarthropods is in surface soil. The response of animals to sudden changes in moisture by moving rapidly from one layer to another will change their impact on the microbial community. Hassal et al. (1986) observed a rapid movement by *Onychiurus subtenuis* from humus into wet litter after rainfall. Field studies showed that this species moved down into the deeper horizons of the litter profile during the dry summer months but that a significant proportion of the population returned to the surface layers within a few hours of the summer rain storms and remained to feed there until the litter dried out again. This migration is related to the presence in the litter layers of microorganisms which have a higher nutritional quality than those in the humus layer.

Seasonal climatic variations affect the microfloral composition and the duration of its activity. The response of the microflora is expressed by its growth rate, while animal response is expressed in changes in grazing pressure and migration. In seasonal grazing interactions, the reduction in microbial biomass caused by mesofaunal grazers is confined to situations where the environmental conditions cause strong feeding pressure and where the microflora is exposed to environmental stress. Elkins and Whitford (1982) found that the number of nematodes associated with litter appeared to be a function of season and not the degree of decomposition. However, a significant change in nematode type occurred which appeared to be a successional phenomenon and not a seasonal one, first with bacterivorous species and then with fungivorous species. Depending upon the type of climate, some periods can be more critical than others. In high latitudes and in areas where there is prolonged snow cover, the activity does not stop during the winter period. Aitchison (1983) identified some winter-active collembola feeding on members of the fungal family Dematiaceae, such as *Cladosporium*. In semi-arid tropical savanna, the monsoon period favours collembola and mites, and decomposition is activated.

THE ECOSYSTEM

The characteristics of the ecosystem determine the type of interactions which underlie the structure of the food web. Animal abundance and diversity is lower in cultivated systems than in natural ecosystems.

Forests, grasslands and arid zones have specific patterns of relationships. In temperate forests, there is some evidence of a link between humus type and animal activity. From the raw humus to the mull, animal activity becomes increasingly diversified. David et al. (1993) compared the soil macrofauna of different types of humus: mesotrophic mull, dystrophic mull, moder and dysmoder. High density and diversity of the saprophagous fauna was found in mesotrophic mull, which was characterised not only by Lumbricidae but also by many other epigeic taxa. In contrast, moder and dysmoder humus types hosted a lower number of taxa, with a predominance of elaterid larvae. In mor and moder types, the dominant animal effect is caused by grazing and the microbial community is dominated by fungi.

In mull humus, the earthworm effect is predominant and the mixture of soil organic matter and mineral soil promotes the formation of soil aggregates where bacterial communities are very active. Herlitzius (1983), in an experiment involving the incubation of litter bags of different meshes over 1 year in a spruce forest, a beech forest and an alluvial forest, obtained a weight loss of 40-44% in the fine-mesh litter bags (44 µm) in all sites, 60% in the alluvial forest, 50% in the beech forest and 44% in the spruce forest for the medium-mesh bags (1.1 mm), and 71% in alluvial and beech forests and 44% in the spruce forest for the coarse-mesh bags (1 cm). This indicates that animal impact is low in raw humus and becomes more important with better humus quality.

In a shortgrass steppe, Moore et al. (1988) estimated that predatory microarthropods obtained 51% of their energy from the bacterial energy pathway, 26% from the fungal pathway and 23% from the root pathway.

For agricultural systems, Hendrix et al. (1986) postulated two different food webs operating in various management regimes of row-crop agriculture. They suggested that conventional tillage systems are more bacteria-dominated, with the Enchytraeidae playing an important role, and that no-tillage systems are more fungi-dominated, with earthworms being predominant.

CONCLUSION

At a regional scale or along a climatic transect from north to south or from high to low altitudes, the trend of biological interactions is linked to a climatic gradient which affects either structural or functional organisation. If moisture is not limiting, the biodiversity and the nutrient turnover increase from cold to warm climates. In cool temperate forest organic layers the fungi represent about 80% of the total microbial biomass. The forest floor faunal community is impoverished, both in species diversity and in number, in comparison with more temperate forest ecosystems. Burrowing inverte-brates are absent and activity is concentrated in spring and eventually in autumn (Parkinson, 1988). In tropical ecosystems, the seasonal trend is less obvious. Earthworms are the most active in humid and subhumid tropics, while termites and ants are more active in semi-arid and arid regions (Lal, 1988). Globally, communities in cold countries have few species, with one or two predominating and with a fungi-dominated food web, while in warm and wet countries have communities of a large number of species and a bacteria-dominated food web.

In the context of the global warming, which predicts an increase in mean annual temperature from 3 to 6°C depending upon the latitude (Mitchell et al., 1990), it can be expected that the responses of organisms will be linked to their generation time and their ability to move from a place to another. At a microbial level, the active part of the populations is adapted to the local climate, but the dormant part has the potential to be activated by environmental changes. A rise of temperature will lead to the development of adapted microorganisms and a change in the microbial communities. For soil fauna, a rise in temperature will not result in an altered composition of the communities. Species from warmer areas are not likely to be able to move. An example of this is provided by earthworms. In northern countries there is low earthworm diversity. South of the limit of the last glaciation, earthworm diversity increases abruptly, showing their low potential to invade new sites (M. Bouché, pers. comm.). From the functional point of view, temperature increase will have two important effects. The first is that food availability will change. The second relates to the period of activity; in cold and wet climates, winter activity will be prolonged, while in warm and dry climates, summer activity will be reduced. These changes will probably alter the balance between species in microbial communities and offer new competitive situations.

References

Aitchison, C.W. 1983. Low temperature and preferred feeding by winter-active Collembola (Insecta: Apterigota). *Pedobiologia* 25: 27-36.

Anderson, J.M. 1988. Spatiotemporal effects of invertebrates on soil processes. *Biology and Fertility of Soils* 6: 216-27.

Anderson, J.M. and Bignell, D.E. 1980. Bacteria in the food, gut contents and faeces of the litter-feeding millipede *Glomeris marginata*. *Soil Biology and Biochemistry* 12: 251-54.

Barois, I. 1992. Mucus production and microbial activity in the gut of two species of *Amynthas* (Megascolecidae) from cold and warm tropical climates. *Soil Biology and Biochemistry* 24: 1507-10.

Bengtsson, G. and Rundgren, S. 1983. Respiration and growth of a fungus, *Mortierella isabellina*, in response to grazing by *Onychiurus armatus* (Collembola). *Soil Biology and Biochemistry* 15: 469-73.

Booth, R.G. and Anderson, J.M. 1979. The influence of fungal food quality on the growth and fecundity of *Folsomia candida* (Collembola: Isotomidae). *Oecologia* 38: 317-23.

Borkott, H. and Insam, H. 1990. Symbiosis with bacteria enhances the use of chitin by the springtail, *Folsomia candida* (Collembola). *Biology and Fertility of Soils* 9: 126-29.

Chakraborty, S., Theodorou, C. and Bowen, G.D. 1985. The reduction of root colonisation by mycorrhizal fungi by mycophagous amoebae. *Canadian J. Microbiology* 31: 295-97.

Clarholm, M. 1985. Interactions of bacteria, protozoa and plants leading to mineralisation of soil nitrogen. *Soil Biology and Biochemistry* 17: 181-87.

Coleman, D.C., Reid, C.P.P. and Cole, C.V. 1983. Biological strategies of nutrient cycling in soil systems. In MacFadyen, A. and Ford, E.D. (eds) *Advances in Ecological Research*. London, UK: Academic Press.

Coûteaux, M.M. 1976. Etude quantitative des Thécamoebiens d'une savane à *Hyparrhenia* à Lamto (Côte d'Ivoire). *Protistologica* 12: 563-70.

Coûteaux, M.M. 1978. Etude quantitative des Thécamoebiens édaphiques dans une savane à *Loudetia* à Lamto (Côte d'Ivoire). *Revue d'Ecologie et de Biologie du Sol* 15: 401-12.

Coûteaux, M.M. and Dévaux, J. 1983. Effet d'un enrichissement en champignons sur la dynamique d'un peuplement thécamoebien d'un humus. *Revue d'Ecologie et de Biologie du Sol* 20: 519-44.

Coûteaux, M.M., Faurie, G., Palka, L. and Steinberg, C. 1988. La relation prédateur-proie (protozoaires-bactéries) dans les sols: Rôle dans la régulation des populations et conséquences sur les cycles du carbone et de l'azote. *Revue d'Ecologie et de Biologie du Sol* 25: 1-31.

Coûteaux, M.M., Mousseau, M., Célérier, M.L. and Bottner, P. 1991a. Atmospheric CO_2 increase and litter quality: Decomposition of sweet chestnut leaf litter under different animal food web complexity. *Oikos* 61: 54-64.

Coûteaux, M.M., Bottner, P., Rouhier, H. and Billès, G. 1991b. Atmospheric CO_2 increase and plant material quality: Production, nitrogen allocation and litter decomposition of sweet chestnut. In Teller, A., Mathy, P. and Jeffers, J.N.R. (eds) *Responses of Forest Ecosystems to Environmental Changes*. London, UK: Elsevier.

Cromack, K. Jr, Fichter, B.L., Moldenne, A.M., Entry, J.A. and Ingham, E.R. 1988. Interactions between soil animals and ectomycorrhizal fungal mats. In Edwards, C.A., Stinner, B.R., Stinner, D. and Rabatin, S. (eds) *Biological Interactions in Soil*. Amsterdam, Netherlands: Elsevier.

Darbyshire, J.F., Elston, D.A., Sompson, A.E.F., Robertson, M.D. and Seaton, A. 1992. Motility of a common soil flagellate *Cercomonas* sp. in the presence of aqueous infusions of fungal spores. *Soil Biology and Biochemistry* 24: 827-31.

David, J.F. 1987. Consommation annuelle d'une litière de chêne par une population adulte du diplopode *Cylindroiulus nitidus*. *Pedobiologia* 30: 299-310.

David, J.F., Ponge, J.F. and Delecour, F. 1993. The saprophagous macrofauna of different types of humus in beech forests of Ardenne (Belgium). *Pedobiologia* 37: 49-56.

Dietz, S. and Bottner, P. 1979. Etude par autoradiographie de l'enfouissement d'une litière marquée au ^{14}C en milieu herbacé. In *Migrations Organo-Minérales dans les Sols Tempérés*. Colloques Internationaux du CNRS No. 303. Montpellier, France: CNRS.

Elkins, N.Z. and Whitford, W.G. 1982. The role of microarthropods and nematodes in decomposition in semi-arid ecosystems. *Oecologia* 55: 303-10.

Faber, J.H., Teuben, A., Berg, M.P. and Doelman, P. 1992. Microbial biomass and activity in pine litter in the presence of *Tomocerus minor* (Insecta, Collembola). *Biology and Fertility of Soils* 12: 233-40.

Fogel, R. 1988. Interactions among soil biota in coniferous ecosystems. In Edwards, C.A., Stinner, B.R., Stinner, D. and Rabatin, S. (eds) *Biological Interactions in Soil*. Amsterdam, Netherlands: Elsevier.

Hanlon, R.D.G. 1981. Some factors influencing microbial growth on soil animal faeces. I. Bacterial and fungal growth on particulate leaf litter. *Pedobiologia* 21: 257-63.

Hanlon, R.D.G. and Anderson, J.M. 1979. The effects of Collembola grazing on microbial activity in decomposing leaf litter. *Oecologia* 38: 93-100.

Hanlon, R.D.G. and Anderson, J.M. 1980. Influence of macroarthropod feeding activities on microflora in decomposing oak leaves. *Soil Biology and Biochemistry* 12: 255-61.

Hassal, M., Parkinson, D. and Visser, S. 1986. Effects of the collembolan *Onychiurus subtenuis* on decomposition of *Populus tremuloides* leaf litter. *Pedobiologia* 29: 219-25.

Hedlund, K., Boddy, L. and Preston, C.M. 1991. Mycelial responses of the soil fungus, *Mortierella isabellina*, to grazing by *Onychiurus armatus* (Collembola). *Soil Biology and Biochemistry* 23: 361-66.

Hendrix, P.F., Parmelee, R.W., Crossley, D.A., Coleman, D.C., Odum, E.P. and Groffman, P.M. 1986. Detritus food webs in conventional and no-tillage agroecosystems. *BioScience* 36: 374-80.

Henkinet, R., Coûteaux, M.M., Billès, G., Bottner, P. and Palka, L. 1990. Accélération du turnover du carbone et stimulation du priming effect par la prédation dans un humus forestier. *Soil Biology and Biochemistry* 22: 555-61.

Herlitzius, H. 1983. Biological decomposition efficiency in different woodland soils. *Oecologia* 5: 78-79.

Ineson, P. and Anderson, J.M. 1985. Aerobically isolated bacteria associated with the gut and faeces of the litter feeding macroarthropod *Oniscus asellus* and *Glomeris marginata*. *Soil Biology and Biochemistry* 17: 843-49.

Ingham, R.E., Trofymow, J.A., Ingham, E.R. and Coleman, D.C. 1985. Interactions of bacteria, fungi, and their nematode grazers: Effects on nutrient cycling and plant growth. *Ecological Monographs* 55: 119-40.

Jost, J., Drake, J.F. and Tsuchiya, HM, 1973. Interactions of *Tetrahymena pyriformis*, *Escherichia coli*, *Azotobacter vinelandii*, and glucose in a minimal medium. *J. Bacteriology* 113: 708-14.

Kheirallah, A.M. 1990. Fragmentation of leaf litter by a natural population of the millipede *Julus scandinavius* (Latzel 1884). *Biology and Fertility of Soils* 10: 202-06.

Kristufek, V., Ravasz, K. and Pizl, V. 1992. Changes in densities of bacteria and microfungi during gut transit in *Lumbricus rubellus* and *Aporrectodea caliginosa* (Oligochaeta: Lumbricidae). *Soil Biology and Biochemistry* 24: 1499-1500.

Lal, R. 1988. Effects of macrofauna on soil properties in tropical ecosystems. In Edwards, C.A., Stinner, B.R., Stinner, D. and Rabatin, S. (eds) *Biological Interactions in Soil*. Amsterdam, Netherlands: Elsevier.

Lavelle, P. and Martin, A. 1992. Small-scale and large-scale effects of endogeic earthworms on soil organic matter dynamics in soils of the humid tropics. *Soil Biology and Biochemistry* 24: 1491-98

Lavelle, P., Melendez, G., Pashanasi, B. and Schaefer, R. 1992. Nitrogen mineralisation and reorganisation in casts of the geophagous tropical earthworm *Pontoscolex corethrurus* (Glossoscolecidae). *Biology and Fertility of Soils* 14: 49-53.

Leetham, J.W. and Milchunas, D.G. 1985. The composition and distribution of soil microarthropods in the shortgrass steppe in relation to soil water, root biomass, and grazing by cattle. *Pedobiologia* 28: 311-25.

Leonard, M.A. 1984. Observations on the influence of culture conditions on the fungal feeding preference of *Folsomia candida* (Collembola: Isotomidae). *Pedobiologia* 26: 361-67.

Levrat, P., Pussard, M. and Alabouvette, C. 1992. Enhanced bacterial metabolism of a *Pseudomonas* strain in response to the addition of culture filtrate of a bacteriophagous amoeba. *European J. Protistology* 28: 79-84

Mitchell, J.F.B., Manabe, S., Meleshko, V. and Tokioka, T. 1990. Equilibrium climate change — and its implications for the future. In Houghton, J.T, Jenkins, G.J. and Ephraums, J.J. (eds) *Climate Change, The IPCC Scientific Assessment*. Cambridge, UK: Cambridge University Press.

Moore, J.C., Ingham, E.R. and Coleman, D.C. 1987. Inter- and intraspecific feeding selectivity of *Folsomia candida* (Collembola: Isotomidae) on fungi. *Biology and Fertility of Soils* 5: 6-12.

Moore, J.C., John, T.V. and Coleman, D.C. 1985. Ingestion of vesicular-arbuscular mycorrhizal hyphae and spores by soil microarthropods. *Ecology* 66: 1979-81.

Moore, J.C., Walter, D.E. and Hunt, H.W. 1988. Arthropod regulation of micro- and mesobiota in below-ground detrital food webs. *Annual Review of Entomology* 33: 419-39.

Newell, K. 1984. Interaction between two decomposer basidiomycetes and collembola under *Stika* spruce: Grazing and its potential effects on fungal distribution and litter decomposition. *Soil Biology and Biochemistry* 16: 235-39.

Palka, L. 1988. Rôle des protozaires bactériophages du sol dans la minéralisation de l'azote en conditions gnotobiotiques. Thèse de Doctorat, Université Blaise Pascal — Clermont II, France.

Parkinson, D. 1988. Linkages between resource availability, microorganisms and soil invertebrates. In Edwards, C.A., Stinner, B.R., Stinner, D. and Rabatin, S. (eds) *Biological Interactions in Soil.* Amsterdam, Netherlands: Elsevier.

Parkinson, D., Visser, S. and Whittaker, J.B. 1979. Effect of collembolan grazing on fungal colonisation of leaf litter. *Soil Biology and Biochemistry* 11: 529-37.

Rabatin, S.C. and Stinner, B.R. 1988. Indirect effects of interactions between VAM fungi and soil-inhabiting invertebrates on plant processes. In Edwards, C.A., Stinner, B.R., Stinner, D. and Rabatin, S. (eds) *Biological Interactions in Soil.* Amsterdam, Netherlands: Elsevier.

Reddel, P. and Spain, A.V. 1991. Earthworms as vectors of viable propagules of mycorrhizal fungi. *Soil Biology and Biochemistry* 23: 767-74.

Rouelle, J. and Pussard, M. 1985. Microflore, protozoaires et lombriciens (*Oligochètes terrestres*). Relations trophiques et stimulations réciproques. *J. Protozoology* 31: 76A.

Scheimer, F. 1983. Comparative aspects of food dependence and energetics of free living nematodes. *Oikos* 41: 32-42.

Scheu, S. 1987. Microbial activity and nutrient dynamics in earthworm casts (Lumbricidae). *Biology and Fertility of Soils* 5: 230-34.

Scheu, S. 1992. Automated measurement of the respiratory response of soil microcompartments: Active microbial biomass in earthworm faeces. *Soil Biology and Biochemistry* 24: 1113-18.

Scholle, G., Wolters, V. and Joergensen, R.G. 1992. Effects of mesofauna exclusion on the microbial biomass in two moder profiles. *Biology and Fertility of Soils* 12: 253-60.

Seastedt, T.R. 1984. The role of microarthropods in decomposition and mineralisation processes. *Annual Review of Entomology* 29: 25-46.

Stefaniak, O. and Seniczak, S. 1981. The effect of fungal diet on the development of *Oppia nitens* (Acari, Oribatei) and on the microflora of its alimentary tract. *Pedobiologia* 21: 202-10.

Steinberg, C. Faurie G., Zegerman M. and Pavé A. 1987. Regulation par les protozoaires d'une population bactérienne introduite dans le sol; modélisation mathématique de la relation prédateur-proie. *Revue d'Ecologie et de Biologie du Sol* 24: 49-62.

Stout, J.D. 1983. Organic matter turnover by earthworms. In Satchell, J.E. (ed) *Earthworm Ecology.* London, UK: Chapman and Hall.

Teuben, A. and Roelofsma, T.A.P.J. 1990. Dynamic interactions between functional groups of soil arthropods and microorganisms during decomposition of coniferous litter in microcosm experiments. *Biology and Fertility of Soils* 9: 145-51.

Teuben, A. and Verhoef, H.A.1992. Direct contribution by soil arthropods to nutrient availability through body and faecal nutrient content. *Biology and Fertility of Soils* 14: 71-75.

Toutain, F. and Villemin, G., Albrecht, A. and Reisinger, O. 1982. Etude ultrastructurale des processus de biodégradation. II. Modèle enchytraeides-litière de feuills. *Pedobiologia* 23: 145-56.

Trofymow, J.A. and Coleman, D.C. 1982. The role of bacterivorous and fungivorous nematodes in cellulose and chitin decomposition in the context of a root/rhizosphere/soil conceptual model. In Frackman, D.W. (ed) *Nematodes in Soil Ecosystems.* Austin, Texas, USA: University of Texas Press.

Volterra, V. 1927. Variations and fluctuations of populations size in coexisting animal species. In Oliviera Pinto, F. and Conolly, B.W. (eds). *Applicable Mathematics of Non Physical Phenomena* (1982 edn). New York, USA: Ellis Horwood .

Warnock, A.J., Fitter, A.H. and Usher, M.B. 1982. The influence of the springtail *Folsomia candida* on the mycorrhizal association of leek *Allium porum* and the vesicular-arbuscular mycorrhizal endophyte *Glomus fasciculatus. New Phytologist* 90: 285-92.

Beyond the Biomass
Edited by K. Ritz, J. Dighton and K.E. Giller
© 1994 British Society of Soil Science (BSSS)
A Wiley-Sayce Publication

CHAPTER 18

Priming effects of macroorganisms on microflora: A key process of soil function?

P. Lavelle and C. Gilot

Bingeman et al. (1953) described the priming effect as the stimulation of soil organic matter decomposition by the addition of fresh organic material. Jenkinson (1966) described it in more general terms as any positive or negative change in the decomposition rate of soil organic matter caused by the addition of fresh organic matter. He illustrated this using a simple experiment in which ryegrass foliage uniformly labelled with ^{14}C was mixed with soil, and carbon dioxide evolution was monitored during an incubation period of 78 days. The production of unlabelled carbon dioxide resulting from soil organic carbon mineralisation was greater in the mixture than in the control soil. The difference was the result of a positive priming action.

Most examples of priming effects described in the literature are positive (reviewed by Jenkinson et al., 1985). They indicate that the addition of fresh organic matter as green manure to soils or of mineral nitrogen as fertiliser often stimulates mineralisation of soil organic matter. However, some authors have observed negative effects. For example, Bingeman et al. (1953) reported such effects in the first few days following the addition of glucose to an organic soil. The observed priming effects may be apparent or real. Apart from experimental factors (such as the exchange of labelled for unlabelled carbon in calcareous soils, or errors resulting from heterogeneous labelling of introduced material), some observed priming effects may be attributable to changes in pH or oxygen supply following the introduction of organic material (Parr and Reuszer, 1959; Barrow, 1960). In experiments where labelled material is added, apparent priming effects may also result from the release of unlabelled material turned over in the microbial biomass (Dalenberg and Jager, 1981).

Real priming effects may stem from three factors:

- the germination of spores, which increases overall microbial activity

- interactions between compounds derived from the added and 'native' organic matter, which render the latter more labile (Mandl and Neuberg 1956)

- an increase in the concentration of extra-cellular enzymes produced by microorganisms, resulting in accelerated decomposition of soil organic matter

In this chapter, the focus is on the third factor, triggered in the rhizosphere and drilosphere (the part of the soil and microflora affected by earthworm activities) by the addition to the soil of specific organic substrates produced by roots or earthworms. A significant proportion of carbon assimilated by plants is translocated directly to the rhizosphere soil as 'rhizodeposition', a mixture of water-soluble exudates and secretions, mucilage and sloughed cells from the root epidermis and cortex (Rovira et al., 1979; Hale et al., 1981). They represent 7-20% of the carbon fixed by photosynthesis, depending upon the plant and soil conditions (Lespinat and Berlier, 1975; Martin, 1977; Haller and Stolp, 1985; Milchunas et al., 1985; Heulin et al., 1987; Trofymow et al., 1987).

Earthworms also produce large amounts of mucus in the anterior gut and add it to the ingested soil. Concentrations vary from 5-7% of the dry weight of the ingested soil in native species of African savannas (Martin et al., 1987) to 15-18% in species with a wide pan-tropical distribution (Barois, 1992) and up to 42% in the Lumbricidae in northern Spain (Trigo et al., 1993). *In vitro* incubations of root mucilage and earthworm intestinal mucus have been conducted to compare the kinetics of microbial responses to the addition of these substrates and assess the priming effects on soil organic matter (Gilot, 1990; Mary et al., 1992). The significance of these mechanisms in soil function is discussed here.

EFFECT OF ROOT MUCILAGE IN THE RHIZOSPHERE

The stimulation of microbial activities by rhizodeposition has been fairly well documented (for example, Samtsevich 1971; Stanghellini and Hancock, 1971; Short and Lacy, 1974; Trofymow and Coleman, 1982; Clarholm, 1985; Guckert, 1985; Billes et al., 1990). Mucilage and root litter are the main sources of carbon in the rhizosphere. In an attempt to assess their mineralisation patterns in the soil, the decomposition of mucilage and fresh maize roots was observed over a 50-day period of laboratory incubation and compared with a control supplemented with equivalent amounts of glucose (Mary et al., 1992) These substrates had been mixed with a sterile sand which had been inoculated with a soil extract to provide microbial populations. The kinetics of mineralisation of the mucilage were clearly different from those of the other two substrates. The decomposition of the maize roots and glucose was rapid during the first 7 days and then proceeded more slowly. The decomposition of the mucilage was slow during the first few days and then accelerated sharply. By the end of the experiment, 81% of the mucilage and 70% of the roots had been mineralised. In a parallel experiment over the same period, 89% of the glucose had been mineralised (*see* Figure 18.1).

An incubation experiment was conducted over 185 days in the presence of soil (orthic Luvisol with 17% clay) (Mary et al., 1991). The organic substrates were the same as those used in the experiment outlined above but their ^{13}C composition differed from the soil carbon; it was possible to discern which parts of the total CO_2 evolved came from the decomposing substrate and from the soil organic matter. All three substrates differed slightly in decomposition patterns. The apparent mineralisation was larger with the addition of glucose than for the other two substrates; after 180 days, the total CO_2-C evolved was equivalent to 67%, 74% and 92% of the carbon introduced with the substrate for roots, mucilage and glucose, respectively. In the first few days of the experiment, the priming effects on soil organic matter were similar, representing 10-15% of the overall mineralisation. After 40% of the substrates had decomposed, significant differences appeared and, after 185 days, the priming effect represented 14%, 19% and 31% of the carbon incorporated as roots, mucilage and glucose, respectively.

In the experiment reported by Mary et al. (1991), no priming effect on nitrogen was observed. Nonetheless, the maximum amount of nitrogen immobilised during the decomposition of the substrates was higher for mucilage (88% of added carbon) than for roots (66%) and glucose (61%). The incubated soil was fumigated to kill the microbial biomass and then further incubated to quantify

Figure 18.1   Kinetics of carbon mineralisation of three substrates added to sand inoculated with a soil extract

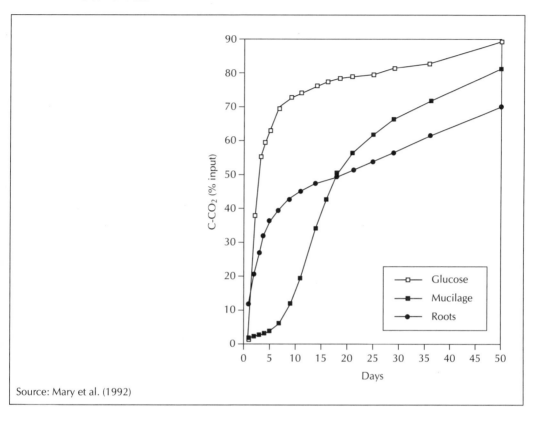

Source: Mary et al. (1992)

the amounts of nitrogen and carbon present as microbial biomass. The N/C ratio of the mineral nitrogen and carbon produced was higher in the mucilage (0.28) than in the roots (0.15) and glucose (0.13) treatments. This suggests that mucilage stimulates bacteria rather than fungi, as bacteria have a much higher relative concentration of nitrogen in their biomass than fungi.

EFFECT OF EARTHWORM INTESTINAL MUCUS IN THE DRILOSPHERE

Some authors have suggested a mutualist relationship between tropical endogeic earthworms and soil microflora in the exploitation of soil organic matter (Lavelle et al., 1980; Barois and Lavelle, 1986; Trigo et al., 1993). The conditions in the anterior part of the earthworm gut suit the activities of soil-free microorganisms: high water content (100-150% of the dry weight of soil), neutral pH and, above all, high concentrations of readily assimilable organic matter as intestinal mucus (5-18% of the dry weight of soil, depending upon the species). This mucus is a mixture of low-molecular-weight (about 200 Da) amino acids and sugars and a glycoprotein of 40 000-60 000 Da (Martin et al., 1987).

Short-term incubations of intestinal mucus were carried out to investigate the response of soil microflora to the addition of these substrates at the same concentration as that observed in the gut (7% for *Millsonia anomala*) (Gilot, 1990) (*see* Figure 18.2 *overleaf*). The intestinal mucus was extracted with water from the anterior gut of *M. anomala* and freeze-dried. It was then incubated for 3 hours in the chambers of a microrespirometer designed by Verdier (1983). Microbial activity increased sharply, reaching a maximum after 45 minutes and then falling to values similar to those observed for the control

Figure 18.2   Changes over time of oxygen absorption in an African Alfisol supplemented with (a) 7% glucose and (b) 7% intestinal mucus of *Millsonia anomala*

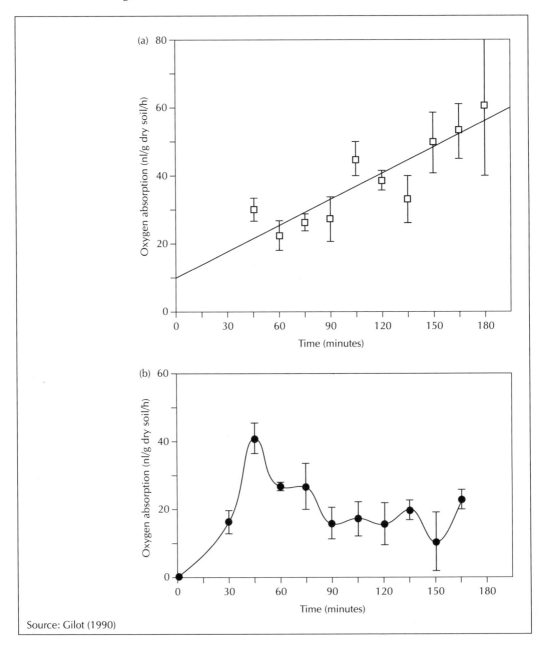

Source: Gilot (1990)

after a further 60 minutes. In the soil supplemented with 7% glucose, a different pattern was observed: respiratory activity increased over the whole period of the experiment (120 minutes).

These results indicate that, in the middle part of the earthworm gut, microorganisms which initially increase their metabolic activity sharply by feeding on the mucus are able to digest soil organic matter at a far higher rate than in bulk soil. Microbial activity at 28°C was 6-10 times higher than in the control (a dry soil sieved at 2 mm, moistened to field capacity) and probably up to 30 times higher than in undisturbed field conditions. The product of the external digestion of microorganisms would be partly reabsorbed with water in the posterior gut to feed the worm. Thus, some 3-19% of the soil organic matter could be assimilated during transit through the gut, which may last from 30 minutes to 2-4 hours (Lavelle, 1978; Barois, 1987; Martin, 1989). Interestingly, when temperature is reduced to 15°C, the increase of microbial respiratory activity in the gut is limited to twice the rate in the control soil.

Zhang et al. (1992) showed that cellulase and mannanase found in the gut content of the earthworm *Pontoscolex corethrurus* were produced by the ingested microflora since they were not found in isolated gut tissue cultures. It is possible, therefore, that in terms of the digestion system there is a mutualist relationship between endogeic earthworms and the ingested soil microflora. It is likely that the efficiency of this digestion system is highly dependent upon temperature. With lower temperatures, the reduction in efficiency would be compensated for by the ingestion of a higher-quality feeding resource and the production of higher amounts of intestinal mucus in the anterior part of the gut to accelerate the activation of the ingested dormant microflora. This may explain why earthworms tend to feed more on litter (easier to digest than soil organic matter) when the average temperature decreases (Lavelle, 1983). On the other hand, Trigo et al. (1993) have found concentrations of intestinal mucus of up to 42% in the anterior gut of temperate earthworm species from northern Spain, much higher than the observed values of 5-18% in tropical species.

Mucus in the drilosphere has a similar effect to that of root exudates in the rhizosphere, or any easily assimilable substrate introduced in the soil.

DISCUSSION

Roots and earthworms produce high amounts of readily assimilable organic substrates in the rhizosphere and drilosphere, respectively. Rhizodeposition may represent up to 10-20% of the total carbon fixed by photosynthesis and earthworms add 5-42% of the dry weight of the ingested soil as intestinal mucus in the anterior part of their gut. In the middle and posterior gut, this mucus is mostly absent. Part of it is metabolised by the ingested soil microflora (Martin et al., 1987); another part is probably reabsorbed by the gut wall and then recycled or simply recirculated in the anterior gut.

Like any readily assimilable resource, root mucilage and earthworm mucus increase microbial activities. Interestingly, microbial response to the addition of these substrates differs from that which occurs after the addition of glucose. The observed priming effect is lower for mucilage than for glucose but the kinetics of microbial response differ. Mary et al. (1992) observed that, after the addition of mucilage, there was a time lag of about 5 days before mineralisation rates started to increase, reaching values which were lower than those obtained with glucose and dead roots. The response of microflora to the addition of earthworm mucus also differed from that obtained with glucose. During the first 45 minutes after the addition of earthworm mucus, respiratory activity increased rapidly, reaching a maximum value of more than 50% higher than that obtained with glucose; there followed a rapid decrease and, after 2 hours, the respiratory activity was the same as that of the control.

There is some evidence that priming effects result from the addition of rhizodeposition to the soil. Sallih et al. (1987) observed the priming effects on nitrogen in a 700-day pot experiment. In pots supporting the continuous activity of live roots, there was a 25% increase of nitrogen mobilised as mineral nitrogen and root and microbial-biomass nitrogen, compared with a control without plants. A large microbial biomass was maintained by the living plant, clear evidence that soil organic matter mineralisation and the subsequent release of mineral nitrogen was enhanced by the priming effect of root exudates. In contrast, Mary et al. (1991) did not observe a priming effect on nitrogen in laboratory incubations and concluded that the priming effect observed on carbon in the same incubations might only have been an apparent effect as opposed to a real priming effect. In this case, the release of unlabelled microbial carbon by microflora would have resulted from the turnover of biomass and the subsequent replacement of unlabelled material by the labelled carbon introduced in the experiment.

These results indicate that macroorganisms such as roots and earthworms can trigger priming effects on soil microflora in the rhizosphere and drilosphere through the production of assimilable substrates. They also suggest that these substrates have specific effects on microflora. Observations of thin sections in the rhizosphere under the electron microscope have shown that only 30% of the microflora is actually activated by the input of exudates (Foster, 1986). Our results show that the specific composition of these substrates may induce specific temporal patterns of the microbial response. In the drilosphere, the response to the addition of mucus was much faster than with glucose and this pattern may be considered as an adaptation to the specific conditions of earthworm digestion; given that the gut transit in the earthworms used for the experiment generally lasted less than a couple of hours (Lavelle, 1978), a fast activation of microflora is critical for rapidly reaching the level of activity beyond which formerly dormant microorganisms recover their ability to digest soil organic matter. In the rhizosphere, processes leading to the release of assimilable nutrients from soil organic matter are probably much slower. The movements of roots growing into the soil are much slower than soil passing through an earthworm gut. As a result, the soil reached by the growing root tip will be influenced by the production of exudates for a longer period than soil introduced in the highly active microsite represented by the gut of an earthworm. Thus, it may be a selective advantage to produce compounds that will have a prolonged effect on microbial activity and will selectively activate bacteria rather than fungi, as is the case for the root mucilage considered in the experiment of Mary et al. (1992).

We propose calling these substrates 'ecological mediators'. They have a specific role in ecosystem function through activating microflora at specific scales of time and space compatible with the needs of roots or earthworms to reingest the assimilable compounds released by the activated microflora.

These observations lead us to suggest three new research avenues that should be explored:

• Investigate the mechanisms which determine apparent or real priming effects. The metabolic activities of microorganisms involved in this process should be investigated and special attention paid to specific microbial processes such as turnover of microbial biomass and the release of enzymes in the surrounding soil.

• Test the specificity of ecological mediators. This involves studying the assimilable compounds produced by roots, earthworms and other soil invertebrates in terms of their chemical composition and the qualitative and quantitative responses of microorganisms (that is, determine which microflora are activated and the time pattern involved); the implications of different time patterns should be considered. The concepts of synchronisation and synlocalisation of nutrient release for plant uptake should be used in this approach (Swift, 1986; van Noordwijk and de Willigen, 1986).

• Test the functional significance of these processes in terms of strategies for exploiting nutrient resources and the consequences for diversity (for example, Lavelle, 1986). Roots, earthworms and other soil macroorganisms have mutualist associations with microorganisms which allow a better use of soil organic resources by favouring organic matter decomposition and mineralisation and the uptake of the assimilates thus released by both components of the association. At the higher temperatures of tropical soils, the range of organic resources used is likely to be enlarged as a result of a much faster and stronger response by microorganisms to the addition of exudates or mucus. In evolutionary time, such differences may have promoted increased diversity as a result of the broader and more diverse base of the organic resources made available.

Acknowledgements

The authors are grateful to Bruno Mary for useful comments on the interpretation of his results.

References

Barois, I. 1987. *Interactions entre les Vers de Terre (Oligochaeta) Tropicaux Géophages et la Microflore pour l'Exploitation de la Matière Organique du Sol.* Paris, France: Laboratoire de Zoologie de l'ENS.

Barois, I. 1992. Mucus production and microbial activity in the gut of two species of *Amynthas* (Megascolecidae) from cold and warm tropical climates. *Soil Biology and Biochemistry* 24: 1507-10.

Barois, I. and Lavelle, P. 1986. Changes in respiration rate and some physicochemical properties of a tropical soil during transit through *Pontoscolex corethrurus* (Glossoscoleciae, Oligochaeta). *Soil Biology and Biochemistry* 18 (5): 539-41.

Barrow, N.J. 1960. Stimulated decomposition of soil organic matter during the decomposition of added organic materials. *Australian J. Agricultural Research* 11: 331-38.

Billes, G., Bottner, P. and Texier, M. 1990. *Régulation du Cycle de l'Azote au Niveau de la Rhizosphère de Plantes Non Fixatrices d'Azote. Cas du Blé.* Paris, France: Ministère de l'Environnement DRAEI.

Bingeman, C.W., Varner J.E. and Martin W.P. 1953. The effect of the addition of organic materials on the decomposition of an organic soil. *Soil Science Society of America Proc.* 17: 34-38.

Bottner, P., Sallih, Z. and Billes, G. 1988. Root activity and carbon metabolism in soils. *Biology and Fertility of Soils* 7: 71-78.

Clarholm, M. 1985. Interactions of bacteria, protozoa and plant leading to mineralisation of soil nitrogen. *Soil Biology and Biochemistry* 17: 181-87.

Dalenberg, J.W. and Jager, G. 1981. Priming effect of small glucose additions to ^{14}C-labelled soil. *Soil Biology and Biochemistry* 13: 219-23.

Foster, R.C. 1986. The ultrastructure of the rhizoplane and rhizosphere. *Annual Review of Phytopathology* 24: 211-34.

Gilot, C. 1990. *Bilan de Carbone du Ver de Terre Geophage Tropical* Millsonia anomala *(Megascolecidae).* Mémoire de DEA 'Production végétale et écologie générale'. Paris-Grignon, France: Institut National Agronomique.

Guckert, A. 1985. Root exudation and microbial activity at the soil-root interface. Paper presented at IPI-ISS Workshop on Potassium, Nanjing, China, September 1985.

Hale, M.G., Moore, L.D.N. and Griffin, G.J. 1981. Root exudates and exudation. In Dommergues, Y.R. and Krupa, S.V. (eds) *Interactions between Non-Pathogenic Soil Microorganisms and Plant.* Amsterdam, Netherlands: Elsevier.

Haller, T. and Stolp, H. 1985. Quantitative estimation of root exudation of maize plants. *Plant and Soil* 86: 207-16.

Heulin, T., Guckert, A. and Balandreau, J. 1987. Stimulation of root exudation of rice seedlings by *Azospirillum* strains: Carbon budget under gnotobiotic conditions. *Biology and Fertility of Soils* 4: 9-14.

Jenkinson, D.S. 1966. The priming action. *J. Applied Radiation Isotopes (Supplement)* 199-208.

Jenkinson, D.S., Fox, R.H. and Rayner, J.H. 1985. Interactions between fertilisers, nitrogen and soil nitrogen: The so-called 'priming-effect'. *J. Soil Science* 36: 425-44.

Lavelle, P. 1983. The structure of earthworm communities. In Satchell, J.E. (ed) *Earthworm Ecology: From Darwin to Vermiculture*. London, UK: Chapman and Hall.

Lavelle, P. 1978. Les vers de terre de la savane de Lamto (Côte d'Ivoire): Peuplements, populations et fonctions dans l'écosystème. Thèse d'Etat, Paris VI, France.

Lavelle, P. 1986. Associations mutualistes avec la microflore du sol et richesse spécifique sous les tropiques: L'hypothèse du premier maillon. *Compte-Rendu de l'Académie des Sciences de Paris* 302: 11-14.

Lavelle, P., Sow, B. and Schaefer, R. 1980. The geophagous earthworm community in the Lamto savanna (Ivory Coast): Niche partitioning and utilisation of soil nutritives resources. In Dindal, D. (ed) *Soil Biology as Related to Land Use Practices*. Washington DC, USA: EPA.

Lespinat, P.A. and Berlier, Y. 1975. Les facteurs externes agissant sur l'excrétion racinaire. *Société Botanique de France. Colloque Rhizosphère* 21-30.

Mandl, I. and Neuberg, C. 1956. Solubilisation, migration and utilisation of insoluble matter in nature. *Advances in Enzymology* 17: 253-59.

Martin, A. 1989. Effets des vers de terre tropicaux géophages sur la dynamique de la matière organique du sol dans les savanes tropicales humides. Thèse de l'Université, Paris XI, France.

Martin, A., Cortez, J., Barois, I. and Lavelle, P. 1987. Les mucus intestinaux de ver de terre, moteur de leurs interactions avec la microflore. *Revue d'Ecologie et Biologie du Sol* 24: 549-58.

Martin, J.A. 1977. Factors influencing the loss of organic carbon from wheat roots. *Soil Biology and Biochemistry* 9: 1-7.

Mary, B., Fresnau, C., Morel, J.L., Mariotti, A. and Guckert, A. 1991. *Decomposition du Mucilage Racinaire de Maïs et Effet sur le Cycle Interne de l'Azote dans le Sol*. Paris, France: Ministère de l'Environnement.

Mary, B., Mariotti, A. and Morel, J.L. 1992. Use of ^{13}C variation at natural abundance for studying the biodegradation of root mucilage, roots and glucose in soil. *Soil Biology and Biochemistry* 24: 1065-72.

Milchunas, D.G., Lauenroth, W.K., Singh, J.S., Cole, C.V. and Hunt, H.W. 1985. Root turnover and production by ^{14}C dilution: Implications of carbon partitioning in plants. *Plant and Soil* 88: 353-65.

Parr, J.F. and Reuszer, H.W. 1959. Organic matter decomposition as influenced by oxygen level and method of application to soil. *Soil Science of America Proc.* 23: 214-22.

Sallih, Z., Bottner, P., Billès, G. and Soto, P. 1987. Interactions racines-microorganismes: Carbone et azote de la biomasse microbien ne développée en présence de racines. *Revue d'Ecologie et Biologie du Sol* 24: 459-71.

Samtsevich, S.A. 1971. Root excretions of plants. An important source of humus formation in the soil. *Humus et Planta* (Prague) 147-54.

Short, G.E. and Lacy, M.L. 1974. Germination of *Fusarium solani* f.sp. *pisi* chlamydospores in the spermosphere of pea. *Phytopathology* 64: 558-62.

Stanghellini, M.E. and Hancock, J.G. 1971. Radial extent of the bean spermosphere and its relation to the behavior of *Pythium ultimum*. *Phytopathology* 61: 165-68.

Swift, M.J. (ed) 1986. *Tropical Soil Biology and Fertility: Interregional Research Planning Workshop*. Special Issue 13, *Biology International*. Paris, France: IUBS.

Trigo, D., Martin, A. and Lavelle, P. 1993. A mutualist system of digestion in temperate earthworms, *Allolobophora molleri* and *Octolasium lacteum*. *Acta Zoologica Fennica* (in press).

Trofymow, J.A. and Coleman, D.C. 1982. The role of bacterivorous and fungivorous nematodes in cellulose and chitin decomposition. In Freckman, D.W. (ed) *Nematodes in Soil Ecosystems*. Austin, Texas, USA: University of Texas Press.

Trofymow, J.A., Coleman, D.C. and Cambardella, C. 1987. Rates of rhizodeposition and ammonium depletion in the rhizospere of anexic oat roots. *Plant and Soil* 97: 333-44.

van Noordwijk, M. and de Willigen, P. 1986. Quantitative root ecology as element of soil fertility theory. *Netherlands J. Agricultural Science* 34: 273-81.

Verdier, B. 1983. Le respiromètre à pression et volume variables. Une technique simple et sensible pour l'étude écophysiologique des animaux du sol. In Lebrun, A.H., de Medts, P., Grégoire-Wibo, C. and Wauthy, G. (eds) *New Trends in Soil Biology*. Louvain-la-Neuve, Belgium: Dieu-Brichart.

Zhang, B.G., Rouland, C., Lattaud, C. and Lavelle, P. 1992. Origin and activity of enzymes found in the gut content of the tropical earthworm *Pontoscolex corethrurus* Müller. *European J. Soil Biology* 29: 7-11.

Beyond the Biomass
Edited by K. Ritz, J. Dighton and K.E. Giller
© 1994 British Society of Soil Science (BSSS)
A Wiley-Sayce Publication

CHAPTER 19

Effects of mesofaunal exclusion on microbial biomass and enzymatic activities in field mesocosms

E. KANDELER, B. WINTER, C. KAMPICHLER and A. BRUCKNER

Soil animals contribute little to soil respiration in comparison to microorganisms (Huhta and Koskenniemi, 1975). However, microcosm studies show that fauna can influence microbial respiration and various soil processes. Soil mesofauna has been considered to enhance decomposition and nutrient cycling rates either indirectly, by affecting the structure and activity of microbial communities, or directly by comminuting litter and excreting nutrients into the soil solution (reviewed by Seastedt, 1984; Ingham et al., 1985; Coleman, 1986; Verhoef and Brussaard, 1990; Lussenhop, 1992). The quantification of grazing effects on microbial metabolism has produced different results, depending upon the interaction studied (Verhoef and Brussaard, 1990). Microbial grazers may feed selectively on senescent colonies and induce a net release of nutrients which stimulates the growth of micro–organisms (Hanlon and Anderson, 1979). In addition, the faunal-microbial interactions depend upon substrate quality, grazing pressure and the temperature and moisture conditions of the soil (Hanlon, 1981; Ineson et al., 1982; Teuben and Roelofsma, 1990).

The evidence exists, therefore, of positive and negative feedback between mesofauna and microorganisms according to culture conditions. Only a few studies of these interactions have been performed in structurally complex systems (Faber and Verhoef, 1991; Leonard and Anderson, 1991; Teuben, 1991), although soil structure is recognised as a key factor in the control of biological interactions (van Veen and van Elsas, 1986).

The use of mesocosms is a promising approach for examining mesofaunal-microbial interactions in structurally complex systems under field conditions (Kampichler et al., 1993). According to Odum (1984), mesocosms are bounded and partially enclosed outdoor experimental set-ups which may bridge the gap between laboratory and ecosystem studies. We used this approach to address two questions regarding the litter layer of a spruce forest. Can soil mesofauna alter the microbial biomass pool under field conditions? Does mesofauna influence the activity of various soil enzymes involved in carbon and nitrogen cycling?

MATERIALS AND METHODS

The study was conducted in a 40-year-old spruce forest 1.5 km south-east of Raumberg in Styria, Austria (47°29'N, 14°7'E, National Grid Reference BMN 5702-0860-4b, 750 m a.s.l., mean annual temperature 6.8°C, mean annual precipitation 1013 mm). The forest is composed exclusively of *Picea abies* Karst. There is no undergrowth and the trees are all about the same age. The soil profile is a Dystic Cambisol (FAO classification) over quarternary sediments with a raw humus litter (Eisenhut, pers. comm.). The chemical properties of the L/F and H horizons are given in Table 19.1.

Table 19.1   Chemical properties of a spruce forest, with the results expressed as the range of five replicated samples

Soil horizon	C_{org} (%)	N_t (%)	pH (0.10 M KCl)	Ca	Mg	K
				(m equiv/100 g soil dry matter)		
L/F	42.2-48.9	1.45-1.79	3.0-3.4	9.9-16.7	4.5-5.8	1.0-1.3
H	27.5-45.9	1.16-1.82	2.7-3.1	2.7-5.1	1.8-6.3	0.4-0.7

For the preparation of the mesocosms, two chromium steel frames which fitted tightly together were used. The inner frame (25 x 25 x 22 cm) was used to cut out the monoliths by driving it into the soil in a confined area of the investigation site. The bottom of each monolith was cut with a metallic plate, and roots were cut with scissors. The outer frame was used to dig randomly distributed cavities in the investigation site, into which the monoliths were lowered after manipulation. A detailed description of the method and technical equipment is given by Bruckner et al. (1994).

Thirty monoliths were defaunated by deep-freezing with solid CO_2. The monoliths were cooled down to -15°C. A previous test had indicated that the whole soil fauna is killed off at this temperature (probably with the exception of protozoa and nematodes). This defaunation technique was preferred to other methods because of its minimal side-effects on soil properties (Huhta et al., 1989). Depending upon the treatments, monoliths were wrapped in nets of monolen polyester. Each net enclosed the entire monolith and prevented the growth of plant roots in the block. The treatments established were:

Treatment 1   Ten blocks were wrapped in fine nets (35 µm mesh) to prevent re-immigration of mesofauna and macrofauna

Treatment 2   Ten blocks were wrapped in coarse nets (1 mm mesh) to allow immigration of mesofauna only

Treatment 3   Ten blocks remained without nets to allow immigration by both mesofauna and macrofauna

Treatment 4   Ten areas of the same size (25 x 25 cm) were chosen as control plots and were left intact; these plots included soil microorganisms, entire soil fauna and living roots

This experimental design should allow for the distinction of the effects of mesofaunal exclusion from the effects of macrofaunal exclusion and manipulation (changes in physical and chemical properties by freezing; cutting of living roots) (*see* Figure 19.1).

Figure 19.1   Experimental design and observed treatments effects

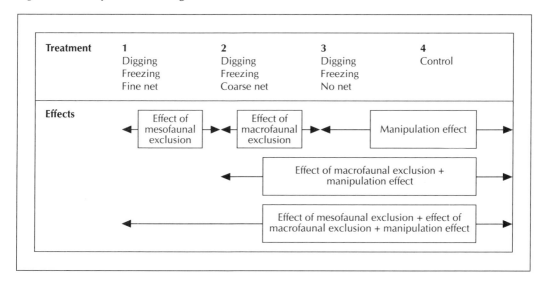

The mesocosms were exposed from October 1991 to June 1992. After this period, they were destructively sampled. Two cores (7.0 cm diameter) were collected in each mesocosm to a depth of 10.0 cm for zoological analysis. Samples for microbiological analysis were taken separately from the litter horizon (L and F layers, approximately 3 cm thick) and from the humus horizon. Results will be given for the litter horizon only. Field moist litter samples were stored in plastic bags at -20°C. After the storage period, samples were allowed to thaw at 4°C for about 3 days. The litter samples were sieved (< 5 mm) and stored in plastic bags at 4°C; microbial analysis started within 2 weeks.

Soil microbial analysis

Microbial biomass

Microbial biomass was determined using both the physiological method of Anderson and Domsch (1978) and the fumigation-extraction method of Amato and Ladd (1988). Substrate saturation and maximum initial respiration response were obtained with an amendment rate of 8.0 mg glucose /g litter dry matter. CO_2 evolved was trapped in 0.05 M NaOH for a 4-hour incubation at 25°C and measured by titration (Jäggi, 1976). For the estimation of biomass nitrogen, litter and soil samples were fumigated with chloroform for 24 hours and extracted with 2 M potassium-chloride. Ninhydrin-reactive nitrogen was determined using a colorimetric procedure (Amato and Ladd, 1988).

Enzyme assays

The measurement of enzyme activity was based on the release and quantitative determination of the product in the reaction mixture. Litter samples were incubated with their respective substrate and buffer or aqueous solution.

Protease activity was estimated using the methods described by Ladd and Butler (1972). The method involves estimating the tyrosine released after an incubation period of 2 hours at 50°C with a buffered caseine solution (pH 8.1). The method used for estimating urease activity involves incubation of the soil with an aqueous urea solution (for 2 hours at 37°C), extraction of ammonium with 1 M KCl and 0.01 M HCl, and colorimetric NH_4^+ determination by a modified indophenol reaction (Kandeler and Gerber, 1988). Soil deaminase activity was measured colorimetrically by deamination of an aqueous arginine solution (incubation for 2 hours at 37°C), as described by Alef and Kleiner (1986). The activities of xylanase and cellulase were determined using the approach described by Schinner et al. (1993). These methods involve the estimation of glucose released after a 24-hour incubation period at 50°C with a buffered xylan solution or carboxymethylcellulose (CMC) solution (pH 5.5). Samples of litter were assayed for beta-glucosidase activity at 37°C for 3 hours, using a buffered salicin solution (pH 6.2).

Analytical methods and statistical procedure

All analytical results were calculated on the basis of oven-dry (105°C) weight of soil. Microbial biomass and soil enzyme activities were determined in duplicate.

Variables of microbial biomass and enzyme activity were tested for normality (Kolmogorov-Smirnov) and homogeneity of variances (Cochran's C). If necessary, data were log-transformed prior to analysis. Differences between mean values of the data were inspected by a one-way analysis of variance (ANOVA), followed by Tukey's honestly significant difference.

RESULTS AND DISCUSSION

The zoological analysis revealed that deep freezing reduced soil mesofauna considerably and that fine nets prevented re-immigration (*see* Figure 19.2). The abundance of Enchytraeidae and Collembola in treatments 2 and 3 resembles the undisturbed field situation. Mites in treatments 2 and 3 did not attain the abundance evident in the undisturbed situation. Since the influence of fauna on microbial processes is not independent of the community structure (combinations of functional groups, including competitive and predator-prey relationships) (Setälä et al., 1991), the observed effects of mesofauna on microflora can be extrapolated only tentatively for the undisturbed situation. Effects of nets on soil properties were considered negligible because no difference in water content between treatments 1, 2 and 3 could be detected (Bruckner et al., 1994), and no root growth could be detected in treatments 1, 2 and 3. In our opinion, this justifies the interpretation of observed effects in treatment 1 to 3 as real faunal effects.

The microbiological analysis of the litter layer showed that soil mesofauna did not significantly influence substrate-induced respiration (SIR) and biomass nitrogen (*see* Figure 19.3 *overleaf*). Blocks into which mesofauna and macrofauna had re-immigrated (treatment 3) showed an increase in SIR. Earlier studies on the influence of mesofauna on microbial biomass did not yield consistent results. A laboratory microcosm study carried out by Faber et al. (1992) revealed that collembola biomass did not contribute significantly to the variation of the fungal standing crop and eukaryotic and prokaryotic SIR. Detailed analysis of their data showed some short-term effects in the animal treatment and fluctuations over time. The time factor in microbial development might be characterised by a succession of species related to substrate quality. Similar results were obtained by Andren and

Figure 19.2 **Abundance of mesofauna (individuals/core) in field mesocosms after an exposure time of 8 months**

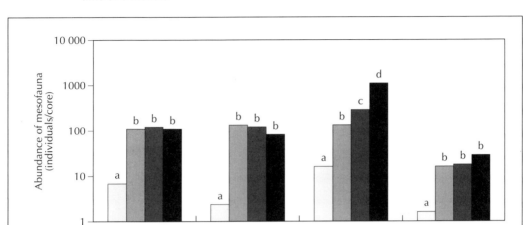

Source: Bruckner et al. (1993)

Schnürer (1985), who did not find any significant effect on respiration and microbial biomass in the presence of collembola. In contrast, the exclusion of mesofauna from litter bags reduced microbial carbon in the organic layer of a moder under beech forest (Scholle et al., 1992). Our results indicate that mesofauna did not affect the pool size of microbial biomass under field conditions after an exposure time of 8 months. During the study, the mesofauna may have stimulated and/or inhibited the bacterial and fungal standing crop, but the net effect on pool size of microbial biomass was negligible. We consider that the exclusion of mesofauna may have caused changes in the community structure of microorganisms. If such changes did occur, however, they exerted no influence on the total microbial biomass.

Soil enzyme activities involved in the carbon cycling (xylanase, cellulase and beta-glucosidase activities) were not affected by exclusion of mesofauna or mesofauna + macrofauna (*see* Figure 19.4 *overleaf*; xylanase activity not shown). In the microcosm study conducted by Teuben and Roelofsma (1990), the introduction of collembola into F1 litter material slightly stimulated cellulase activity. The authors concluded that grazing led to increased microbial activity, resulting in increased enzyme production. Cellulase and xylanase are produced mainly by saprophytic fungi in aerobic environments (Alexander, 1977). As cellulase, xylanase and beta-glucosidase activities were not influenced by the various treatments in our study, it is unlikely that the exclusion of mesofauna shifted the balance of the soil microbial community from fungi to bacteria in the litter layer.

Figure 19.3   Microbial biomass nitrogen (a) and substrate-induced respiration (b) in field mesocosms after an exposure time of 8 months

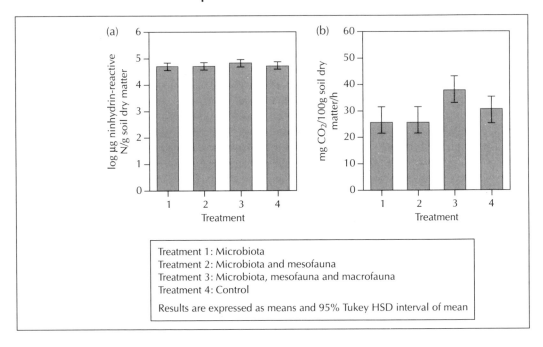

Treatment 1: Microbiota
Treatment 2: Microbiota and mesofauna
Treatment 3: Microbiota, mesofauna and macrofauna
Treatment 4: Control

Results are expressed as means and 95% Tukey HSD interval of mean

Figure 19.4   Cellulase (a) and beta-glucosidase (b) in field mesocosms after an exposure time of 8 months

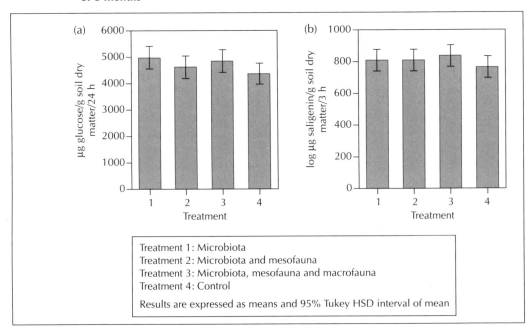

Treatment 1: Microbiota
Treatment 2: Microbiota and mesofauna
Treatment 3: Microbiota, mesofauna and macrofauna
Treatment 4: Control

Results are expressed as means and 95% Tukey HSD interval of mean

Figure 19.5 shows the effect of mesofaunal exclusion on enzymes involved in the nitrogen cycling. Urease and arginine deaminase activity showed similar trends, which are illustrated by the data for arginine deaminase activity. Both enzyme activities were unaffected by soil animals, but protease activity increased with increasing complexity of the fauna. Grazing by *Onychiurus armatus* (Collembola) induced changes in hyphal morphology and extracellular protease production of the fungus *Mortierella isabellina* (Hedlund et al., 1991). This additional protease production was localised in morphologically changed mycelium and the enzyme was classified as a serine protease. In our study, the mesofauna did not influence the size of microbial biomass and arginine deaminase activity, which is bound mainly to the microbial biomass, but increased protease activity. Compensatory growth of grazed fungi, therefore, may buffer the biomass pool. The enhanced production of protease appears to be a direct physiological response to grazing.

Figure 19.5 **Arginine deaminase (a) and protease (b) in field mesocosms after an exposure time of 8 months**

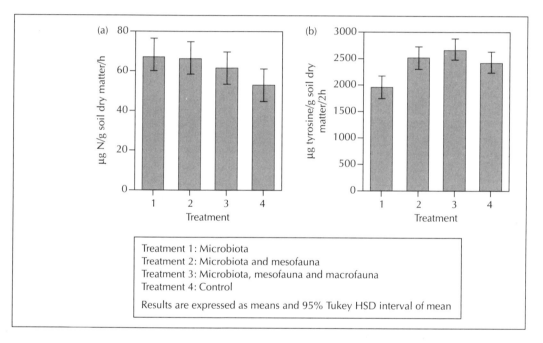

Treatment 1: Microbiota
Treatment 2: Microbiota and mesofauna
Treatment 3: Microbiota, mesofauna and macrofauna
Treatment 4: Control

Results are expressed as means and 95% Tukey HSD interval of mean

Protease activity in control blocks was lower than the activities in treatments 2 and 3 (*see* Figure 19.5). Water uptake by roots is probably responsible for the lower water content in these blocks (data not shown). It is likely that this accounts for the lower protease activity, as proteases are known to be sensitive to desiccation.

In conclusion, we consider that the mesocosm approach is a useful tool in efforts to answer questions about functional relationships under field conditions. Soil enzyme activities are suitable parameters for the indication of faunal-microbial interactions, particularly from a multivariate perspective. Further studies should clarify the temporal course of mesofaunal-microbial interactions.

Acknowledgements

We wish to thank Dr G.Eder for his support of our field work and A.Hofrichter and E.Kohlmann for their technical assistance in the laboratory. We are grateful to Dipl. Ing. G. Hofer for critically reviewing the manuscript and to Dr J. Plant for improvements of the language of the manuscript. Financial support was obtained from the Ministry of Science and Research, Austria.

References

Alef, K. and Kleiner, D. 1986. Arginine ammonification, a simple method to estimate microbial activity potentials in soils. *Soil Biology and Biochemistry* 18: 233-35.

Alexander, M. 1977. *Introduction to Soil Microbiology*. (2nd edn). New York, USA: John Wiley.

Amato, M. and Ladd, J.N. 1988. Assay for microbial biomass based on ninhydrin-reactive nitrogen in extracts of fumigated soils. *Soil Biology and Biochemistry* 20: 107-14.

Anderson, J.P.E. and Domsch, K.H. 1978. A physiological method for quantitative measurement of microbial biomass in soils. *Soil Biology and Biochemistry* 10: 215-21.

Andrén, O. and Schnürer, J. 1985. Barley straw decomposition with varied levels of microbial grazing by *Folsomia fimetaria* (L.) (Collembola, Isotomidae). *Oecologia* 68: 57-62.

Bruckner, A., Kampichler, C., Wright, J., Bauer, R. and Kandeler, E. 1993. Using mesocosms to investigate mesofaunal-microbial interactions in soil: Re-immigration of fauna to defaunated monoliths. *Mitteilungen der Deutschen Bodenkundlichen Gesellschaft* 69: 151-54.

Bruckner, A., Wright, J., Kampichler, C., Winter, B. and Kandeler, E. 1994. Using mesocosms to assess biotic processes in soil: Experiences with a new method. *Pedobiologia*. (submitted).

Coleman, D.C. 1986. The role of microfloral and faunal interactions in affecting soil processes. In Mitchell, M.J. and Nakas, J.P. (eds) *Microfloral and Faunal Interactions in Natural and Agro-Ecosystems*. Dordrecht, Netherlands: Martinus Nijhoff/DrW.Junk.

Faber, J.H. and Verhoef, H.A. 1991. Functional differences between closely related soil arthropods with respect to decomposition processes in the presence or absence of pine tree roots. *Soil Biology and Biochemstry* 23:15-23.

Faber, J.H., Teuben, A., Berg, M.P. and Doelman P. 1992. Microbial biomass and activity in pine litter in the presence of *Tomocerus minor* (Insecta, Collembola). *Biology and Fertility of Soils* 12: 233-40.

Hanlon, R.D.G. 1981. Influence of grazing by collembola on the activity of senescent fungal colonies on media of different nutrient concentration. *Oikos* 36: 362-67.

Hanlon, R.D.G. and Anderson, J.M. 1979. The effects of collembola grazing on microbial activity in decomposing leaf litter. *Oecologia* 38: 93-99.

Hedlund K., Boddy, L. and Preston, C.M. 1991. Mycelial responses of the soil fungus, *Mortierella isabellina*, to grazing by *Onychiurus armatus* (Collembola). *Soil Biology and Biochemistry* 23: 361-66.

Huhta, V. and Koskenniemi, A. 1975. Numbers, biomass and community respiration of soil invertebrates in spruce forests at two latitudes in Finland. *Annales Zoologici Fennici* 12:164-82.

Huhta, V., Wright, D.H. and Coleman D.C. 1989. Characteristics of defaunated soil. I. A comparison of three techniques applied to two different forest soils. *Pedobiologia* 33:417-26.

Ineson, P., Leonard, M.A. and Anderson, J.M. 1982. Effect of collembolan grazing upon nitrogen and cation leaching from decomposing litter. *Soil Biology and Biochemistry* 14: 601-05.

Ingham, R.E., Trofymow, J.A. and Ingham, E.R. 1985. Interactions of bacteria, fungi, and their nematode grazers: Effects on nutrient cycling and plant growth. *Ecological Monographs* 55:119-40.

Jäggi, W. 1976. Die Bestimmung der CO_2-Bildung als Mass der bodenbiologischen Aktivität. *Schweizer Landwirtschaftliche Forschung* 15: 371-80.

Kampichler, C., Bruckner, A., Kandeler, E., Bauer, R. and Wright, J. 1993. A mesocosm study using undisturbed soil monoliths. *Acta Zoologica Fennica* (in press).

Kandeler, E. and Gerber, H. 1988. Short-term assay of soil urease activity using colorimetric determination of ammonium. *Biology and Fertility of Soils* 6: 68-72.

Ladd, J.N. and Butler, J.H.A. 1972. Short-term assays of soil proteolytic enzyme activities using proteins and dipeptide derivatives as substrates. *Soil Biology and Biochemistry* 4: 19-30.

Leonard, M.A. and Anderson, J.M. 1991. Growth dynamics of Collembola (*Folsomia candida*) and a fungus (*Mucor plubeus*) in relation to nitrogen availability in spatial simple and complex laboratory systems. *Pedobiologia* 35: 163-73.

Lussenhop, J. 1992. Mechanisms of microarthropod-microbial interactions in soil. *Advances in Ecological Research* 23: 1-33.

Odum, E.P. 1984. The mesocosm. *BioScience* 34: 558-62.

Schinner, F., Ohlinger, R., Kandeler, E. and Margesin, R. 1993. *Bodenbiologische Arbeitsmethoden.* (2nd edn). Berlin, Germany: Springer-Verlag.

Scholle, G., Wolters, U. and Jörgensen, R.G. 1992. Effects of mesofauna exclusion on the microbial biomass in two moder profiles. *Biology and Fertilty of Soils* 12: 253-60.

Seastedt, T.R. 1984. The role of microarthropods in decomposition and mineralization processes. *Annual Review of Entomology* 29:25-46.

Setälä, H., Tyynismaa, M., Martikainen, E. and Huhta, V. 1991. Mineralisation of C, N and P in relation to decomposer community structure in coniferous forest soil. *Pedobiologia* 35: 285-96.

Teuben, A. 1991. Nutrient availability and interactions between soil arthropods and microorganisms during decompostion of coniferous litter: A mesocosm study. *Biology and Fertility of Soils* 10: 256-66.

Teuben, A. and Roelofsma, T.A.P.J. 1990. Dynamic interactions between functional groups of soil arthropods and microorganisms during decompostion of coniferous litter in microcosm experiments. *Biology and Fertility of Soils* 9: 145-51.

van Veen, J.A. and van Elsas, J.D. 1986. Impact of soil structure and texture on the activity and dynamics of the soil microbial population. In Megujar, F. and Gantar, M. (eds) *Perspectives in Microbial Ecology. Proc. Fourth International Symposium on Microbial Ecology.* Ljubljana, Yugoslavia: Slovene Society for Microbiology.

Verhoef H.A. and Brussaard L. 1990. Decomposition and nitrogen mineralization in natural and agroecosystems: The contribution of soil animals. *Biogeochemistry* 11: 175-211.

CHAPTER 20

Analysis of fungal communities on decomposing beech litter

A. Kjøller and S. Struwe

Two major microbial populations in the decomposer system are bacteria and microfungi. Microfungi are the most active decomposers of plant material. Bacteria play a secondary role despite their high numbers; they have particular functions in the mineralisation of nitrogen compounds. Previous studies have tended to focus on the isolation and identification of soil microfungi rather than on the functional and taxonomic issues needed to provide a fuller understanding of the decomposer community. In a recent publication, Newell (1992) stated that 'if one has sound data for microbial biomass dynamics in decaying litter, along with identities of the decomposer fungi, then one can make useful projections and hypotheses regarding nutrient flow (and) environmental controls on decay rates ... founded upon basic knowledge of the properties of fungal species.'

Total microbial biomass has been determined in various ecosystems mainly by the use of the fumigation technique (Jenkinson and Powlson, 1976). Figures varying from 100 to 1200 μg C_{mic}/g soil, for a variety of agricultural soils in Europe and Canada, have been given in reviews by Anderson and Domsch (1989) and Insam et al. (1989). Comparable figures from litter systems are not available as the fumigation technique cannot be applied to litter (Jenkinson and Ladd, 1981).

Another method based on function is the substrate-induced respiration (SIR) technique, recommended by Newell (1992). Combined with selective inhibition by antibiotics, it is possible using the SIR technique to determine the fungal/bacterial ratio in soil samples (for example, Anderson and Domsch, 1973, 1975, 1978). The technique has also been improved for use in litter samples (Beare et al., 1990; Neely et al., 1991).

We have conducted detailed studies on fungal succession and decomposer patterns in beech (*Fagus silvatica*) litter (Kjøller and Struwe, 1989, 1990). This work showed that, over a period of 2 years of decomposition, the diversity changed and the population consisted of at least three fungal communities. Simultaneous with the taxonomic succession, a functional succession was also identified, initially dominated by pectinolytic fungi, later by cellulolytic fungi and, during the last year of decomposition, by chitinolytic fungi. Lignin-utilising fungi persisted throughout the decomposition period. The aim of the study reported here was to analyse fungal abundance and activity during the decomposition of beech litter, focusing on community similarities and differences and applying the SIR technique in beech litter from forest floors with varying acidity (pH 4.5 and 7.0). The study comprised the following steps:

- isolating fungi by different methods and on different media

- testing the applicability of the SIR technique in litter systems

- identifying the fungal contribution to SIR by selective inhibition

MATERIALS AND METHODS

Beech litter and soil were collected in two forests in Denmark (Strødam, pH 4.5; Allindelille, pH 7.0) at different times during one year. Most of the litter was processed on the same day that it was collected; occasionally it was kept at 5°C for short periods.

The spread plate and soil washing techniques were used to enumerate colony-forming units (CFUs) from litter and soil and for isolating microfungi. In the spread plate technique, 10 g of litter (or soil) were blended for 2-3 minutes in sterile Winogradsky solution. Appropriate tenfold dilutions were plated on soil extract agar SEA (glucose 1.0 g; peptone 1.0 g; yeast extract 1.0 g; K_2HPO_4 1.0 g; soil extract 400 ml; agar 20 g; distilled water 600 ml) with two pH values (5 and 7) and the addition of either 100 mg penicillin and 150 g streptomycin or 50 mg cycloheximide per litre substrate to prevent bacterial or fungal growth. The spread plates were incubated for 2-4 weeks at ambient temperature and colonies of fungi and bacteria were counted. One hundred randomly selected fungal colonies were subcultured and then isolated again until pure cultures were obtained.

In the soil washing technique, blended litter (or soil) was washed through a system of sieves with decreasing mesh sizes, the final sieve having a mesh size of 0.5 mm. One hundred washed particles were placed on separate plates with the addition of penicillin and streptomycin (pH 5 or 7); in one experiment, the soil extract medium diluted 100 times was used as the isolation substrate. The incubation period was 2-4 weeks, at ambient field temperature. Fungal isolates from spread plates and soil washing plates were generally identified to generic level only.

The SIR technique was carried out as described by Neely et al. (1991). Preliminary experiments were conducted to adjust the concentration of glucose and antibiotics and the duration of incubation time, in order to obtain the maximum inhibition and reproducible results. Three concentrations of glucose (0.015, 0.08 and 0.4 mg/g dry litter) were applied to determine the concentration which had the highest initial respiration response. Three concentrations of the two antibiotics, streptomycin (5, 16 and 50 mg/g dry litter) and cycloheximide (25, 80 and 100 mg/g dry litter), applied separately or together, were tested to determine the concentrations when the sum of the effect of each antibiotic added separately was the same as the effect of both antibiotics together. Different incubation periods were used (1, 2.5, 5 and 24 hours). The selected procedure involved the application of 16 mg/g dry litter streptomycin, 80 mg/g dry litter cycloheximide, 0.08 mg/g dry litter glucose and an incubation period of 2.5 hours.

The litter was chopped in a blender; 1 g (dry weight equivalent) was placed in 20 ml vials and kept at 5°C overnight with appropriate antibiotic additions. After the addition of glucose, the vials were incubated for 2.5 hours at room temperature and the CO_2 evolved was measured in a Microlab GC with TC detector (60°C; oven 30°C; 60°C injector; H_2 as carrier gas). The experiments were conducted in triplicate.

RESULTS AND DISCUSSION

This study was a continuation of our work on litter decomposition in deciduous forests in which we attempted to find ways of improving understanding of fungal presence and potential activity of key strains (Kjøller and Struwe, 1989, 1990). One of the aims of the present study was to establish whether the same fungal community would develop on the same litter type in different forests, exemplified by beech litter with low and high pH values and using different isolation media and methods.

The result of the analysis of the fungal flora on decomposing beech litter with low (Strødam) and high (Allindelille) pH values are presented in Tables 20.1 to 20.4. The occurrence of each fungal

Table 20.1   Occurrence of fungi (expressed as a percentage of the total number of isolates) in two beech forests in Denmark (Strødam and Allindelille) with low and high pH values[a]

| | % occurrence | | | | | |
| | Strødam (pH 4.5) | | | Allindelille (pH 7.0) | | |
Isolates	Upper litter layer	Bottom litter layer	Soil	Upper litter layer	Bottom litter layer	Soil
Acremonium spp.	22.0	1.1	1.1	6.1	1.1	9.7
Cladosporium herbarum	19.5	5.5	—	13.1	8.4	3.2
Mortierella spp.	14.6	29.7	19.3	2.0	—	3.2
Penicillium spp.	2.4	16.5	5.7	3.0	20.0	25.0
Mucor sp.	1.2	9.9	11.4	—	5.3	4.8
Cephalosporiopsis sp.	4.9	28.6	—	1.1	4.2	11.3
Aureobasidium pullulans	11.0	—	—	2.0	1.1	—
Chalaropsis sp.	1.2	—	—	3.0	—	1.5
Cylindrocladium sp.	2.4	—	—	1.0	—	—
Sterile mycelia	—	3.3	26.1	11.1	10.5	19.4
Trichoderma viride	—	3.3	2.3	—	—	3.2
Mortierella vinacea	—	1.1	5.7	1.0	—	—
Fusarium sp.	—	1.1	—	—	—	1.5
Yeast II	—	—	18.2	—	1.1	—
Verticillium spp.	—	—	2.3	1.0	1.1	—
Aspergillus sp.	—	—	1.1	—	—	—
Yeast I white	3.7	—	6.8	—	—	—
Menisporella sp.	12.2	—	—	—	—	—
Scopulariopsis sp.	1.2	—	—	—	—	—
Atractium sp.	1.2	—	—	—	—	—
Phialophora sp.	1.2	—	—	—	—	—
Diplodia sp.	1.2	—	—	—	—	—
Sterile, yeast-like	—	—	—	46.5	44.2	12.9
Diplorhinotrichum sp.	—	—	—	2.0	1.1	1.5
Paecilomyces sp.	—	—	—	1.0	1.1	—
Alternaria tenuis	—	—	—	4.0	—	—
Geotrichum sp.	—	—	—	1.0	—	—
Sporotrix sp.	—	—	—	1.0	—	—
Chrysosporium sp.	—	—	—	—	1.1	—
Humicola sp.	—	—	—	—	—	1.5

Note: a   The isolation method was soil washing, and soil extract agar with ambient pH species or genus is expressed as a percentage of the total number of isolates. Although the data were not subjected to statistical analysis, some obvious similarities and differences are worth noting.

As shown in Table 20.1, *Mortierella* species, which are common in acidic forest soils, were very frequent in the two litter layers and in the soil in the low pH forest, but they were almost absent in the alkaline soil, where a sterile, yeast-like fungus was dominant. *Penicillium* species were common in the bottom litter layer and soil in the alkaline forest. A higher frequency of *Cladosporium herbarum*, *Mortierella* species and *Aureobasidium pullulans* was found in the acidic litter (and soil) than in the alkaline litter. The same tendency is evident in the data given in Table 20.2. *Cladosporium herbarum* and *Mortierella* species were twice as frequent on low pH agar than on high pH agar in litter and soil from Allindelille but showed the same frequency in the litter and soil from Strødam. *Aureobasidium pullulans* was twice as frequent on low pH agar than on high pH agar when isolated from Allindelille litter.

Cladosporium herbarum and *Aureobasidium pullulans*, along with the *Mortierella* species, were more frequent on soil extract agar than on the diluted soil extract, and *Penicillium* species occurred more often on the diluted medium (*see* Table 20.3). In a separate experiment in which all the strains isolated were tested on the diluted soil extract medium, all of them were able to grow, indicating an ability to grow in environments where nutrients are limited. Data from the experiment in which fungi from the two forests were isolated using the soil washing and spread plate techniques are given in Table 20.4; both techniques allowed a separation of the population of actively growing fungi from fungi existing mainly as conidia and resting structures. In the spread plate technique, there was a dominance

Table 20.2   Occurrence of fungi (expressed as a percentage of the total number of isolates) at two pH values in the isolation media

	% occurrence							
	Strødam (pH 4.5)				Allindelille (pH 7.0)			
	Beech leaves (5-6 months)		Soil		Beech leaves (5-6 months)		Soil	
Isolates	pH 5[a]	pH 7[a]	pH 5	pH 7	pH 5	pH 7	pH 5	pH 7
Sterile mycelia	49	44	22	34	64	69	51	39
Acremonium spp.	5	7	9	11	16	13	15	27
Penicillium spp.	5	4	44	25	1	3	7	9
Cladosporium herbarum	21	22	—	—	13	7	1	3
Yeast	2	6	1	—	3	2	4	8
Aureobasidium pullulans	14	6	—	1	—	3	—	—
Candida sp.	2	1	—	1	—	3	—	—
Trichoderma viride	—	1	3	4	—	—	1	—
Mortierella spp.	—	1	10	11	—	—	18	9
Mucor sp.	—	3	8	11	—	—	—	3
Verticillium sp.	—	—	—	—	2	—	—	—
Piptocephalis sp.	—	—	—	—	2	—	—	—
Alternaria tenuis	—	—	—	—	—	3	—	—
Fusarium sp.	—	—	—	—	—	—	1	—

Note: a   pH in the isolation medium

of two yeasts, which were almost absent after soil washing, and a much lower diversity. The fungal community in litter and in soil were very different, and there was no indication that the soil served to store fungal conidia produced during decomposition of the litter.

Table 20.3   Occurrence of the most frequent fungi (expressed as a percentage of the total number of isolates) isolated on nutrient-rich (SEA) and nutrient-poor (SEA$_0$) media from litter and soil from a beech forest at Strødam, Denmark

Isolates	Upper litter layer SEA	Upper litter layer SEA$_0$	Bottom litter layer SEA	Bottom litter layer SEA$_0$	Soil SEA	Soil SEA$_0$
Acremonium spp.	22.0	23.0	1.1	5.0	1.1	12.5
Cladosporium herbarum	19.5	14.0	5.5	5.0	—	2.5
Mortierella spp.	14.6	11.0	29.7	14.0	19.3	12.5
Penicillium spp.	2.4	—	16.5	23.0	5.7	22.5
Mucor sp.	1.2	4.0	9.9	13.0	11.4	5.0
Aureobasidium pullulans	11.4	4.0	—	—	—	1.3
Cylindrocladium sp.	2.4	3.0	—	5.0	—	—
Sterile mycelia	—	36.0	3.3	10.0	26.1	20.0

Table 20.4   Occurrence of the most frequent fungi (expressed as a percentage of the total number of isolates) isolated from litter and soil from a beech forest at Strødam, Denmark using two isolation methods

Isolates	Soil washing Upper litter layer	Soil washing Bottom litter layer	Soil washing Soil	Spread plate Upper litter layer	Spread plate Bottom litter layer	Spread plate Soil
Sterile mycelia	56.1	25.8	2.1	18.4	39.8	9.0
Acremonium spp.	10.2	11.2	18.9	1.0	1.9	2.5
Yeast I white	1.0	4.5	—	8.2	3.9	38.8
Cladosporium herbarum	11.2	—	4.1	1.5	1.9	10.1
Mortierella spp.	—	2.2	4.1	—	2.9	14.8
Yeast II dark	—	1.1	—	66.8	35.0	12.3
Trichoderma viride	3.1	22.5	6.8	—	—	—
Mucor spp.	2.0	10.1	12.2	—	—	—
Aureobasidium pullulans	7.1	—	1.4	—	—	—
Botrytis cinerea	3.1	—	5.4	—	—	—
Alternaria tenuis	1.0	—	—	—	—	—
Hyalorhinocladiella sp.	—	2.2	—	—	—	—
Verticillium spp.	—	—	4.1	—	—	—
Torula sp.	—	—	1.4	—	—	—

As noted above, the study also sought to examine the difference in fungal SIR between very acidic and alkaline litter, to confirm the finding reported by Neele et al. (1991) that the fungal/bacterial ratio decreased during decomposition. The most conservative approach used in the analysis of microbial communities is a colony-count on traditional media, such as soil extract agar, to enumerate CFUs of bacteria and fungi. High numbers of bacteria were found in the two types of beech litter, the highest in litter from neutral soil (*see* Table 20.5). The same population sizes were observed when media with different pH values (4.5 and 7.0) were used for the enumeration.

Table 20.5   Total number of bacteria and fungi in beech litter and soil

Site		Bacteria (no./g dry weight)	Fungi (no./g dry weight)
Allindelille pH 7			
pH 7[a]	litter (upper layer)	1.1×10^8	1.2×10^6
	litter (bottom layer)	1.9×10^8	2.1×10^6
	soil	6.7×10^6	2.0×10^{5b}
pH 4.5[a]	litter	5.4×10^8	4.6×10^7
	soil	3.1×10^6	6.0×10^5
Strødam pH 4.5			
pH 4.5[a]	litter (upper layer)	7.5×10^6	4.4×10^5
	litter (bottom layer)	1.6×10^{8b}	5.5×10^5
	soil	1.0×10^6	2.9×10^5
pH 7[a]	litter (upper layer)	7.2×10^7	1.8×10^{7b}
	litter (bottom layer)	4.1×10^{7b}	1.6×10^7
	soil	8.6×10^6	1.6×10^{6b}

Note: a   pH in growth medium
        b   The standard error was less than 20% except for these values

CFUs provide only a rough idea of the fungal community in question; they do not provide information on actual growth and activity, only on past activity such as the formation of conidia. To investigate the activity of the two microbial groups, the SIR technique combined with selective inhibition was used. The SIR technique is based on the initial respiratory response of the microbial population to substrate amendments such as the addition of glucose. Antibiotics may be added for the selective measurement of fungal and bacterial respiration by inhibition with streptomycin and cycloheximide.

Most experiences using this technique originate from work with soil samples. In all the forest and agricultural soils investigated, the fungal part of the total respiration was always between 70 and 80% (*see* Table 20.6). A tropical forest soil from China with a pH value ranging between 3.4 and 4.7 (Yang and Insam, 1991) showed low fungal activity in soil from the A horizon and greater fungal activity in deeper horizons. The results of an attempt to adapt the method for litter, mainly readily decomposable

Table 20.6   Quantification of fungal contribution to soil respiration

Soil and litter	Proportion of fungal respiration	References
Quercus soil		Anderson and Domsch (1973, 1975, 1978)
pH 4.2	70-80%	
Beech litter		—
pH 2.9	60%	
Agricultural soil		—
pH 5.4	78%	
pH 5.2-7.5	70%	
Tropical soil		Yang and Insam (1991)
pH 3.4-4.7	25%	
Forest soil		Tate (1991)
pH 3.5		
O horizon	82%	
A horizon	69%	
Plant litter[a]		Beare et al. (1990)
Trifolium incarnatum	70%	
Digitoria sanguinalis	67%	
Vicea villosa	66%	
Secale cereale	61%	
Sorghum bicolor	71%	
Quercus prinus	57%	

Note: a   After 24 days of decomposition

litter, showed that 60-70% of the respiration was fungal after 24 days of decomposition (Beare et al., 1990; Neely et al., 1991). The conclusion of these experiments was that the ratio of fungal to bacterial SIR tended to decrease over time (that is, after 56 and 100 days).

In our work, we have adopted the same procedures as those used by Neely et al. (1991) to determine the fungal/bacterial ratio in the slowly decomposable litter from beech. To date, the experiments have been carried out with litter from the two sites with neutral and low pH in order to test the common assumption that fungi dominate in acidic environments, and therefore that fungal activity would be higher in the litter from acidic soil. Litter from the two beech forests was collected monthly from July 1992 to January-February 1993, representing 7- to 14-month-old litter. Newly shed leaves were sampled in November-December 1992 and January-February 1993.

Respiration figures showed that the fungal contribution to total respiration was over 50%, reaching 80% in 7- to 12-month-old acidic litter (July to December) and falling to about 30% during January and February. In the newly fallen leaves, the proportion of fungal respiration was initially very low; after 1-2 months it increased to about 50%, as shown in Table 20.7 (*overleaf*). The fungal respiration in the alkaline litter showed a wide variation between samples, fluctuating between 7 and 47.5% in the

7- to 12-month-old litter and increasing a month later to 65%. It was not possible to establish any kind of inhibition in the newly fallen neutral leaves (*see* Table 20.7). Alder and ash litter were also examined. The fungal respiration contributed about 50% of the respiration in both litter types, which decomposed within 6 months at neutral pH.

Table 20.7   **Fungal contribution to substrate-induced respiration (SIR) in beech litter from two forests (Strødam and Allindelille) in Denmark**

| | | Strødam (pH 4.5) | | | | Allindelille (pH 7.0) | | | |
		1991 litter	%	1992 litter	%	1991 litter	%	1992 litter	%
1992	July	80.3				July	10.0		
	August	75.5				August	40.8		
	October	72.2				October	7.0		
	November	49.3		November	4.8	November	47.5	November	—[a]
	December	67.0		December	45.0	December	12.1	December	—[a]
1993	January	33.5		January	51.0	January	65.2	January	—[a]
	February	31.0		February	—[a]			February	—[a]

Note: a   No inhibition

Work conducted by Anderson and Domsch (1975) showed that the ratio between fungal and bacterial respiration was not related to variations in soil pH. They investigated six different soils with pH values ranging between 2.9 and 7.5. In all soils, the fungal part of respiration exceeded 60%, but was lowest in the low pH soils. It is not possible to draw a clear conclusion about the degree of fungal dominance and dependency on pH from our monthly determinations. The fluctuation between the monthly samples was large, and there are environmental factors other than pH that determine fungal activity during decomposition. Consequently, it was not possible to confirm the assumption that fungal activity would be higher in the acidic litter.

Other attempts to discriminate between the decomposer groups have been suggested. In this regard, key enzymes would be of interest. Estimations of the relative contributions made by fungi and bacteria to soil cellulase activity have been provided by Rhee et al. (1987). They measured total carboxymethyl-cellulase (CMC) activity and found that the ratio between the CMC of fungal and bacterial origin was 6/1 in soil obtained from a deciduous forest. We are planning to develope the SIR method to include substrates other than glucose (such as cellulose, chitin and pectin) to estimate different proportions of the active fungal population, in addition to the characterisation of functionally distinct microbial communities.

The understanding of the relationship between the diversity and function of microbial (particularly fungal) communities during decomposition needs further research. Most studies reported to date have been either pure culture studies (isolation, identification, investigation of growth rates and enzymatic potentials) or biomass or process studies. In our study, we have determined the taxonomic differences between fungal flora on acidic and alkaline beech litter and the fungal/bacterial ratio during decomposition over various periods of time of these two litter types.

References

Anderson, J.P.E. and Domsch, K.H. 1973. Quantification of bacterial and fungal contributions to soil respiration. *Archiv für Mikrobiologie* 93: 113-27.

Anderson, J.P.E. and Domsch, K.H. 1975. Measurement of bacterial and fungal contributions to respiration of selected agricultural and forest soils. *Canadian J. Microbiology* 21: 314-22.

Anderson, J.P.E. and Domsch, K.H. 1978. A physiological method for the quantitative measurement of microbial biomass in soils. *Soil Biology and Biochemistry* 10: 215-21.

Anderson, T.-H. and Domsch, K.H. 1989. Ratios of microbial biomass carbon to total organic carbon in arable soils. *Soil Biology and Biochemistry* 21: 471-79.

Beare, M.H., Neely, C.L., Coleman, D.C. and Hargrove, W.L. 1990. A substrate-induced respiration (SIR) method for measurement of fungal and bacterial biomass on plant residues. *Soil Biology and Biochemistry* 22: 585-94.

Insam, H., Parkinson, D. and Domsch, K.H. 1989. Influence of macroclimate on soil microbial biomass. *Soil Biology and Biochemistry* 21: 211-21.

Jenkinson, D.S. and Powlson, D.S. 1976. The effect of biocidal treatments on the metabolism in soil. V. A method for measuring soil biomass. *Soil Biology and Biochemistry* 8: 209-13.

Jenkinson, D.S. and Ladd, J.N. 1981. Microbial biomass in soil. Measurement and turnover. In Paul, E.A. and Ladd, J.M. (eds) *Soil Biochemistry* (Vol. 5). New York, USA: Marcel Dekker.

Kjøller, A. and Struwe, S. 1989. Functional groups of fungi during succession. In Hattori, T., Ishida, Y., Maruyama, Y., Morita, R.Y. and Uchida, A. (eds) *Proc. 5th International Symposium on Microbial Ecology: Recent Advances in Microbial Ecology*. Tokyo, Japan: Japan Scientific Societies Press.

Kjøller, A. and Struwe, S. 1990. Decomposition of beech litter. A comparison of fungi isolated on nutrient-rich and nutrient-poor media. *Transactions of the Mycological Society of Japan* 31: 5-16.

Neely, C.L., Beare, M.H., Hargrove, W.L. and Coleman, D.C. 1991. Relationships between fungal and bacterial substrate-induced respiration, biomass and plant residue decomposition. *Soil Biology and Biochemistry* 23: 947-54.

Newell, S.Y. 1992. Estimating fungal biomass and production in decomposing litter. In Carroll, G.C. and Wicklow, D.T. (eds). *The Fungal Community. Its Organization and Role in the Ecosystem*. New York, USA: Marcel Dekker.

Rhee, Y.H., Hah, Y.C. and Hong, S.W. 1987. Relative contributions of fungi and bacteria to soil carboxymethylcellulase activity. *Soil Biology and Biochemistry* 19: 479-81.

Tate, R.L. 1991. Microbial biomass measurement in acidic soil: Effect of fungal/bacterial activity ratios and of soil amendment. *Soil Science* 152: 220-25.

Yang, J.C. and Insam, H.,1991. Microbial biomass and relative contributions of bacteria and fungi in soil beneath tropical rain forest, Hainan Island, China. *J. Tropical Ecology* 7: 385-93.

Beyond the Biomass
Edited by K. Ritz, J. Dighton and K.E. Giller
© 1994 British Society of Soil Science (BSSS)
A Wiley-Sayce Publication

CHAPTER 21

Compositional analysis of microbial communities: Is there room in the middle?

D.C. Coleman

The renowned physicist, Richard Feynman, challenged his colleagues in theoretical physics and engineering with the statement: 'There's room at the bottom' (Feynman, 1961). He was considering the seemingly remote possibility of the development and implementation of molecular machines. Of course, the past 20 years have been replete with numerous examples of the foresightedness of his ideas, and many organisations and researchers have 'made their names' developing machines which are less than one nanometer in diameter.

The aim of the contributions in this book is to go beyond the general concept of microbial biomass *per se* by examining the possibilities of extending the concept across all the major terrestrial ecosystems and biomes of the world. Therefore, it is appropriate here to paraphrase Feynman: 'There's room in the middle.' Thus the purview of this chapter ranges from the 2-90 μm pores and porenecks, microaggregates, fields and regions to the biosphere. This task is daunting, but there are ways to discern patterns across so many levels of organisation (for example, Levin, 1992).

A general concern in studies of soil microbial ecology is to go beyond the static, standing crop information on biomass, and move into the dynamic aspects relating to production and turnover rates (Nannipieri et al., 1990; Smith and Paul, 1990; Tunlid and White, 1992). It is imperative that our studies be embedded in a conceptual framework large enough to be applicable at various field and site levels and yet be transferable across landscapes and regions, so as to interact effectively with the objectives and needs of the ecologists involved in Global Change research (for example, Lubchenco et al., 1991; Klopatek et al., 1992). I offer some comments and examples of studies of soil community structure and function, and suggest some approaches to expand the basic and applied aspects of soil ecology.

SOIL COMMUNITY ORGANISATION

Organisation in space

Soils are organised in many ways and by many agents. We will concentrate on biologically mediated organisation, noting that numerous physico-chemical mechanisms exist (Oades et al., 1989) and that they operate in conjunction with biological ones (Oades and Waters, 1991). A useful introduction is the arrangement of size classes containing all biota, from bacteria to the megafauna (Swift et al., 1979) (*see* Figure 21.1). When considered in this fashion, the biota are arranged according to the types of microclimates which they experience. Thus the microfauna, such as protozoa and small nematodes, are true water-film inhabitants and experience similar types of moisture stresses as those experienced by the microbial populations. The mesofauna inhabit the aerial pores and range between the various pore spaces. The macrofauna are the most conspicuous in size, and also often have extensive burrowing

Figure 21.1   Size classification of soil organisms by body width

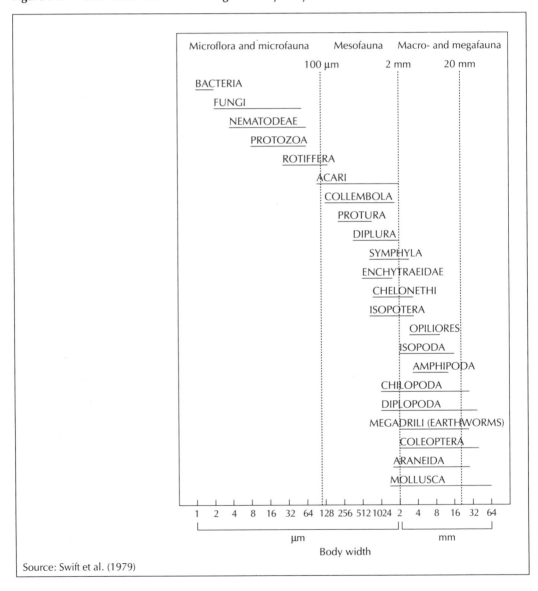

Source: Swift et al. (1979)

and tunnelling activities. Because of the numerous pores which exist in soil (thousands per cubic centimeter), Vannier (1987) termed this milieu the 'porosphere'. It alternates between several states: fully saturated (aquatic); with some water films and open pores (sub-aquatic); and mostly open pores (aerial), as a function of the duration of drying-down after precipitation.

Organisation in space and time

There is an extensive literature on the proximate (Price 1992) and ultimate or evolutionary factors (Price, 1988, 1991) influencing the organisation of invertebrate and soil communities (Margulis and Fester, 1991). I will confine my remarks here to the more short-term (years) results of experimental studies in soil ecology.

One of the aims of ecosystem-level research is to study processes in space and time in the context of various manipulative treatments. This work has been carried out by numerous investigators, several of whom have made use of long-term management regimes; these regimes include those in the Broadbalk plots at Rothamsted, UK (Jenkinson and Rayner, 1977) and the Morrill plots in Illinois and the Great Plains, USA (Haas et al., 1957). For more than a decade, researchers at Colorado State University, USA (Ingham et al., 1986a, b) and the University of Georgia, USA (Beare et al., 1992; Hendrix et al., 1986) have used a 'mesocosm' approach, first described formally by E.P. Odum (1984), to manipulate portions of the soil biota and investigate some of the short- and long-term changes which would appear in soil processes and properties in the course of a multiple-year study.

The general hypothesis tested is that the composition of decomposer communities and their trophic interactions can influence patterns of organic matter transformations and nitrogen dynamics in agroecosystems. At the Horseshoe Bend Experimental area near Athens, Georgia, USA, we used conventional tillage (CT) (moldboard plowing) and no-tillage (NT) (direct-drilled) crop management to test this idea. Cropping sequences were sorghum (*Sorghum bicolor* L. Moench) in summer and rye (*Secale cereale* L.) as a winter cover crop. Biocides were applied to exclude or reduce markedly the populations of bacteria, fungi, microarthropods and earthworms. We measured decomposition rates and nitrogen fluxes in surface and buried rye litter bags in both systems. The abundance and biomass of all microbial and faunal groups were greater on buried litter than on surface litter, with buried litter decay rates (1.4-1.7%/day) some 2.5 times faster than surface litter rates (0.5-0.7%/day) (*see* Figures 21.2 and 21.3 *overleaf*). The ratio of fungal to bacterial biomass and the fauna feeding upon them was 2.5 times higher in the NT surface litter than in the CT buried litter by the end of a 5-month summer study period. The fungicide and bactericide treatments reduced decomposition of the NT surface litter by 36% and 25% of controls, respectively. Where fungivorous microarthropods were experimentally excluded, there was a < 5% reduction in litter decomposition in both tillage treatments, but with fungal densities increasing after the exclusion of microarthropods, there was a 25% greater nitrogen retention in NT when compared with the control after 56 days of decomposition (*see* Figure 21.4 *overleaf*). This indicates the importance of microarthropod regulation of fungal activity, and hence decomposition, under field conditions. No such pattern was observed under CT conditions (*see* Figure 21.5 *overleaf*).

In subsequent experiments we used a similar approach with biocides and measured water-stable aggregates. Plots which received the fungicide captan were shown to have a significantly lower amount of large (< 2000 μm - > 250 μm) macroaggregates, indicating that these aggregates may serve as 'hot spots' of microbial activity. Using wet-sieving (Beare et al., 1993), three primary pools of aggregate-associated soil organic matter were quantified: aggregate-unprotected, macroaggregate-protected and aggregate-resistant carbon and nitrogen.

Figure 21.2   Effects of biotic treatments on percentage dry mass remaining in surface litter under no-tillage land management

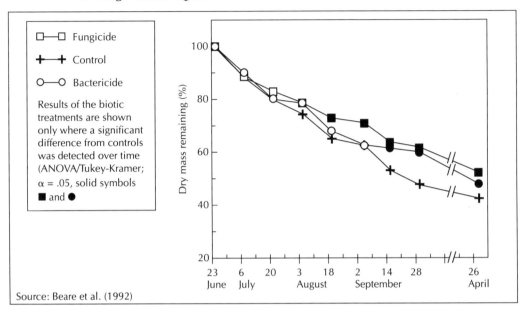

Source: Beare et al. (1992)

Figure 21.3   Effects of biotic treatments on percentage dry mass remaining in buried litter under conventional tillage land management

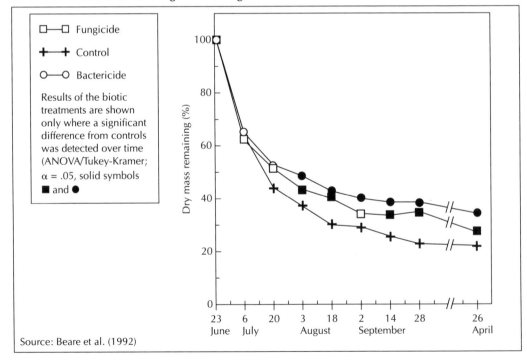

Source: Beare et al. (1992)

Figure 21.4   Effects of biotic treatments on percentage nitrogen remaining in surface litter under no-tillage land management

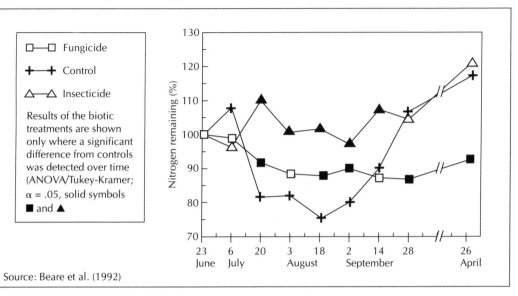

Source: Beare et al. (1992)

Figure 21.5   Effects of biotic treatments on percentage nitrogen remaining in buried litter under conventional tillage land management

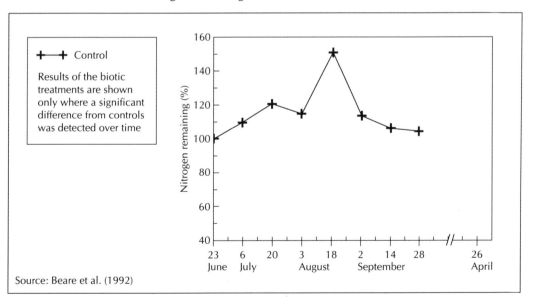

Source: Beare et al. (1992)

Aggregate-unprotected pools of soil organic matter were 21-65% higher in surface soils of NT than CT, with larger differences in the macroaggregate size classes (> 250 μm). Macroaggregate-protected soil organic matter accounted for 23% and 22% of the mineralisable carbon and nitrogen, respectively, in NT (0-15cm) but only 11% and 6% of the total mineralisable carbon and nitrogen in CT, respectively, as determined by laboratory incubations of intact and crushed macro- and microaggregates (*see* Figure 21.6). We concluded from this that macroaggregates in NT soils furnish an important

Figure 21.6 **Cumulative carbon mineralised (C_{min}, g/kg sand-free aggregates) from (a) intact and crushed macroaggregates (> 2000 and 250-2000 μm) and (b) intact microaggregates (106-250 and 53-106 μm) of the 0-5 and 5-15 cm depths of soil under no tillage and conventional tillage**

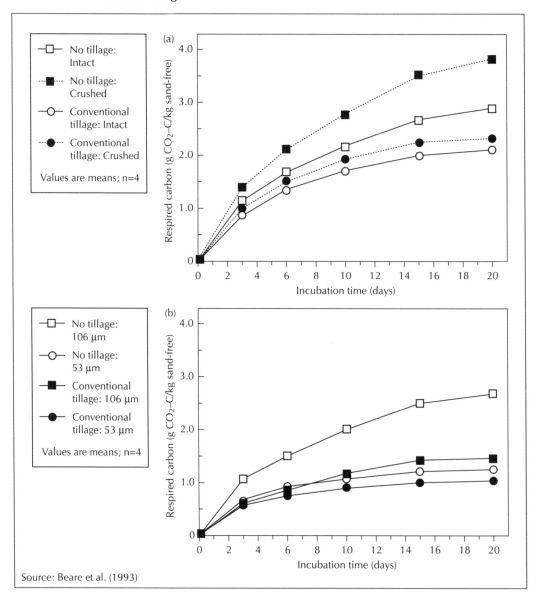

Source: Beare et al. (1993)

mechanism for the protection of soil organic matter that is otherwise mineralised by the frequent disruption of aggregates under CT practices.

Trophic interactions: Loops and webs

There have been some interesting contributions recently to our knowledge of trophic interactions in soil systems (for example, the 'soil microbial loop', Clarholm, *Chapter 22, this volume*). There are, as yet, several unknowns in soil trophic interactions. If sources of reduced carbon and nitrogen are present in the soil matrix, then an organism or organisms will exploit the opportunity to use them as a resource (Barron, 1988, 1992). Barron (1988) observed numerous examples of lignicolous fungi in the genera *Agaricus*, *Coprinus*, *Lepista* and *Pleurotus* which produced coralloid, haustorial hyphae which attacked microcolonies of soil bacteria, principally *Agrobacterium* and *Pseudomonas* (*see* Figure 21.7). This behaviour was limited in species range of predators, with over 100 genera in the Basidiomycotina, Zygomycotina, Oomycetes, Deuteromycotina and Ascomycotina not performing in this fashion. In contrast, there are several genera in the Basidiomycotina, particularly ligninolytic forms, which may form adhesive structures and traps at times of low nitrogen availability to ensnare soil nematodes and, possibly other mesofauna, as suitable nitrogen sources (Barron, 1992).

The wide range of micro- and mesofauna which are attacked by fungi (amoebae, nematodes, rotifers and tardigrades) (Barron, 1977, 1981) suggests that this general predatory pathway needs to be explored in greater detail, particularly in field studies. In the early 1920s several researchers attempted to control plant-parasitic nematodes in Hawaiian pineapple fields, with very little success

Figure 21.7   Microcolonies of the bacterium *Pseudomonas* being attacked by *Lepista nuda,* showing rod-shaped bacterial cells still present around the perimeter of the colonies

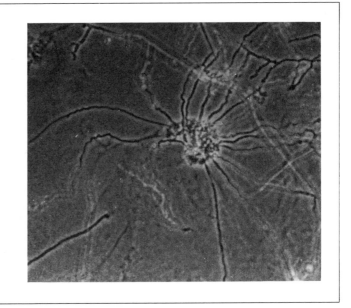

Source: Barron. (1988)

(Mankau, 1980). The nematode-trapping fungi were usually eliminated by other competitors in the usually nitrogen-rich crop/soil environment. Much progress has been made on the biochemistry of trapping mechanisms and general activity of these organisms (Dackman et al., 1992). It will be interesting to see if future studies show that a greater diversity of trapping fungi can be induced into activity by maintaining a deliberately low-nitrogen environment sporadically for the time duration required for adequate control of the pest fauna of concern.

Research on plant pathogens in both natural and man-managed ecosystems needs more emphasis. Studies on plant pathogens such as *Rhizoctonia solani* Kuhn, fed upon by the collembola *Proisotoma minuta* Tullberg and *Onychiurus encarpatus* Denis in the presence of three species of biological control fungi, points to the impact of certain voracious fungivores on the maintenance of crop plants, such as cotton (Curl, 1979; Curl et al., 1988). The pioneering studies conducted by Newell (1984a, b) in which certain collembola altered the species composition of decomposer fungi in conifer needle litter, and hence decomposition kinetics, should be borne in mind, as should the more recent results of decomposition kinetics at various depths in the litter and soil profile (Faber, 1991).

Various whole-system models of food chains and food webs (Hendrix et al., 1986; Hunt et al., 1987; Moore and de Ruiter, 1991; de Ruiter et al., 1993) have been useful, but are much simplified from real-life situations. It is generally agreed, however, that the prevalence of indirect effects (that is, enhance–ment of productivity of exploited populations after being consumed) is extensive above ground (Dyer et al., 1993) and below ground (Moore and de Ruiter, 1991). Omnivory is widespread in desert arthropod-dominated food webs, and in below-ground food webs (Hunt et al., 1987). The implications of these processes for system-level stability are significant and controversial (Oksanen, 1991).

Standing crops, turnover rates and energetics of soil systems

A good general synthesis of biotic interactions must take into account the marked contrast between the theoretical minimum generation times (hours) of microbes and actual net turnover times (a few to several months). In contrast, a much smaller biomass of microbivorous protozoa (chiefly flagellates and amoebae) prey upon primary decomposers and have as much as 10 doublings or turnovers per growing season (Clarholm, 1985; Gupta and Germida, 1988), with a significant impact on plant nitrogen uptake (Elliott et al., 1979; Kuikman and van Veen, 1989). Similar patterns emerge for the entire heterotrophic assemblage of mesofauna and macrofauna (*see* Table 21.1). The consumption and respiration rates and the turnover times of micro- , meso- and macrofauna showed marked differences between theoretical and actual secondary production and turnover rates (Hunt et al., 1987; Parmelee et al., 1990; Beare et al., 1992; Coleman et al., 1993). The ability of most of the soil biota to enter quiescent phases, presumably nutrient or water-stress induced, has a significant influence on the long-term stability of the soil if the organic integrity of the soil matrix is maintained.

ROOT/SOIL INTERACTIONS

In the interactions of roots, soil and microbes, it is necessary to quantify more adequately the contributions made by roots and residues to the soil carbon pools and their turnover rates (Jenkinson, 1966a, b; Jenkinson and Rayner, 1977).

Total rhizosphere respiration of annual plants has been estimated to be in the range of 20-80% of the plant carbon transfer to the below-ground system via roots (Lundegårdh, 1924; Martin and Kemp,

Table 21.1 Average standing crop and energetic parameters of microorganisms, mesofauna and earthworms in a lucerne ley and no-tillage agroecosystem

	Naked amoeba	Flagellates	Ciliates	Bacteria	Fungi
Typical size in soil	30 μm	10 μm	80 μm	0.5-1x 1-2 μm	Ø 2.5 1.0-5.5 μm
Mode of living	in water films on surfaces	free-swimming in water films	free-swimming in water films	on surfaces	free and on surfaces
Biomass (kg dw/ha)	95%	5% 50[a]	<1%	500-750[b]	700-2700[c]
% active	0-100			15-30	2-10
Estimated turnover times/season		10		2-3	0.75
Number of bacteria/ division x 10³	3-8	0.6-1	20-2000		
Minimum generation time in soil (hours)		2-4		0.5	4-8

	Microbivorous nematodes	Collembolans	Mites	Enchytraeids	Earthworms
Typical size in soil	Ø ~ 40 μm	Ø 5000 μm	Ø 1000 μm	Ø 1000 μm	Ø 5000 μm
Mode of living	in water films free and on surfaces	free	free	free	free in soil
Biomass (kg dw/ha)	1.5-4[d]	0.2-0.5[d]	2-8[d]	1-8[d]	25-50[d]
% active	0-100	80-100	80-100	?	0-100
Estimated turnover times/season	2-4	2-3	2-3	?	3
Minimum generation time in soil (hours)	120	720	720	170	720

Note: a Most probable number (MPN) technique
      b Direct counts plus size class estimations
      c Direct estimate of fungal hyphal length and diameter
      d Extractions and sorting
Source: Adapted from Clarholm (1985), Hendrix et al. (1987) and Beare et al. (1992) in Coleman et al. (1993)

1986; Lambers et al., 1990). The functional categories of rhizosphere respiration (root versus microbial) have seldom been studied *in situ* because of the difficulties involved in partitioning respiration between roots and their rhizosphere associates (Helal and Sauerbeck, 1991). Root respiration is a direct cost or tax on the plant, whereas the loss of labile carbon from exudation or exfoliation is passed on to the carbon pool in the soil, where it is presumably rapidly incorporated into the microbial biomass. Numerous researchers have attempted to grow plants in solution cultures or artificial soils under gnotobiotic conditions (Lambers, 1987). Unfortunately, these results bear little relevance to real-life conditions, such as the impact of factors (for example, soil texture) on exudation and respiration patterns.

Several recent studies (Minchin and McNaughton, 1984; Meharg and Killham, 1988; Wang et al., 1989) have employed pulse labelling of plants with $^{14}CO_2$ or $^{11}CO_2$ to investigate rhizosphere respiration, using the translocation of the current photosynthate as the source of the respiratory substrate. We developed a novel procedure to measure root respiration, rhizo-microbial respiration and soluble-carbon concentration simultaneously in the rhizospheres of intact plants using ^{14}C pulse-labelling and tracing (Cheng et al., 1993). Adding ^{12}C-glucose to the rhizosphere immediately before pulse labelling of plant shoots should have the following effects: a dilution of the water-soluble ^{14}C-labelled carbon in the rhizosphere-rhizoplane; and a lower rate of $^{14}CO_2$ output from the root-soil component and a

Figure 21.8   Rates of $^{14}CO_2$ evolution from root-soil columns after 10-minute labelling of shoots with $^{14}CO_2$ in the air

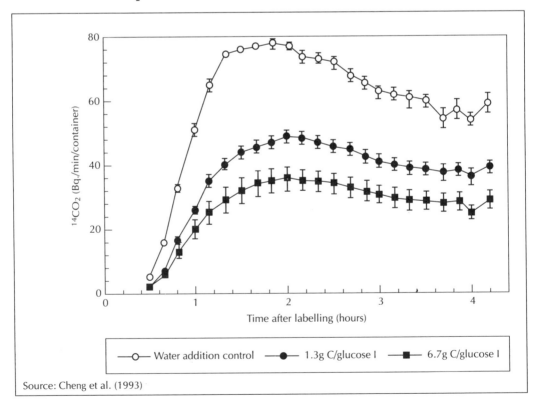

Source: Cheng et al. (1993)

higher concentration of water-soluble ^{14}C-labelled carbon in the rhizosphere and bulk soil than the water addition control during the initial period (4-6 hours) after labelling. This hypothesis was developed from the concept of 'isotopic trapping' (Wolf, 1964).

Soil samples (a sandy clay loam) used in developing this procedure came from the Horseshoe Bend Research Area of the University of Georgia. We set up PVC containers in the laboratory (5 cm internal diameter, 15 cm tall) capped at the lower end, with air inlet tubing at the tip and air outlet tubing at the bottom, for growing wheat (*Triticum aestivum* L.) seedlings. The [14]CO_2 evolved from each container was trapped in ethanolamine scintillation cocktail, with the traps being changed every 10 minutes. Root, shoot and soil samples were quantified and soluble ^{14}C in 20 g fresh root-free soil was extracted in 0.5 M K_2SO_4 solution.

The addition of ^{12}C-glucose solution to the root-soil column 1 hour before pulse labelling of shoots with [14]CO_2 substantially reduced the rates of [14]CO_2 evolution from the root-soil columns and increased the amount of water-soluble ^{14}C in the soil from 3 to 12-fold compared with the water-addition controls (*see* Figure 21.8).

These results confirmed our initial hypothesis and validated the isotopic trapping mechanism. Therefore, rhizo-microbial respiration ([14]CO_2 evolved from microbial utilisation of exudates) must be inversely proportional to the glucose-^{12}C concentration in the rhizosphere. Root respiration remains independent of the glucose-^{12}C concentration in the rhizosphere. The following equation can be written:

$$\frac{^{14}CO_2 \text{ with glucose - root respiration}}{^{14}CO_2 \text{ without glucose - root respiration}} = \frac{\text{Soluble C (g/l)}}{\text{Soluble C (g/l)} + \text{glucose C (g/l)}}$$

Using this equation, we computed root respiration rates, rhizo-microbial respiration rates and soluble carbon concentrations in the rhizosphere (*see* Table 21.2 *overleaf*). Root respiration and rhizo-microbial respiration contributed 40.6% and 59.4% of total rhizosphere respiration, respectively. Soluble carbon concentration in the rhizosphere of 3-week-old wheat plants was 667 mg C/l of soil water.

All these calculations were based on data produced from one short (10 or 20 minutes) pulse of [14]CO_2 assimilated into the plants. Therefore, the values are all related to the current photosynthate and cannot be extrapolated to represent older carbon already incorporated into plant tissues before the experiment or younger carbon assimilated later.

This procedure opens up new opportunities for research in the areas of root respiration physiology, plant-soil interactions and rhizosphere ecology. Using this procedure we can now move some of our studies from being based mainly on solution cultures to being conducted under true soil conditions.

SOIL QUALITY — A QUANTIFIABLE CONCEPT?

The large-scale programmes in several biomes in North America and Antarctica, such as the USA Long-Term Ecological Research (LTER) programme and the Tropical Soil Biology and Fertility Programme (TSBF) involving a 27-site network of research projects on decomposition and nutrient cycling in the context of tropical agroecosystem productivity, are prime candidates for seeking an integration of structure, composition and function in this wide array of ecosystems. As work progresses at the interface between human-influenced and more natural ecosystems, we face an ever-increasing need to be able to provide quantitative information on more qualitative concepts, such as soil quality.

Table 21.2   Cumulative $^{14}CO_2$ evolved (0.5-4.2 hours) in root-soil columns with two levels of
glucose addition and a water-addition control, and calculated root respiration,
rhizo-microbial respiration and concentration of soluble carbon in the rhizosphere
after pulse labelling of shoots (10 minutes) with $^{14}CO_2$

Mean glucose concentration[a]	
(g C/l soil water)	
glucose I	1.19
glucose II	6.78
$^{14}CO_2$ evolved	
(KBq/container) (mean ± 1 SE)	
water only	13.72 ± 0.27
glucose I	8.50 ± 0.34
glucose II	6.30 ± 0.63
Root respiration	
(KBq/container)	5.57
(% water addition)	40.61
Rhizo-microbial respiration	
(KBq/container)	8.15
(% water addition)	59.39
Water soluble C	
(g C/l soil water)	0.667

Note: a   The calculation of mean glucose concentration was based on the amount of glucose added and the
uptake rate of 8.68 mg glucose/container/hour obtained from a separate study with the same soil and
plant in an identical container
Source: Cheng et al. (1993)

In the past few years, there has been increasing interest in finding ways to assess the overall quality
or health of a given soil or ecosystem. Soil quality has been defined as the degree to which a soil can:
promote biological activity (plant, animal, and microbial); mediate water flow through the environment; and maintain environmental quality by acting as a buffer that assimilates organic wastes and
ameliorates the effects of contaminants (Doran and Parkin, 1994; Larson and Pierce, 1994). Recent
reviews on the effects of microorganisms on soil aggregates (Anderson, 1991) and of soil invertebrates
on soil structure (Wolters, 1991) are useful adjuncts to this discussion.

Because it has often been observed that microbial carbon is not readily correlated with total soil
carbon in all types of soils (Sparling, 1992), we need a more robust approach, such as the respiration/
biomass ratio, as noted by Anderson and Domsch (1978), or calculations of the metabolic quotient for
CO_2 (qCO_2), which correlates well with successional status in a number of terrestrial ecosystems
(Insam and Haselwandter, 1989).

Until recently there have been only a few attempts to use bioindicators to evaluate soil quality (for
example, Zlotin, 1985; Foissner, 1987; Paoletti et al., 1991; Stork and Eggleton, 1992). An increasing
number of studies are examining soils in regions of radioactive contaminants, such as at Chernobyl
(Krivolutskii and Pokharzhevskii, 1991), and in areas contaminated with toxic chemicals (Koehler,

1992). Certain species or assemblages of fauna responded to the disturbance in each case, serving as possible bioindicators of that disturbance. However, the rate of dispersal of key organisms involved in the detritus/decomposition pathway, such as oribatid mites as late colonisers, must also be considered (Beckmann, 1988). Bioindicators which are useful in assessing soil quality range from single organisms to biological communities and processes such as decomposition and nutrient turnover (Linden et al., 1994; *see* Table 21.3).

Table 21.3   Properties of soil fauna for use as indicators of soil quality

Organisms and populations
   Individuals
      Behaviour, morphology, physiology

   Populations
      Numbers and biomass
      Rates of growth, mortality and reproduction
      Age distribution

Communities
   Functional groups
      Guilds (e.g., burrowers vs non-burrowers, litter vs soil dwellers, etc.)

   Trophic groups
      Food chains and food webs (microbivores, predators, etc.)

   Biodiversity
      Species richness, dominance, evenness
      Keystone species

Biological processes
   Bioaccumulation
      Heavy metals and organic pollutants

   Decomposition
      Fragmentation of organic matter
      Mineralisation of carbon and nutrients

   Soil structure modification
      Burrowing and biopore formation
      Fecal deposition and soil aggregation
      Mixing and redistribution of organic matter

Source: Linden et al. (1993)

A complete synthesis still eludes us, possibly because many disciplines are required to achieve the synoptic coverage necessary for the successful measurement and prediction of trends. Recent progress has been made by Bentham et al. (1992), who compared the soil microbiological and physico-chemical characteristics of soils restored to grassland or woodland after opencast coal mining in the UK with undisturbed, control sites. They measured microbiological activity by soil dehydrogenase activity, soil adenosine triphosphate (ATP) for biomass and ergosterol contents to estimate the fungal component. A cluster analysis based on the three microbiological indices discriminated between habitat types (*see* Figure 21.9), which was not possible using the soil physico-chemical properties alone. The authors concluded that habitat classification based on 3-dimensional ordination of soil microbiological properties could be used to assess progress in reclamation potential of soils. In addition, the inclusion of soil fauna, such as primary detritivores (for example, microbivorous nematodes; Neher et al., 1993), and fauna which ingest soil, such as enchytraeids and earthworms, would make the assessment more holistic in terms of the effects on soil aeration and porosity (for example, Edwards et al., 1992; Hendrix et al., 1992).

Figure 21.9   **A three-dimensional ordination of soil microbial adenosine triphosphate (ATP) concentration, dehydrogenase activity and ergosterol concentration for each site**

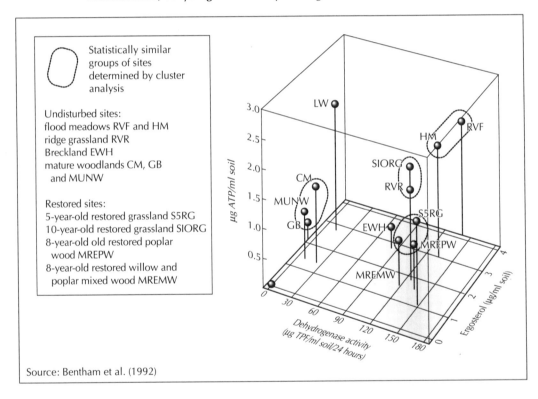

Source: Bentham et al. (1992)

GLOBAL CHANGE AND 'SCALING' PROBLEMS

One of the major concerns of ecologists is how best to integrate our efforts into the large-scale framework of the biospheric processes which are so prominently featured in the popular press (for example, Gore, 1992) and in scientific literature. Several processes in terrestrial ecosystems, namely production of greenhouse gases such as methane and nitrous oxide (Wessman, 1992; Schimel, 1993), or ammonia from urine patches and unmanaged and agricultural soils (Schlesinger and Hartley, 1992), are of great importance in the calculation of global carbon and nitrogen budgets. Although our most common terms of reference are within field or landscape boundaries, it is the movement of substances across the landscape, either aqueous or volatile organics and inorganics, such as ammonia, methane and nitrous oxide, which really act as linking or unifying agents across regions and the biosphere as a whole (Schimel, 1993). Sources of soil respiration, as the summation of major biological processes in terrestrial ecosystems (Raich and Schlesinger, 1992), are controversial with respect to their relative contributions to global warming. Oechel et al. (1993) and Smith and Shugart (1993) measured greater rates of increase of carbon dioxide evolution from tundra ecosystems worldwide than had been previously measured; it has been claimed that the short-term (a few years) rate of output from tropical regions increases significantly with small increases in temperature (Townsend et al., 1992; *see* Figure 21.10). Soil respiration rates in a network of sites from the poles to the equator should be monitored carefully over the next few decades as part of various long-term studies.

Figure 21.10   Effects of Q_{10} on the 100-year integrated flux of carbon (Pg = 1015 g) between soils and the atmosphere for boreal and tropical moist forests

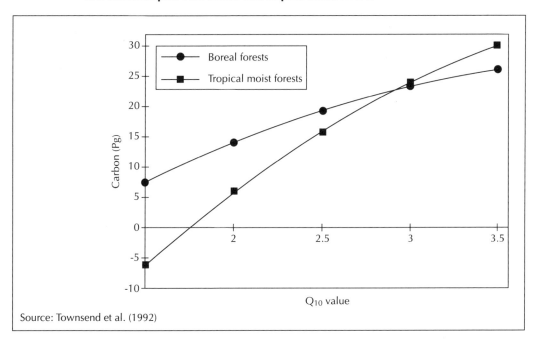

Source: Townsend et al. (1992)

The best way to observe the impact of processes across landscapes is to use an array of sensors to determine the extent of changes and their impact over space and time (Mooney, 1990; Coleman et al., 1992). Hierarchy theory (Allen and Starr, 1982; O'Neill et al., 1986) is an organising concept which is useful in scaling. The scale under observation is affected by the potential behaviour of its components and by the environmental constraints imposed by the higher levels. Lower-level dynamics are too fast to be seen as variables; they are experienced as average or integer values and appear in patterns as a blend. They can be ignored, therefore, given a relatively stable system. Reciprocally, higher-level dynamics are so slow that they are experienced as constants, and larger-scale patterns are seen only as a uniform local condition. However, when the system is disrupted, the response dynamics at the fine scales break up the constraint system and move it into a new configuration (O'Neill, 1988; Wessman, 1992).

This brings us full circle in the ways in which we can scale up and down in ecosystems. As problems such as disturbances and responses to them occur in a given landscape, then we should focus downward, employing specific functional measurements, such as qCO_2 or specific enzyme activities as related to active biomass (Insam and Haselwandter, 1989; Bentham et al., 1992), or faunal activity (Linden et al., 1993), and then focus outward to the larger-scale problems under investigation.

CONCLUSION

Some large surprises undoubtedly await us as we continue to discover how soil systems function under the increasing onslaught of human and human-induced activities in all the ecosystems of the world. The magnitude of the impact on soil ecosystems from land-use change will probably continue to be the dominant force of change, as contrasted to the more indirect ones of global change, driven by increasing concentrations of greenhouse gases. As noted by Levin (1992), 'the key is to separate the components of variability into those that inhibit persistence and coexistence, those that promote these, and those that are noise.' Indeed, our principal concerns, going over more than nine orders of magnitude just within a soil alone (Oades and Waters, 1991), must be to assess the problems of aggregation and simplification, which has been termed 'the problem of determining minimal sufficient detail' (Levin, 1991). We have a 'sufficiency', indeed, to carry our research on soil community processes well into the new millennium.

Acknowledgements

The author wishes to thank Drs M.H. Beare, D.A. Crossley, Jr., W. Cheng and P.F. Hendrix for their comments on the manuscript. Support for the development of ideas and research results was provided by the Coweeta LTER and Georgia Agroecosystem Grants, funded by the USA National Science Foundation, and the Andrew W. Mellon Foundation.

References

Allen, T.F.H. and Starr, T.B. 1982. *Hierarchy. Perspectives for Ecological Complexity*. Chicago, Illinois, USA: University of Chicago Press.
Anderson, J.P.E. and Domsch, K.H. 1978. A physiological method for the quantitative measurement of microbial biomass in soils. *Soil Biology and Biochemistry* 10: 215-21.
Anderson, T.-H. 1991. Bedeutung der mikroorganismen für die bildung von aggregaten im boden. *Zeitschrift für Pflanzenernährung und Bodenkunde* 154: 409-16.
Barron, G.L. 1977. *The Nematode-Destroying Fungi*. Guelph, Canada: Canadian Biological Publications.

Barron, G.L. 1981. Predators and parasites of microscopic animals. In Cole, G. and Kendrick, B. (eds) *Biology of Conidial Fungi* (Vol. 2). New York, USA: Academic Press.

Barron, G.L. 1988. Microcolonies of bacteria as a nutrient source for lignicolous and other fungi. *Canadian J. Botany* 66: 2505-10.

Barron, G.L. 1989. New species and new records of fungi that attack microscopic animals. *Canadian J. Botany* 67: 267-71.

Barron, G.L. 1992. Lignolytic and cellulolytic fungi as predators and parasites. In Carroll, G. and Wicklow, D.T. (eds) *The Fungal Community* (2nd edn). New York, USA: Marcel Dekker.

Beare, M.H., Parmelee, R.W., Hendrix, P.F., Cheng, W., Coleman, D.C. and Crossley, D.A., Jr. 1992. Microbial and faunal interactions and effects on litter nitrogen and decomposition in agroecosystems. *Ecological Monographs* 62: 569-91.

Beare, M. H., Cabrera, M.L., Hendrix, P.F. and Coleman, D.C. 1993. Aggregate-protected and unprotected pools of organic matter in conventional and no-tillage Ultisols. *Soil Science Society of America J.* (in press).

Beckmann, M. 1988. The development of soil mesofauna in a ruderal ecosystem as influenced by reclamation measures. I. Oribatei (Acari). *Pedobiologia* 31: 391-408.

Bentham, H., Harris, J.A., Birch, P. and Short, K.C. 1992. Habitat classification and soil restoration assessment using analysis of soil microbiological and physico-chemical characteristics. *J. Applied Ecology* 29: 711-18.

Cheng, W., Coleman, D.C., Carroll, C.R. and Hoffman, C.A. 1993. *In situ* measurement of root respiration and soluble carbon concentrations in the rhizosphere. *Soil Biology and Biochemistry* 25: 1189-96.

Clarholm, M. 1985. Possible roles for roots, bacteria, protozoa and fungi in supplying nitrogen to plants. In Fitter, A.H., Atkinson, D., Read, D.J. and Usher, M.B. (eds) *Ecological Interations in Soil: Plants, Microbes and Animals*. London, UK: Blackwell.

Coleman, D.C., Odum, E.P. and Crossley, D.A., Jr. 1992. Soil biology, soil ecology, and global change. *Biology and Fertility of Soils* 14: 104-11.

Coleman, D.C., Hendrix, P.F., Beare, M.H., Cheng, W.X. and Crossley, D.A., Jr. 1993. Microbial and faunal dynamics as they affect soil organic matter dynamics in subtropical agroecosystems. In Paoletti, M.G., Foissner, W. and Coleman, D.C. (eds) *Agroecology and Conservation Issues in Temperate and Tropical Regions. Vol. 3: Soil Biota, Nutrient Cycling, and Farming Systems*. Chelsea, Michigan, USA: Lewis Publishing.

Curl, E.A. 1979. Effects of mycophagous collembola on *Rhizoctonia solani* and cotton-seedling disease. In Schippers, B. and Gams, W. (eds) *Soil-Borne Plant Pathogens*. London, UK: Academic Press.

Curl, E.A., Lartey, R. and Peterson, C.M. 1988. Interactions between root pathogens and soil microarthropods. *Agriculture, Ecosystems and Environment* 24: 249-61.

Dackman, C., Jansson, H.-B. and Nordbring-Hertz, B. 1992. Nematophagous fungi and their activities in soil. In Stotzky, G. and Bollag, J.-M. (eds) *Soil Biochemistry* (Vol. 7). New York, USA: Marcel Dekker.

De Ruiter, P. C., Moore, J.C., Zwart, K.B., Bouwman, L.A., Hassink, J., Bloem, J., De Vos, J.A., Marinissen, J.C.Y., Didden, W.A.M., Lebbink, G. and Brussaard, L. 1993. Simulation of nitrogen mineralisation in the below-ground food webs of two winter wheat fields. *J. Applied Ecology* 30: 95-106.

Doran, J.W. and Parkin, T.B. 1994. Defining and assessing soil quality. In Doran, J.W., Bezdicek, D.F. and Coleman, D.C. (eds) *Defining Soil Quality for a Sustainable Environment*. SSSA Special Publication. Madison, Wisconsin, USA: American Society of Agronomy (in press).

Dyer, M.I., Coleman, D.C., Freckman, D.W. and McNaughton, S.J. 1993. A cross-experiment analysis of source-sink relationships in *Panicum coloratum* L. using the ^{11}C technology. *Ecological Applications* 3: 654-65.

Edwards, W.M., Shipitalo, M.J., Traina, S.J., Edwards, C.A. and Owens, L.B. 1992. Role of *Lumbricus terrestris* (L.) burrows on quality of infiltrating water. *Soil Biology and Biochemistry* 24: 1555-61.

Elliott, E.T., Coleman, D.C. and Cole, C.V. 1979. The influence of amoebae on the uptake of nitrogen by plants in gnotobiotic soil. In Harley, J.L. and Russell, R.S. (eds) *The Soil-Root Interface*. London, UK: Academic Press.

Faber, J.H. 1991. Functional classification of soil fauna: A new approach. *Oikos* 62: 110-17.

Feynman, R.P. 1961. There's room at the bottom. In Gilbert, H.D. (ed) *Miniaturisation*. New York, USA: Reinhold.

Foissner, W. 1987. Soil protozoa: Fundamental problems, ecological significance, adaptations in ciliates and testaceans, bioindicators, and guide to the literature. *Progress in Protistology* 2: 69-212.

Gore, A.J. 1992. *Earth in the Balance*. Cambridge, Massachusetts, USA: Houghton Mifflin.

Gupta, V.V.S.R. and Germida, J.J. 1988. Populations of predatory protozoa in field soils after 5 years of elemental S fertiliser application. *Soil Biology and Biochemistry* 20: 787-91.

Haas, H.J., Evans, C.E. and Miles, E.F. 1957. *Nitrogen and Carbon Changes in Great Plains Soils as Influenced by Cropping and Soil Treatments*. USDA Technical Bulletin No. 1164. Washington DC, USA: USDA.

Helal, H.M. and Sauerbeck, D. 1991. Short-term determination of the actual respiration rate of intact plant roots. In Michael, B.L. and Persson, H. (eds) *Plant Roots and Their Environment*. Amsterdam, Netherlands: Elsevier.

Hendrix, P.F., Parmelee, R.V., D.A. Crossley, J., Coleman, D.C., Odum, E.P. and Groffman, P. 1986. Detritus food webs in conventional and no-tillage agroecosystems. *Bioscience* 36: 374-80.

Hendrix, P.F., Crossley, D.A., Jr., Coleman, D.C., Parmelee, R.W. and Beare, M.H. 1987. Carbon dynamics in soil microbes and fauna in conventional and no-tillage agroecosystems. *INTECOL Bulletin* 15: 59-63.

Hendrix, P.F., Mueller, B.R., Bruce, R.R., Langdale, G.R. and Parmelee, R.W. 1992. Abundance and distribution of earthworms in relation to landscape factors on the Georgia Piedmont, USA. *Soil Biology and Biochemistry* 24: 1357-61.

Hunt, H.W., Coleman, D.C., Ingham, E.R., Ingham, R.E., Elliott, E.T., Moore, J.C., Rose, S.L., Reid, C.P.P. and Morley, C.R. 1987. The detrital food web in a shortgrass prairie. *Biology and Fertility of Soils* 3: 57-68.

Ingham, E.R., Trofymow, J.A., Ames, R.N., Hunt, H.W., Morley, C.R., Moore, J.C. and Coleman, D.C. 1986a. Trophic interactions and nitrogen cycling in a semi-arid grassland soil. I. Seasonal dynamics of the natural populations, their interactions and effects on nitrogen cycling. *J. Applied Ecology* 23: 597-614.

Ingham, E.R., Trofymow, J.A., Ames, R.N., Hunt, H.W., Morley, C.R., Moore, J.C. and Coleman, D.C. 1986b. Trophic interactions and nitrogen cycling in a semiarid grassland soil. II. System responses to removal of different groups of soil microbes or fauna. *J. Applied Ecology* 23: 615-30.

Insam, H. and Haselwandter, K. 1989. Metabolic quotient of the soil microflora in relation to plant succession. *Oecologia* 79: 174-78.

Jenkinson, D.S. 1966a. The turnover of organic matter in soil. In Silow, R.A. (ed) *The Use of Isotopes in Soil Organic Matter Studies*. London, UK: Pergamon Press.

Jenkinson, D.S. 1966b. The priming action. In Silow, R.A. (ed) *The Use of Isotopes in Soil Organic Matter Studies*. London, UK: Pergamon Press.

Jenkinson, D.S. and Rayner, J.H. 1977. The turnover of soil organic matter in some of the Rothamsted classical experiments. *Soil Science* 123: 298-305.

Klopatek, C.C., O'Neill, E.G., Freckman, D.W., Bledsoe, C.S., Coleman, D.C., Crossley, D.A., Jr., Ingham, E.R., Parkinson, D. and Klopatek, D. 1992. The sustainable biosphere initiative: A commentary from the US Soil Ecology Society. *Bulletin of the Ecological Society of America* 73: 223-28.

Koehler, H.H. 1992. The use of soil mesofauna for the judgement of chemical impact on ecosystems. *Agriculture, Ecosystems and Environment* 40: 193-205.

Krivolutzkii, D.A. and Pokarzhevskii, A.D. 1991. Soil fauna as bioindicators of biological after-effects of the Chernobyl atomic power station accident. In Jeffrey, D.W. and Modden, B. (eds) *Bioindicators and Environmental Management*. London, UK: Academic Press.

Kuikman, P. and van Veen, J.A. 1989. The impact of protozoa on the availability of bacterial nitrogen to plants. *Biology and Fertility of Soils* 8: 13-18.

Lambers, H. 1987. Growth, respiration, exudation and symbiotic associations: The fate of carbon translocated to the roots. In Gregory, P.J., Lake, J.V. and Rose, D.A. (eds) *Root Development and Function*. Cambridge, UK: Cambridge University Press.

Lambers, H., van der Werf, A. and Konings, H. 1990. Respiratory patterns in roots in relation to their functioning. In Waisel, Y., Eshel, A. and Kafkafi, U. (eds) *Plant Roots: The Hidden Half.* New York, USA: Marcel Dekker.

Larson, W.E. and Pierce, F.J. 1994. The dynamics of soil quality as a measure of sustainable management. In Doran, J.W., Bezdicek, D.F. and Coleman, D.C. (eds) *Defining Soil Quality for a Sustainable Environment.* SSSA Special Publication. Madison, Wisconsin, USA: American Society of Agronomy (in press).

Levin, S.A. 1991. The problem of relevant detail. In Busenberg, S. and Martelli, M. (eds) *Differential Equations Models in Biology, Epidemiology and Ecology.* Lecture Notes in Biomathematics 92. Berlin, Germany: Springer-Verlag.

Levin, S.A. 1992. The problem of pattern and scale in ecology. *Ecology* 73: 1943-67.

Linden, D.R., Hendrix, P.F., Coleman, D.C. and van Vliet, P.C.J. 1994. Faunal indicators of soil quality. In Doran, J. W., Bezdicek, D.F. and Coleman, D.C. (eds) *Defining Soil Quality for a Sustainable Environment.* SSSA Special Publication. Madison, Wisconsin, USA: American Society of Agronomy (in press).

Lubchenco, J., Olson, A.M., Brubaker, L.B., Carpenter, S.R., Holland, M.M., Hubbel, S.P., Levin, S.A., MacMahon, J.A., Matson, P.A., Melillo, J.M., Mooney, H.A., Peterson, C.H., Pulliam, H.R., Real, L.A., Regal, P.J. and Risser, P.G. 1991. The sustainable biosphere initiative: An ecological research agenda. *Ecology 72: 371-*412.

Lundegårdh, H. 1924. *Der Kreislauf der Kohlensäure in der Natur.* Jena, Germany: G. Fisher.

Mankau, R. 1980. Biological control of nematode pests by natural enemies. *Annual Review of Phytopathology* 18: 415-40.

Margulis, L. and Fester, R. (eds) 1991. *Symbiosis as a Source of Evolutionary Innovation.* Cambridge, Massachusetts, USA: MIT Press.

Martin, J.K. and Kemp, J.R. 1986. The measurement of carbon transfers within the rhizosphere of wheat grown in field plots. *Soil Biology and Biochemistry* 18: 103-07.

Meharg, A.A. and Killham, K. 1988. A comparison of carbon flow from pre-labelled and pulse-labelled plants. *Plant and Soil* 112: 225-31.

Minchin, P.E.H. and McNaughton, G.S. 1984. Exudation of recently fixed carbon by non-sterile roots. *J. Experimental Biology* 35: 74-82.

Mooney, H.A. (Chmn). 1990. *Research Strategies for the US Global Change Research Program.* National Research Council. Washington DC, USA: National Academy Press.

Moore, J.C. and de Ruiter, P.C. 1991. Temporal and spatial heterogeneity of trophic interactions in below-ground food webs. In Crossley, D.A., Coleman, D.C., Hendrix, P.F., Cheng, W., Wright, D.H., Beare, M.H. and Edwards, C.A. (eds) *Modern Techniques in Soil Ecology.* Amsterdam, The Netherlands: Elsevier.

Nannipieri, P., Grego, S. and Ceccanti, B. 1990. Ecological significance of the biological activity in soil. In Bollag, J.-M. and Stotzky, G. (eds) *Soil Biochemistry* (Vol 6). New York, USA: Marcel Dekker.

Neher, D.A., Peck, S.L., Rawlings, J.O. and Campbell, C.L. 1993. Variability of plant parasitic and free-living nematode communities within and between fields and geographic regions of North Carolina. *J. Nematology* (in press).

Newell, K. 1984a. Interaction between two decomposer basidiomycetes and a collembolan under *Sitka* spruce: Distribution, abundance and selective grazing. *Soil Biology and Biochemistry* 16: 227-33.

Newell, K. 1984b. Interaction between two basidiomycetes and Collembola under *Sitka* spruce: Grazing and its potential effects on fungal distribution and litter decomposition. *Soil Biology and Biochemistry* 16: 235-40.

O'Neill, R.V. 1988. Hierarchy theory and global change. In Rosswall, T., Woodmansee, R.G.and Risser, P.G. (eds) *Scales and Global Change.* Chichester, UK: John Wiley.

O'Neill, R.V., DeAngelis, D.L., Waide, J.B. and Allen, T.F.H. 1986. *A Hierarchical Concept of Ecosystems.* Princeton, New Jersey, USA: Princeton University Press.

Oades, J.M. and Waters, A.G. 1991. Aggregate hierarchy in soils. *Australian J. Soil Research* 29: 815-28.

Oades, J.M., Gillman, G.P. and Uehara, G. 1989. Interactions of soil organic matter and variable-charge clays. In Coleman, D.C., Oades, J.M. and Uehara, G. (eds) *Dynamics of Soil Organic Matter in Tropical Ecosystems.* Honolulu, Hawaii, USA: University of Hawaii Press.

Odum, E.P. 1984. The mesocosm. *BioScience* 34: 558-62.

Oechel, W.C., Hastings, S.J., Vourlitis, G., Jenkins, M., Riechers, G. and Grulke, N. 1993. Recent change of Arctic tundra ecosystems from a net carbon dioxide sink to a source. *Nature* 361: 520-23.

Oksanen, L. 1991. Trophic levels and trophic dynamics: A consensus emerging? *Trends in Research in Ecology and Evolution* 6: 58-60.

Paoletti, M.G., Favretto, M.R., Stinner, B.R., Purrington, F.F. and Bater, J.E. 1991. Invertebrates as bioindicators of soil use. *Agriculture, Ecosystems and Environment* 34: 341-62.

Parmelee, R.W., Beare, M.H., Cheng, W., Hendrix, P.F., Rider, S.J., D.A. Crossley, J. and Coleman, D.C. 1990. Earthworms and enchytraeids in conventional and no-tillage agroecosystems: A biocide approach to assess their role in organic matter breakdown. *Biology and Fertility of Soils* 10: 1-10.

Price, P.W. 1988. An overview of organismal interaction in ecosystems in evolutionary and ecological time. *Agriculture, Ecosystems and Environment* 24: 369-77.

Price, P.W. 1991. The web of life: Development over 3.8 billion years of trophic relationships. In Margulis, L. and Fester, R. (eds) *Symbiosis as a Source of Evolutionary Innovation.* Cambridge, Massachusetts, USA: MIT Press.

Price, P.W. 1992. The resource-based organisation of communities. *Biotropica* 24: 273-82.

Raich, J.W. and Schlesinger, W.H. 1992. The global carbon dioxide flux in soil respiration and its relationship to vegetation and climate. *Tellus* 44B: 81-89.

Schimel, D.S. 1993. Population and community processes in the response of terrestrial ecosystems to global change. In Kareiva, P.M., Kingsolver, J.G. and Huey, R.B. (eds) *Biotic Interactions and Global Change.* Sunderland, Massachusetts, USA: Sinauer Associates.

Schlesinger, W.H. and Hartley, A.E. 1992. A global budget for atmospheric NH_3. *Biogeochemistry* 15: 191-211.

Smith, J.L. and Paul, E.A. 1990. The significance of soil microbial biomass estimations. In Bollag, J.-M. and Stotzky, G. (eds) *Soil Biochemistry* (Vol. 6). New York, USA: Marcel Dekker.

Smith, T.M. and Shugart, H.H. 1993. The transient response of terrestrial carbon storage to a perturbed climate. *Nature* 361: 523-26.

Sparling, G.P. 1992. Ratio of microbial biomass carbon to soil organic carbon as a sensitive indicator of changes in soil organic matter. *Australian J. Soil Research* 30: 195-207.

Stork, N.E. and Eggleton, P. 1992. Invertebrates as determinants and indicators of soil quality. *American J. Alternative Agriculture* 7: 38-47.

Swift, M.J., Heal, O.W. and Anderson, J.M. 1979. *Decomposition in Terrestrial Ecosystems.* Berkeley, California, USA: University of California Press.

Townsend, A.R., Vitousek, P.M. and Holland, E.A. 1992. Tropical soils could dominate the short-term carbon cycle feedbacks to increased global temperatures. *Climatic Change* 22: 293-303.

Tunlid, A. and White, D.C. 1992. Biochemical analysis of biomass, community structure, nutritional status, and metabolic activity of microbial communities in soil. In Stotzky, G. and Bollag, J.-M. (eds) *Soil Biochemistry* (Vol. 7). New York, USA: Marcel Dekker.

Vannier, G. 1987. The porosphere as an ecological medium emphasised in Professor Ghilarov's work on soil animal adaptations. *Biology and Fertility of Soils* 3: 39-44.

Wang, G.M., Coleman, D.C., Freckman, D.W., Dyer, M.I., McNaughton, S.J., Acra, M.A. and Goeschl, J.D. 1989. Carbon partitioning patterns of mycorrhizal versus non-mycorrhizal plants. Real-time dynamic measurements using $^{11}CO_2$. *New Phytologist* 112: 489-93.

Wessman, C.A. 1992. Spatial scales and global change: Bridging the gap from plots to GCM grid cells. *Annual Review of Ecology and Systematics* 23: 175-200.

Wolf, G. 1964. *Isotopes in Biology.* New York, USA: Academic Press.

Wolters, V. 1991. Soil invertebrates — effects on nutrient turnover and soil structure: A review. *Zeitschrift für Pflanzenernährung und Bodenkunde* 154: 389-402.

Zlotin, R.I. 1985. Role of soil fauna in the formation of soil properties. *Soil Ecology and Management* 12: 39-47.

Beyond the Biomass
Edited by K. Ritz, J. Dighton and K.E. Giller
© 1994 British Society of Soil Science (BSSS)
A Wiley-Sayce Publication

CHAPTER 22

The microbial loop in soil

M. CLARHOLM

The importance of protozoan feeding on bacteria in nutrient regeneration has been acknowledged in the marine environment since the mid-1960s (Johannes, 1965; Pomeroy, 1970). Later, Williams (1981) demonstrated that, in contrast to earlier beliefs, most of the carbon fixed by minute green algae in the open waters was not channelled along the grazer food chain via zooplankton to fish, but rapidly recycled to CO_2 by bacteria and bacteria-feeding protozoa. This fast turnover of carbon and nutrients through bacteria and protozoa was later termed 'the microbial loop' (Azam et al., 1983).

In soil, the microbial loop has been described as 'the fast flora' around roots (Coleman, 1976; Coleman et al., 1988). Bacteria, utilising root-derived carbon, bind inorganic nutrients transported towards the roots by mass flow and diffusion into bacterial organic matter. The nutrients are thus transported to the root surface, but are inaccessible for root uptake. Grazing by protozoa, mainly naked amoebae, will release about one third of the bacterial nitrogen as ammonium available for uptake by roots. One third is incorporated into the protozoan biomass, and one third is egested by the protozoa in organic forms as parts of cell walls and organelles (Fenchel, 1982a).

This chapter examines the conditions for bacterial consumption prevailing for different groups of protozoa and the importance of the microbial loop as a nutrient regenerator for primary producers in water, in soil under forest and in cultivated soil. It argues that naked amoebae are the most important bacterial grazers in soil and that the importance of the microbial loop in nutrient regeneration will decrease in favour of fungal activities in the order: water > cultivated soil > soil under permanent natural vegetation.

BACTERIAL-PROTOZOAN MIGRATION FROM WATER TO LAND

Unicellular, eukaryotic protozoa were the first bacterial feeders to develop on Earth (Fenchel and Blackburn, 1979). Protozoa accelerated nutrient turnover in the aquatic environment, and this increased nutrient availabity made possible the evolution of larger, multicellular organisms. The bacterial-protozoan complex developed in water and migrated onto land. Both bacteria and protozoa can survive dry conditions in resting states, but they need water to be active. On land, water availability is often suboptimal for biological processes. To cope with this situation, water transport systems such as the vascular system in plants, fungal hyphae and filamentous bacteria evolved. In unfertilised soil with a permanent vegetation cover, levels of inorganic nutrients are always low, even including periods of high plant uptake. Together with the mycorrhizae, saprophytic fungi close the circulation of nutrients. This leads to a tighter transfer of nutrients within undisturbed terrestrial ecosystems than in cultivated soils and water, where, for example, nitrogen is periodically lost through leaching and denitrification. There are even situations reported where ions were not part of the soil solution, but moved directly from fallen leaves (Herrera et al., 1978) and dying roots (Ritz and Newman, 1985) to living plants, via the mycorrhizae.

Plants on land need more stabilising tissues than plants in water. The supporting tissues are made of lignin, cellulose and other substances, which are more difficult to decompose than the algal cell walls made by polysaccharide polymers. Consequently, saprophytic fungi with enzymes able to release nutrients and carbon from more complex organic substances dominate decomposition in soil, whereas they have a low impact in the aquatic environment.

BACTERIA IN SOIL

Although many common soil bacteria are motile, most bacterial cells in soil are located at, and grow on, surfaces (Clark, 1967; Lynch, 1988). More bacteria are associated with organic surfaces than with surfaces of mineral particles (Marshall, 1976; Lynch, 1988). Most bacterial taxa common in soil have rod-shaped cells as a species characteristic and only a small fraction consists of spherical true cocci (Clark, 1967). However, rod-shaped cells do not make up a large part of the total bacterial community, and their existence is transient. The rods will continue to divide until their food sources are exhausted, whereupon they again become small and rounded (Schaechter, 1968). The rod shape is thus an indication that the cells are actively growing. When viewed by direct microscopy, most bacterial cells in soil are small and rounded. Judging from their shape and size, they are thus mainly in a dormant state. Rod-shaped bacteria in soil are found on fresh litter, on root surfaces and around hyphae.

Drying followed by rewetting triggers bacterial growth activities. On the first day after rain, Clarholm and Rosswall (1980) found that 7% of the total bacterial population in the humus layer of a pine forest consisted of long rods; on the second day this had fallen to 4%. The long rods divided and there was an increase in medium-sized rods which made up 7%, 10% and 11% of the total number of bacterial cells on days 3, 4 and 5, respectively. During the same period, the proportion of long rods reverted to the usually recorded 1-2%. This series of observations indicated a short burst of bacterial production after rain. Bacterial production in soil thus seems to be concentrated in both time and space, leaving large spaces and long periods without production (Nedwell and Gray, 1988).

PROTOZOA AS BACTERIAL CONSUMERS

Based on their mode of locomotion, protozoa are commonly divided into ciliates, flagellates and amoebae. The first two groups are free-swimming and catch bacteria with their cilia or flagella, while amoebae crawl slowly on surfaces, engulfing bacteria by surrounding them with their pseudopodia. With free access to bacteria, many species in all three groups have the same growth rates as bacteria, with divisions every 2-4 hours at room temperature, depending upon the species (Clarholm, 1985b). They are therefore the most potent bacterial feeders in soil (Sandon, 1932). Most soil bacteria are found attached to the soil material. Therefore, naked amoebae feeding on surfaces ought to be more favoured as bacterial consumers in soil than flagellates and ciliates, which are adapted to feeding in free water. In both forest humus and arable soil, naked amoebae have been shown to cause a decline of elevated bacterial biomass within 1-2 days as a result of grazing (Clarholm, 1981). Ciliates generally need larger bodies of water than the other groups to be active, but some ciliates are better adapted to life in soil. In soil samples investigated in water in the laboratory, small, hypotrich ciliates can be seen 'kicking' off bacteria from a surface with their leg-like ventral cirri, making it possible to catch them with their cilia. Similarly, members of the genus *Colpoda* can grow in small amounts of water with high bacteria concentrations (for example, along a streak of bacteria on an agar plate; pers. obs.).

If changes in numbers of protozoa are to be used to evaluate the potential grazing pressure, then the capacity of individuals to consume bacteria must be known. Flagellates need only a couple of hundred bacteria to divide, while naked amoebae and ciliates need numbers an order of magnitude higher to multiply (Clarholm, 1985b). Each naked amoeba or ciliate thus represents a grazing pressure at least 10 times greater than that represented by a flagellate. The relative importance of the three groups as bacterial consumers in the aquatic and terrestrial environments can be predicted from their modes of feeding. In open waters, where the bacterial concentration is too low to support multiplication of ciliates, flagellates are the dominant consumers (Fenchel, 1982b). Ciliates dominate in more productive aquatic situations near the sediment surface, where they tend to filter-feed on bacteria, but some also ingest flagellates and small ciliates. Amoebae are scarce because of the absence of solid surfaces on which to crawl. Similarly, in soil, flagellates dominate protozoan activities where bacterial production is low (for example, in soil without plants, and at the start of a flush of bacterial activity after rain). Where there is a sizable bacterial production, naked amoebae are the dominant consumers.

The relationship between naked amoebae and flagellates can be demonstrated with data from the humus layer of a pine forest (*see* Figure 22.1 *overleaf*), where bacteria and flagellates started to increase immediately after rain and peaked after 2-3 days; flagellates then decreased 1 day after the bacteria had started to decrease. The numbers of naked amoebae increased more slowly and peaked at day 4-5 after rain (*see* Figure 22.1). The same relationship between bacteria, flagellates and naked amoebae was reported in a 1-year study in which daily estimates of protozoa and bacteria were made at Rothamsted, UK in a long-term, manured, arable field kept under fallow (Cutler et al., 1922). An inverse relationship between numbers of bacteria and naked amoebae was found in 86% of the daily observations (Cutler et al., 1922). In the same study, flagellates were observed to be dependent upon a bacterial production to sustain themselves, but they did not reduce the numbers of bacteria (Darbyshire et al., 1993).

MICROCOSMS AND THE REAL WORLD

One way used to study the effects of microbial interactions has been to build gradually more complex microcosms by adding organism groups to sterilised soil. Increased amounts of inorganic nitrogen have been reported in soil in microcosms with both bacteria and protozoa, compared with soil containing only bacteria (Coleman et al., 1977; Kuikman and van Veen, 1989). These results are presumably correct for the situation investigated, but how do the created situations relate to the real world? Obvious differences are the absence of fungi and larger animals and the presence of a new pool of organic matter — the necromass of the organisms killed by sterilisation.

Increased nitrogen mineralisation after the addition of flagellates only to sterilised soil reinoculated with bacteria has been reported (Gerhardson and Clarholm, 1986; F. Verhagen, pers. comm.). When enumerating bacteria in reinoculated soil using direct microsocopy (Clarholm, 1985a), it was clear that the reintroduced bacteria had not managed to colonise the soil particles in the same way as the native bacteria had in unsterilised soil. The bacteria were larger and fewer than normally observed in the same

Figure 22.1   Daily estimates of numbers of bacteria, naked amoebae and flagellates in the humus layer of a pine forest after rain

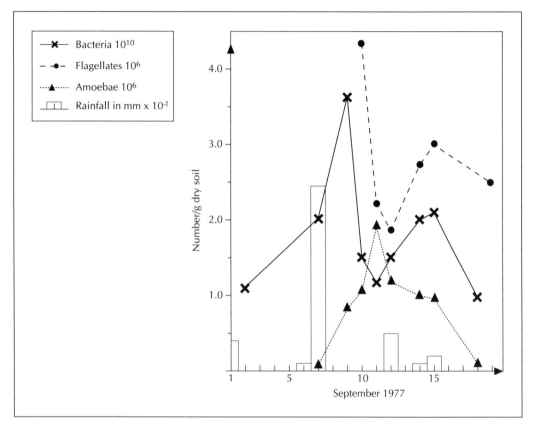

soil when not sterilised. Furthermore, they tended to float about, rather than being attached to surfaces (Clarholm, 1985a). Reinoculation with bacteria thus created an aquatic-like situation in the soil. In this situation, flagellates should be more favoured as bacterial grazers than in unsterilised soil, where most bacteria are attached to surfaces.

Microcosm experiments performed to date have shown that:

- bacteria exceeding the carrying capacity of the soil (van Veen et al., 1985) will be quickly consumed by the grazers most suited to the situation

- nitrogen and other nutrients are released during this process

- a carbon source for bacteria, such as roots, must be present for protozoa to be continously active (Kuikman et al., 1990)

However, an assessment of the importance of these interactions for nutrient regeneration in the field cannot yet be made because too few relevant studies have been conducted. In a barley field, short-term decreases in numbers of bacteria corresponded to 10% and 17% of the plant uptake of nitrogen over the same 2-day periods, if one assumed that one third of the bacterial nitrogen became available for plant uptake (Clarholm, 1989), the other sources being fertiliser and nitrogen released by decomposer fungi. A model of nitrogen circulation in the same treatment indicated that 14% of the harvested nitrogen was made available through faunal activities, mainly by protozoa (Paustian et al., 1990).

THE IMPORTANCE OF THE MICROBIAL LOOP IN WATER AND ON LAND

In the aquatic environment, where fungi are of little importance, bacteria and protozoa play the dominating role in decomposition and nutrient mineralisation (Azam et al., 1983; Wangersky, 1984). The traditional view of the situation in the sea, where nutrients were thought to be regenerated mainly in the sediment, has gradually changed with the recognition of the activities on organic particles and in secondarily formed aggregates floating in the free water. These particles are biological hot spots and it is estimated that only 5% of the organic matter produced will eventually reach the bottom of the sea (Wangersky, 1984). Thus, most of the nutrients are regenerated within the water column via the microbial loop (Azam et al., 1983; Münster and Chróst, 1990)

A natural soil undisturbed by humans contains mycorrhizal roots and heterotrophic decomposer fungi in addition to bacteria and protozoa. The fungi form two types of pipeline systems, which differ in their sources of energy. Mycorrhizae are supplied from above-ground plant parts, while decomposer fungi obtain their energy mainly through the decomposition of dead plant parts of different ages. The energy sources for bacteria in soil are less clear. Their physiology indicates a dependence upon low-molecular-weight substrates. The most obvious sources are fresh above-ground litter and root-derived carbon. A third source can be envisaged around decomposer fungi when enzymes are secreted from the hyphae to act on the organic matter. The resulting smaller molecules may be utilised by both the decomposer fungi and bacteria.

SPATIAL CONSTRAINTS

Fungal hyphae can grow past air pockets and nutrients are transported within hyphae to support fungal growth on suitable carbon sources. The possibility of nitrogen translocation to suitable carbon sources is an important factor in coniferous forests, where the C/N ratios of the fresh organic matter are much wider than in water and arable soil and where there are no burrowing animals mixing the substrates. The importance of upward translocation of nutrients by fungi was demonstrated in two coniferous forest ecosystems where the transport of nitrogen from the mineral soil to the pine needle litter on the forest floor was estimated to be 9 kg N/ha (Hart and Firestone, 1991) and 4 kg kg N/ha (Fahey et al., 1985), representing about half the plant uptake.

Bacteria are attached to surfaces and caught in isolated bodies of water, the sizes of which are determined by the structure of the soil (Foster, 1988). The heterogeneous distribution of organic matter and available nutrients in soil will cause problems for bacteria in terms of the supply of suitable energy, resulting in long periods of inactivity.

THE RELATIONSHIP BETWEEN THE MICROBIAL LOOP AND FUNGI

Few parts of the fine-root system of coniferous plants are non-mycorrhizal. However, little information on mycorrhiza-derived carbon supporting bacterial growth is available (Ingham and Molina, 1991).

Significantly increased numbers of naked amoebae and ciliates within mats of the ectomycorrhizal basidiomycete, *Hysterangium setchellii* (Cromack et al., 1988), indicated the presence of an available carbon source for bacteria. Mycorrhizal fungi may be better decomposers of organic material than has been assumed (Trojanowski et al., 1984; Dighton et al., 1987) (*see* Figure 22.2a). There is also the

Figure 22.2   Possible interactions of decomposer fungi, mycorrhizal fungi and the 'microbial loop' in nutrient mineralisation from organic matter and uptake by the mycorrhizae

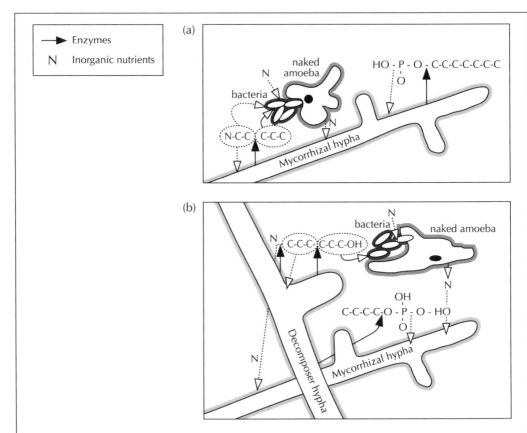

(a)   The mycorrhizal fungus will obtain its carbon from the plant. Its hypha is able to produce catalytic enzymes, such as phosphatases, which will be used to release nutrients to be taken up by the mycorrhiza and forwarded to the plant. Carbon in smaller units than existed before the fungal attack will be used by bacteria for growth and nutrient immobilisation. Protozoa will release part of the immobilised nutrients through grazing.

b)   If the mycorrhizal fungus functions only as a thin root without catalytic enzymes, then it has to rely on the activities of a decomposer fungus to obtain inorganic nutrients. The decomposer fungus produces extracellular enzymes to obtain a carbon source and nutrients for its own growth and metabolic activities. In that process, inorganic nutrients will also be liberated for mycorrhizal uptake. Bacteria will scavenge for carbon substrates around the decomposer fungus and compete with the mycorrhiza for nutrients. Protozoan grazing of bacteria will release nutrients, otherwise locked up in bacteria, for uptake by the mycorrhiza.

possiblity of a four-partner situation whereby saprophytic decomposer fungi are intermingled within the mycorrhizal mat (*see* Figure 22.2b). The decomposer fungus would release enzymes to cut long carbon chains, providing substrate both for itself and for scavenging bacteria, close to the enzyme-producing hypha. At the same time, nutrients released by enzymes from the decomposer hyphae would be taken up by the mycorrhizae but also by the bacteria present. Protozoan grazing on bacteria would again release part of the nitrogen for mycorrhizal uptake. But what might have caused the mycorrhizal mat to develop where it did? Substrate quality of the organic matter is probaby involved, but which organisms released the energy for the bacteria and the nutrients for the mycorrhiza? A root with a mycorrhiza with decomposer abilities might be the most efficient solution for plants in the terrestrial habitat. The bacteria-protozoa complex in soil would then be a relict with regard to the decomposition of organic matter.

In cultivated soils, humans have changed the naturally stable terrestrial situation through frequent mixing of the soil, large inputs and exports of nutrients, and monocultural crop management practices which include long periods during which fertile soils are left with no plant cover and thus no root uptake. In relation to the importance of bacterial-protozoan interactions, this is a situation which lies between the aquatic and the natural, terrestrial situation, since mixing the soil increases the possibilities for bacteria to reach suitable substrates for growth. Decreased mixing of soil in experiments involving minimum tillage has also been reported to increase the proportion of fungi in relation to bacteria (Beare et al., 1992).

Mycorrhizae play an insignificant role in conventionally fertilised agricultural soils. Decomposer fungi are generally more important, but their relative contribution to the microbial biomass seems to vary. In arable soils derived from forests, which are common in, for example, Sweden and Germany, the dominance of fungi is strong in the microbial biomass (Anderson and Domsch, 1973; Schnürer et al., 1985), while bacteria have been reported to dominate the microbial biomass in arable soils in the Netherlands (Bloem et al., 1992). The arable soils in the Netherlands are derived mainly from the sea or lakes, and at least part of the organic matter has an aquatic history. Another difference is the common use of fumigation in the Netherlands to fight pathogens. Nitrogen additions to boreal forest soils are reported to change the size and activity of the microbial biomass; long-lasting reductions in the fungal-dominated microbial biomass, as well as a reduced CO_2 evolution, have been reported in boreal forest soils after nitrogen fertilisation (Söderström et al., 1983; Nohrstedt et al., 1989).

THE IMPORTANCE OF THE MICROBIAL LOOP IN MINERALISATION

In nature, bacteria and protozoa generally occur together and they should be viewed as a unit in relation to nutrient turnover. The degree of importance of the microbial loop in terms of the amount of nutrients mineralised depends upon the amount of bacteria produced. In this respect, the amount of bacterial biomass exceeding carrying capacity, which is immediately turned over, constitutes the interesting part of the bacteria, since the remainder is protected from grazing. It is almost impossible to measure the bacterial production above carrying capacity because it is unevenly distributed and rapidly turned over. If the amount of carbon available for bacteria could be estimated, indirect calculations could be made. It is, however, only the input of above-ground litter that is known in most ecosystems. Too little is known at present about the amounts of root-derived carbon and how the activities of saprophytic fungi contribute to the pool of carbon available for bacteria to determine their absolute contribution to nurient regeneration.

Acknowledgements

The author wishes to thank O. Andrén, J. Bengtsson, J.-E. Nylund and T. Persson for their constructive comments on the manuscript. The work described in this chapter was supported by a grant from the National Board for Industrial and Technical Development, Stockholm, Sweden.

References

Anderson, J.P.E. and Domsch, K.H. 1973. Quantification of bacterial and fungal contributions to soil respiration. *Archiv für Mikrobiologie* 93: 113-27.

Azam, F., Fenchel ,T., Field, J., Gray, J., Meyer-Reil, L. and Thingstad, F. 1983. The ecological role of water-column microbes in the sea. *Marine Ecology Progress Series* 10: 257-63.

Beare, M.H., Parmelee, R.W., Hendrix, P.W., Cheng, W., Coleman, D.C. and Crossley, D.A. 1992. Microbial and faunal interactions and effects on litter nitrogen and decomposition in agroecosystems. *Ecological Monographs* 62: 569-91.

Bloem, J., De Ruiter, P.C., Koopman, G.J., Lebbink, G. and Brussard, L. 1992. Microbial numbers and activity in dried and rewetted arable soil under integrated and conventional management. *Soil Biology and Biochemistry* 24: 655-65.

Clarholm, M. 1981. Protozoan grazing of bacteria in soils: Impact and importance. *Microbial Ecology* 7: 343-50.

Clarholm, M. 1985a. Interactions of bacteria, protozoa and plants leading to mineralisation of soil nitrogen. *Soil Biology and Biochemistry* 17: 181-87.

Clarholm, M. 1985b. Possible role for roots, bacteria, protozoa and fungi in supplying nitrogen to plants. In Fitter, A.H. and Usher, M. (eds) *Ecological Interactions in Soil*. Oxford, UK: Blackwell .

Clarholm, M. 1989. Effects of plant-bacterial-amoeban interactions on plant uptake of nitrogen under field conditions. *Biology and Fertility of Soils* 8: 373-78.

Clarholm, M. and Rosswall T. 1980. Biomass and turnover of bacteria in a forest soil and a peat. *Soil Biology and Biochemistry* 12: 49-57.

Clark, F.E. 1967. Bacteria in soil. In Burges, A. and Raw, F. (eds) *Soil Biology*. London, UK: Academic Press.

Coleman, D.C. 1976. A review of root processes and their influence on soil biota in terrestrial ecosystems. In Anderson, J.M. and Macfayden, A. (eds) *The Role of Terrestrial and Aquatic Organisms in Decomposition Processes*. Oxford, UK: Blackwell.

Coleman, D.C., Cole, C.V., Anderson, R.V., Blaha, M., Campion, M.K., Clarholm, M., Elliott, E.T., Hunt, H.W., Shaefer, B. and Sinclair, J. 1977. An analysis of rhizosphere-saprophytic interactions in terrestrial ecosystems. *Ecological Bulletins (Stockholm)* 25: 299-309.

Coleman, D.C., Crossley, D.A.J., Beare, M.H. and Hendrix, P.F. 1988. Interactions of organisms at root/soil and litter/soil interfaces in terrestrial systems. In Edwards, C.A., Stinner, B.R., Stinner, D. and Rabatin, S. (eds) *Biological Interactions in Soil*. Amsterdam, Netherlands: Elsevier.

Cromack, K.J., Fichter, B.L., Moldenke, A.M., Entry, J.A. and Ingham, E.R. 1988. Interactions between soil animals and ectomycorrhizal fungal mats. *Agriculture, Ecosystems and Environment* 24: 161-68.

Cutler, D.W., Crump, L.M. and Sandon, H. 1922. A quantitative investigation of the bacterial and protozoan population of the soil, with an account of the protozoan fauna. *Philosophical Transactions of the Royal Society of London (Series B)* 211: 317-50.

Darbyshire, J.F., Zwart, K.B.and Elson, D.A. 1993. Growth and nitrogeneous excretion of a common soil flagellate, *Cercomonas* sp. *Soil Biology and Biochemistry* (in press).

Dighton, J., Thomas, E.D. and Latter, P.M. 1987. Interactions between tree roots, mycorrhizas, a saprophytic fungus and the decomposition of organic substrates in a microcosm. *Biology and Fertility of Soils* 4: 145-50.

Fahey, T.J., Yavitt, J.B., Pearson, J.A. and Knight, D.H. 1985. The nitrogen cycle in lodgepole pine forests, southeastern Wyoming. *Biogeochemistry* 1: 257-75.

Fenchel, T. 1982a. Ecology of heterotrophic microflagellates. II. Bioenergetics and growth. *Marine Ecology Progress Series* 8: 225-31.

Fenchel, T. 1982b. Ecology of hetertrophic microflagellates. IV. Quantitative occurrence and importance as bacterial consumers. *Marine Ecology Progress Series* 9: 35-42.

Fenchel, T. and Blackburn, T.H. 1979. *Bacteria and Mineral Cycling.* London, UK: Academic Press.

Foster, R.C. 1988. Microenvironments of soil microorganisms. *Biology and Fertility of Soils* 6: 189-203.

Gerhardson, B. and Clarholm, M. 1986. Microbial communities on plant roots. In Jensen, V., Kjøller, A. and Sørensen, L.H. (eds) *Microbial Communities in Soil.* Amsterdam, Netherlands: Elsevier.

Hart, S. and Firestone, M.K. 1991. Forest floor-mineral soil interactions in the internal nitrogen cycle of an old-growth forest. *Biogeochemistry* 12: 103-27.

Herrera, R., Merida, T., Stark N. and Jordan C. 1978. Direct phosphorus transfer from leaf litter to roots. *Naturwissenschafter* 65: 208-09.

Ingham, E.R. and Molina, R. 1991. Interactions among mycorrhizal fungi, rhizosphere organisms, and plants. In Barbosa, P., Krischik, V. and Jones, C.G. (eds) *Microbial Mediation of Plant-Herbivore Interactions.* Chichester, UK: John Wiley

Johannes, R.E. 1965. Influence of marine protozoa on nutrient regeneration. *Limnology Oceanography* 10: 434-42.

Kuikman, P.J., van Elsas, J.D., Jansen , A.G., Burgers, S.L.G.E. and van Veen, J.A. 1990. Population dynamics and activity of bacteria and protozoa in relation to their spatial distribution in soil. *Soil Biology and Biochemistry* 22: 1063-73.

Kuikman, P.J. and van Veen, J.A. 1989. The impact of protozoa on the availability of bacterial nitrogen to plants. *Biology and Fertility of Soils* 8: 13-18.

Lynch, J.M. 1988. The terrestrial environment. In Lynch, J. and Hobbie, J. (eds) *Microorganisms in Action: Concepts and Applications in Microbial Ecology.* Oxford, UK: Blackwell .

Marshall, K.C. 1976. *Interfaces in Microbial Ecology.* Cambridge, Massachusetts, USA: Harvard University Press.

Münster, O. and Chróst, R.J 1990. Origin, composition, and microbial utilisation of dissolved organic matter. In Overbeck, J. and Chróst, R.J. (eds) *Aquatic Microbial Ecology: Biochemical and Molecular Approaches.* New York, USA: Brock/Springer.

Nedwell, F.B. and Gray, T.R.G. 1988. Soils and sediments as matrices for microbial growth. In Fletcher, M., Gray, T.R.G. and Jones, J.G. (eds) *Ecology of Microbial Communities.* Cambridge, UK: Cambridge University Press.

Nohrstedt, H.-O., Arnebrant, K., Bååth, E. and Söderström, B. 1989. Changes in carbon content, respiration rate, ATP content and microbial biomass in nitrogen-fertilised pine forest soils in Sweden. *Canadian J. Forestry Research* 19: 323-28.

Paustian, K., Andrén, O., Clarholm, M., Hansson, A.-C., Lagerlöf, J., Lindberg, T., Pettersson, R. and Sohlenius, B. 1990. Carbon and nitrogen budgets of four agro-ecosystems with annual and perennial crops, with and without N fertilisation. *J. Applied Ecology* 27: 60-84.

Pomeroy, L.R. 1970. The strategy of mineral cycling. *Annual Review of Ecological Systems* 1: 171-90.

Ritz, K. and Newman, E. 1985. Evidence for rapid cycling of phosphorus from dying roots to live plants. *Oikos* 45: 174-80.

Sandon, H. 1932. *The Food of Protozoa.* Cairo, Egypt: MIRS-Sokkar Press.

Schaechter, M. 1968. Growth: Cells and populations. In Mandelstam, J. and McQuillen, K. (eds) *Biochemistry of Bacterial Growth.* Oxford, UK: Blackwell.

Schnürer, J,. Clarholm, M. and Rosswall, T. 1985. Microbial biomass and activity in an agricultural soil with different organic matter contents. *Soil Biology and Biochemistry* 17: 611-18.

Söderström, B., Bååth, E. and Lundgren, B. 1983. Decrease in soil microbial activity and biomasses owing to nitrogen amendments. *Canadian J. Microbiology* 29: 1500-06.

Trojanowski, J,. Haider, K. and Hutterman, A. 1984. Decompostion of ^{14}C-labelled lignin, holocellulose and lignocellulose by mycorrhizal fungi. *Archiv für Mikrobiologie* 139: 202-06.

van Veen, J., Ladd J. and Amato, M. 1985. Turnover of carbon and nitrogen through the microbial biomass in a sandy loam and a clay soil incubated with [^{14}C(U)]glucose and [^{15}N](NH_4)$_2SO_4$ under different moisture regimes. *Soil Biology and Biochemistry* 17: 747-56.

Wangersky, P.J .1984. Organic particles and bacteria in the ocean. In Hobbie, J.E and Williams, P.J. Le B. (eds) *Hetertrophic Activity in the Sea*. New York, USA: Plenum Press

Williams, P.J . Le B. 1981. Incorporation of microheterotrophic processes into the classic paradigm of the planctonic food web. *Kieler Meerforschung, Sonderheft* 5: 1-28.

Beyond the Biomass
Edited by K. Ritz, J. Dighton and K.E. Giller
© 1994 British Society of Soil Science (BSSS)
A Wiley-Sayce Publication

CHAPTER 23

Holistic approaches to the study of populations, nutrient pools and fluxes: Limits and future research needs

P. NANNIPIERI, L. BADALUCCO and L. LANDI

Holism is a theory which views reality as being made up of 'wholes' that are more important in terms of functioning than the simple sum of their parts. This theory has been extended to various disciplines; in soil science, the holistic approach has been used to investigate nutrient transformations (Stevenson, 1986). Using the techniques which are currently available, it is impossible to monitor each inorganic and organic compound and to quantify the rates of each abiotic and biotic reaction in a complex system such as soil. Thus, the system has been partitioned into pools which have a functional meaning; fluxes among pools represent physical processes (for example, leaching and volatilisation) or abiotic and/or biotic transformations. This has allowed a better mathematical description of the system (Stevenson, 1986).

The major limitations to the discrete pool approach is that the system is likely to vary continously and that the models based on pools will never completely reflect biological, chemical or physical reality. Current techniques allow some pools and fluxes to be evaluated with a reasonable degree of accuracy, whereas this is not the case with other pools and fluxes described by mathematical models. Further insights in nutrient cycling require: a reconciliation of differences among conceptual, mathematical and operational definitions of pools; a more detailed description of systems involving a larger number of pools and fluxes than those currently determined; and a direct estimation of functional pools and fluxes identified in conceptual models.

Nutrient cycling is rather complex from a biological, chemical and physical perspective. In view of the limited space, it will not be possible to discuss here the major research needs of nutrient cycling. The discussion focuses on nutrient mineralisation and immobilisation processes, whose complexity makes them an excellent testing ground for assessing new methods and their application to mathematical modelling.

Particular attention is given to conceptual and experimental procedures for determining new abiotic and biotic pools and respective fluxes with a functional meaning. The current model describing ition and immobilisation of nutrients in soil organic matter is based on the following basic

* microbial biomass is a significant source and sink of nutrients

* microbial decomposition of organic matter and microbial synthesis occur simultaneously in soil

* nutrients in the organic matter pool are heterogeneous in terms of biological activity (a fraction cycles very rapidly but some components cycle very slowly)

DEFINITION AND DETERMINATION OF NEW POOLS AND FLUXES IN MINERALISATION AND IMMOBILISATION PROCESSES

Microbial pools

Microbial biomass has been considered as an undifferentiated whole in studies on mineral cycling and energy flow in soil. This approach has been made possible by the availability of the fumigation technique which allows a direct and accurate estimation of relative carbon, nitrogen, phosphorus and sulphur content (Jenkinson and Ladd, 1981; Jenkinson, 1988). Other methods based on the extraction of specific components of living microbial cells, such as the adenosine triphosphate (ATP) method, or on the physiological response of microrganisms to an exogenous substrate, such as the substrate-induced respiration (SIR) method, are as easy to use and not as time-consuming as the $CHCl_3$ fumigation-extraction or fumigation-incubation methods. However, these methods permit only the indirect determination of the nutrient content of microbial biomass (Anderson and Domsch, 1978; Jenkinson, 1988). Thus, the risk of error is higher in indirect than direct methods. New methods for characterising microbial subpools should permit the direct determination of the nutrient content.

The use of molecular biology techniques, such as DNA extraction, purification, cleaving by restriction endonucleases and probing using DNA/DNA hybridisation techniques, has the potential to provide a better definition of the composition and ecology of soil microflora. However, its contribution to a better characterisation of nutrient cycling may be irrelevant. The number of a microbial species is not related to the activity of this species. Many factors, all resulting from the peculiarity of the soil environment, are responsible for the lack of a relationship between microbial numbers and activity (Nannipieri et al., 1990). The extraction and purification procedures in soil molecular biology techniques could be developed for estimating the nutrient content of microbial subpools with a functional role in nutrient cycling. Molecular probing using DNA/DNA hybridisation requires extraction of DNA from soil. This is usually carried out in two ways: microorganisms may be lysed directly in soil prior to extraction; or microbial cells may be separated from soil and lysed to recover DNA (Sayler et al., 1992; Trevors, 1992). The latter technique could be developed and used to obtain discrete microbial subpools such as bacterial, fungal or protozoan fractions with a functional role in nutrient cycling. The separation of cells from soil allows the determination of the nutrient content if the extraction and purification procedures are sufficiently effective. Obviously, the technique should be easy to apply and not time-consuming.

Another possible approach for partitioning microbial biomass in subpools is the discrimination between fungi and bacteria by specific inhibitors. Commonly used antibiotics, such as the bactericide streptomycin and the fungicide cycloheximide, specifically inhibit protein synthesis acting on bacterial (70s) and fungal (80s) ribosomes, respectively (Lukens, 1971; Kucers and Bennet, 1979;

Landi et al., 1993). The selective inhibition of the respiratory responses of bacteria and fungi to glucose has been used to determine the bacteria/fungi ratio in soil (Anderson and Domsch, 1973). No conclusive results were obtained when both inhibitors were used singly or in combination to separate the contribution of fungi and bactera to soil respiration and net nitrogen mineralisation rates (Landi et al., 1993). The efficacy of an inhibitor in soil depends upon biotic and abiotic processes (*see* Table 23.1). Interactions between the antibiotic molecule and inorganic colloids or its association with biologically resistant humic substrates may render the inhibitor inaccessible to either decomposer cells or target microorganisms. An antibiotic may be inactivated or used as a carbon and nitrogen source by microorganisms in the soil. Usually, long periods of biocide residence in soil result in microbial community changes and allow indirect or non-target effects to occur (Ingham and Coleman, 1984).

Table 23.1   Some problems in the use of specific microbial inhibitors, such as cycloheximide and streptomycin, to discriminate between the contribution of fungi and bacteria in the soil

1.   Target and non-target effects on soil microflora
2.   The sensitive/insensitive ratio component of soil microflora may vary according to soil type
3.   Microbial inactivitation of the inhibitor
4.   Use of the inhibitor as a source of carbon and nitrogen by soil microorganisms
5.   Adsorption of the inhibitor by soil colloids

The determination of specific biomarkers may be used to discriminate narrower classes of soil microorganisms; for example, ergosterol exists only in fungi, muramic acid has been found in bacteria, and lipopolysaccharides (LPS) are specific for Gram-negative bacteria (Tunlid and White, 1992).

It has been hypothesised that a better characterisation of microorganisms in nutrient cycling may be obtained by discriminating between the active and inactive pools. Turnover times estimated by modellers for microbial biomass range from 1 to 3 years, while values for non-living organic pools ranged from 25 to thousands of years (Parton et al., 1987). In short-term studies, the partitioning of microbial biomass into subpools characterised by different turnover times may make an important contribution to a better knowledge of nutrient cycling. In long-term studies, however, this partitioning may be less feasible and rewarding than the subdivision of the non-living organic pool.

In essence, the partioning of microbial biomass into subpools is required if we are to improve our understanding of nutrient cycling in soil. The use of antibiotics does not seem to solve the problem, especially with long periods of biocide residence in soil. The approach based on the extraction of microbial groups from soil, followed by the determination of the relative nutrient content, seems promising but an appropriate technique has to be devised.

Non-living organic pools

Current models describe soil organic matter quality and nutrient availability in terms of discrete organic matter fractions, or pools, which vary in decomposition activity. Conceptually useful models should include, in addition to microbial biomass, at least pools for non-living but available or active nutrients and stabilised or recalcitrant forms of nutrients. Indeed, isotopic techniques have shown that a fraction of soil organic matter and immobilised nutrients can be remineralised very rapidly (Duxbury et al., 1989), while ^{14}C dating has shown that some soil organic carbon, and presumably also the nutrients associated with it, have a turnover time of 1000 years or more (Campbell et al., 1967). Paul and Juma (1981) fractionated organic nitrogen into microbial biomass (24-week relative half-life), active non-microbial biomass (77-week half-life), stabilised organic matter (27-year half-life) and old organic matter (600-year half-life). Biomass nitrogen was determined directly by the fumigation-incubation method. Total active nitrogen was determined from total nitrogen and ^{15}N mineralised during laboratory incubation, on the assumption that recently immobilised ^{15}N had uniformily mixed with the active nitrogen pool but entered no other fractions. An old carbon fraction was estimated from ^{14}C dating and a C/N ratio for this fraction was assumed; the stabilised nitrogen was that which remained. Although several assumptions on which these calculations were based are questionable, this remains one of the few experimental determinations of conceptually meaningful organic nitrogen fractions.

Generally, incubations under controlled physical conditions are used to estimate active pools of any nutrient; the size of these fractions does not define an absolute amount of a specific class molecule but instead provides a relative index of biological nutrient availability under a particular set of conditions. The aerobic nitrogen mineralisation procedure described by Stanford and Smith (1972) is the most commonly used measurement of an active nutrient fraction.

Organic matter mineralisation can be studied by considering the elemental bond classes of nutrients; five classes of chemical bonds (C-C, N-C, S-C, S-O-C and P-O-C) were distinguished by Hunt et al. (1986). The phosphorus bonded to carbon was neglected because it was much smaller than the phosphorus linked to oxygen. The sulphur bonded to carbon was considered relatively inert in terms of the microbial attack. Nitrogen was considered to be stabilised due to the bonding with carbon and probably mineralised by organisms oxidising carbon to provide energy; in contrast, sulphur and phosphorus atoms present as esters were considered to be mineralised by the action of extracellular hydrolases according to the need of the element (McGill and Cole, 1981). In the simulation model, based on the concepts developed by Hunt et al. (1986), the activity of sulphate in the soil solution was simulated fairly well, considering the action of sulphohydrolases and the repressive effect of high soil solution sulphate concentration on enyzme production.

Modellers use different terms to name the discrete pools of organic matter. For example, Duxbury et al. (1989) proposed three pools of soil organic matter: labile organic matter, colloidally and chemically protected organic matter, and physically protected organic matter (see Table 23.2). The terminology is not well defined because the physical and chemical significance of the various organic fractions, as well as their functioning, is rather vague. It is logical that physico-chemical interactions and protection, such as chemical structures, would also be continuosly, rather than discretely, variable. The very old organic fraction may be protected by chemical interactions with mineral colloids; some authors call this 'chemical protection', while others see it as 'physical protection'. Generally, physical protection is considered for the fraction of intermediate age rather than for the very old fraction. The conceptual or experimental basis for this distinction is not clear. The chemical structure of organic molecules, in itself, is insufficient to account for the extreme variation in age and turnover time. Although humic molecules are more resistant to microbial degradation than biopolymers, their intrinsic chemical recalcitrance is much smaller than the observed stabilisation in soil (Stevenson et al., 1989). The half-life of 1000-year-old humic fractions is in the order of weeks when extracted and fractionated fractions are added to unextracted soil.

A better understanding of the stability of soil organic matter to mineralisation will be obtained by improving the knowledge of the chemical structure and the synthesis of extracellular organic nutrient molecules in soil. Current information on organic nitrogen in soil comes from the acidic hydrolysing

Table 23.2   Pools of soil organic matter and nutrients under temperate conditions

Pool	Turnover time (years)	Pool size controls
Microbial biomass	2.5	Substrate availability
Labile organic matter	20.0	Residue inputs, climate
Colloidally and chemically protected organic matter	1000.0	Soil mineralogy, texture
Physically protected organic matter distribution	Depends upon physical distrubances	Tillage and aggregate disruption, soil particle size

Source: Adapted from Duxbury et al. (1989)

procedure which gives amino acids and amino sugars as the main identifiable organic nitrogen compounds in the soil hydrolysate (Stevenson, 1982). The soluble nitrogen not accounted for as NH_3 or known compounds is the hydrolysable unknown fraction (10-20% of soil nitrogen); 20-35% of the total is not solubilised by acid hydrolysis and is termed 'acid insoluble nitrogen', the chemical structure of which is unknown. It is almost impossible to formulate plausible hypotheses on the presence and experimental procedures for determining organic nitrogen pools if the chemical structure of most of the organic nitrogen in soil is not known. Such knowledge is important but not conclusive for understanding the stability of organic nitrogen to mineralisation; we need to understand how the adsorption of organic nitrogen molecules by soil colloids affects the degradation of organic nitrogen.

Mineralisation and immobilisation processes

Isotope dilution techniques have been used to estimate the gross rates of mineralisation and immobilisation processes. This has involved, for example, adding $^{15}NH_4^+$ and observing the rate at which the atom % ^{15}N enrichment of the NH_4^+ pool declines as a result of mineralisation of organic nitrogen. Consumption of NH_4^+ does not change the ^{15}N enrichment of the pool; thus, it is possible to calculate the gross mineralisation rate from the rate of dilution of pool enrichments. On the other hand, gross rates of NH_4^+ consumption (microbial assimilation and any other consumptive process) are calculated from the disappearance of the ^{15}N label. According to Kirkham and Bartholomew (1954), these calculations can be carried out if three assumptions are valid: microorganisms do not discriminate between ^{14}N and ^{15}N; considered rates of nitrogen processess remain constant during the incubation period; and ^{15}N assimilated during the incubation period is not remineralised. It is well established that, in some processes, discrimination between ^{14}N and ^{15}N occurs (Delwiche and Steyn, 1970) but it is probably not important for transformations of enriched samples over a few days. The error caused by assuming rate is overshadowed by the experimental error with the exclusion of some situations such as when dry soil is re-wetted for incubation or when the soil has been fumigated prior to incubation (Bjarnason, 1988). Remineralisation of immobilised ^{15}N occurs after 1 week (Bristow et al., 1987; Bjarnason, 1988) and thus both the first and third assumptions are valid within the first few days of incubation.

In the holistic approach, mineralisation and immobilisation are considered as single-step processes. Taking nitrogen mineralisation as an example, many metabolic steps, each catalysed by a specific enzyme, are involved in the conversion of organic nitrogen to NH_4^+-N; in the case of proteins, proteases and peptidases are involved in the release of amino acids, and eventually amino acid dehydrogenases and amino acid oxidases catalyse the conversion of amino acids to ammonia (Ladd and Jackson, 1982; Nannipieri et al., 1990). An interesting task would be to determine the critical rate-limiting step of the entire reaction sequence; the kinetics of the overall process could then be defined by those of the rate-limiting reaction (Nannipieri et al., 1990). No marked differences may exist between pure culture and soil in terms of the rate of intracellular metabolic sequences. The rates of extracellular reactions (for example, proteases and peptidases) may be markedly different in soil compared with those of the same reactions in solution; a substrate, rapidly transformed in solution by an enzyme, may be adsorbed in soil by colloids, becoming less accessible to the enzyme. Thus, the reaction rate is retarded and the metabolic step may become the critical rate-limiting step of the overall reaction sequence in soil.

The use of a specific enzyme inhibitor could be used to stop biotic reactions in soil. In this way, it may be possible to block the nitrogen immobilisation so as to permit the determination of the gross nitrogen mineralisation rates. In most bacteria, cellular nitrogen for the synthesis of nitrogen molecules is derived from the amido glutamine group, the amino glutamate group or the amido asparagine group (Reitzer and Magasanik, 1987). When the ammonium ion concentration of the growth medium is greater than 1 mM, ammonia is incorporated directly into glutamate, glutamine and asparagine. When ammonium ion concentration is lower than 0.1 mM, ammonia is incorporated into glutamine only according to the reaction catalysed by glutamine synthetase (GS):

$$NH_3 + glutamate + ATP \text{ ------> } glutamine + ADP + Pi$$
$$Mg^{2+}$$

This occurs because the K_m of glutamine synthetase is lower than the K_m of glutamate dehydrogenase (GHD) which catalyses the reductive amination of 2-oxoglutarate with the formation of glutamate according to the following reaction:

$$NH_3 + 2\text{-ketoglutarate} + NADPH \text{ ------> } glutamate + NADP^+$$

Methionine sulphoximine (MSX), a specific inhibitor of GS activity, reduced the ammonia incorporation by 24% and 13% in the O2 and O3 subhorizons, respectively, of a forest soil (Schimel and Firestone, 1989); about 20% of ammonia incorporation occurred through abiotic processes probably involving reactions of NH_3 with activated phenols or quinones. The incorporation was studied in soil slurries, which permit better homogeneous conditions than aerated soils, but these conditions differ from those occurring *in situ*. The inhibitor depressed urease production when 5 g of oven-dried soil were treated with 1.5 ml water containing 1000 μm glucose-carbon and 0 or 250 μm of nitrogen as $(NH_4)_2SO_4$ or KNO_3 (McCarty et al., 1992). We have used the inhibitor in soils incubated at 50% of the water-holding capacity to inhibit the NH_4^+ assimilation and thus separate nitrogen mineralisation from nitrogen immobilisation (Landi et al., 1994). An increase in ammonia concentration occurred in the presence of methionine sulphoximine. Further experiments are needed to ascertain the degree of inhibition of ammonia assimilation by methionine sulphoximine. These experiments should be short term so as to minimise problems related to microbial degradation of the inhibitor.

There is a need to find an accurate, less time-consuming method for determining gross mineralisation and immobilisation rates. The use of specific enzyme inhibitors, blocking one of the two processes, seems promising. However, further research is needed to ascertain the validity of this approach.

CONCLUSION

A more detailed description of nutrient cycling involving a larger number of pools and fluxes than those currently determined is required, particularly for the non-living organic fraction. Analytical methods should be available for determining functional pools and fluxes identified in conceptual models.

There is an urgent need to link chemical and biological studies. Chemists tend to fractionate and study components of soil organic matter without regard to the biological significance of the fractions. Biologists and ecologists have developed concepts and models without fully considering the chemical and physical aspects of pools of different biological stability. It is also necessary to relate studies on mechanisms and processes at the microenvironment level to the possible implications in nutrient cycling, and to improve our knowledge of the chemical nature of associations among organic molecules and between organic and inorganic constituents. However, these studies should not be limited to the chemical aspects; they must also take into account the effects of different types of associations on the stability of soil organic matter components during biological decomposition. Eventually, studies need to be done to determine how the overall architecture of soil affects decomposition processes.

In this review we have emphasised the role of microorganisms as a source and a sink of nutrients. However, we do not want to neglect the importance of acquiring a better knowledge of microbial physiology *in situ*. For example, in order to understand the effect of climatic conditions on nutrient cycling we need to know more about how microorganisms respond physiologically to changes in temperature and moisture. Although there is an extensive bibliography on single-factor response studies (for example, the effect of temperature or of soil moisture on various processes), there is still confusion as to how to integrate these studies in order to predict their quantitative effects on nutrient cycling (Duxbury et al., 1989). Microbial uptake systems are well characterised in pure culture but their significance in natural systems has rarely been considered. In the case of nitrogen, it is generally accepted that microorganisms prefer NH_4^+ and that the decay of organic nitrogen always results in nitrogen mineralisation and the release of NH_4^+ into the inorganic pool. It seems that substrate-specific populations incorporate low-molecular-weight organic nitrogen compounds released from the breakdown of organic matter, and directly assimilate the amino groups (Hadas et al., 1992). The quantitative contribution of this process to nitrogen immobilisation in soil needs to be ascertained.

References

Anderson, J.P.E. and Domsch, K.H. 1973. Quantification of bacterial and fungal contribution to soil respiration. *Archives of Microbiology* 93: 113-17.

Anderson, J.P.E. and Domsch, K.H. 1978. A physiological method for quantitative measurements of microbial biomass in soils. *Soil Biology and Biochemistry* 10: 215-21.

Bjarnason, S. 1988. Calculations of gross nitrogen immobilisation and mineralisation in soil. *J. Soil Science* 39: 393-406.

Bristow, A.W., Ryden, J.C. and Whitehead, D.C. 1987. The fate of several time intervals of ^{15}N-labelled ammonium nitrate applied to an established grass sward. *J. Soil Science* 38: 245-54.

Campbell, C.A., Paul, E.A., Rennie, D.A. and McCallum, K.J. 1967. Applicability of the carbon-dating method of analysis to soil humus studies. *Soil Science* 104: 217-24.

Delwiche, C.C. and Steyn, P.L. 1970. Nitrogen isotope fractionation in soils and microbial reactions. *Environmental Science Technology* 4: 929-35.

Duxbury, J.M., Scott Smith, M., Doran, J.W., Jordan, C., Szott, L. and Vance, E. 1989. Soil organic matter as a source and a sink of plant nutrients. In Coleman, D.J., Oades, J.M. and Uehara, G. (eds) *Dynamics of Soil Organic Matter in Tropical Ecosystems*. Honolulu, Hawaii, USA: NifTAL Project.

Hadas, A., Safer, M., Molina, J.A.E., Barak, P. and Clapp, C.E. 1992. Assimilation of nitrogen by soil microbial population: NH_4^+ versus organic N. *Soil Biology and Biochemistry* 24: 137-43.

Hunt, H.W., Stewart, J.W.B. and Cole, C.V. 1986. Concepts of sulfur, carbon and nitrogen transformations in soil: Evaluation of simulating modeling. *Biogeochemistry* 2: 163-77.

Ingham, E.R. and Coleman, D.C. 1984. Effects of streptomycin, cycloheximide, fungizone, captan, carbofuran, cygon and PCNB on soil microbes populations and nutrient cycling. *Microbial Ecology* 10: 345-57.

Jenkinson, D.S. 1988. The determination of microbial biomass carbon and nitrogen in soil. In Wilson, J.R. (ed) *Advances in Nitrogen Cycling in Agricultural Ecosystems*. Wallingford, UK: CAB International.

Jenkinson, D.S. and Ladd, J.N. 1981. Microbial biomass in soil, measurement and turnover. In Paul, E.A. and Ladd, J.N. (eds) *Soil Biochemistry* (Vol. 5). New York, USA: Marcel Dekker.

Kirkham, D. and Bartholomew, W.V. 1954. Equations for following nutrient transformations in soil, utilising tracer data. *Soil Science Society of American Proc.* 18: 33-34.

Kucers, A. and Bennett, N. M. 1979. *The Use of Antibiotics*. London, UK: Heineman.

Ladd, J.N. and Jackson, R.B. 1982. Biochemistry of ammonification. In Stevenson, F.J. (ed) *Nitrogen in Agricultural Soils*. Madison, Wisconsin, USA: American Society of Agronomy.

Landi, L., Badalucco, L., and Nannipieri, P. 1994. Effectiveness of antibiotics to distinguish the contribution of fungi and bacteria to net nitrogen mineralisation, nitrification and respiration. *Soil Biology and Biochemistry* (in press).

Lukens, R.J. 1971. *Chemistry of Fungicidal Action*. New York, USA: Springer-Verlag.

McCarty, G.W., Shogren, D.R. and Bremner, J.M. 1992. Regulation of urease production in soil by microbial assimilation of nitrogen. *Biology and Fertility of Soils* 12: 261-64.

McGill, W.B. and Cole, C.V. 1981. Comparative aspects of cycling of organic C, N, S and P through soil organic matter. *Geoderma* 26: 267-86.

Nannipieri, P., Ceccanti, B. and Grego, S. 1990. Ecological significance of the biological activity. In Bollag, J.-M. and Stotzky, G. (eds) *Soil Biochemistry* (Vol. 6). New York, USA: Marcel Dekker.

Parton, W.J., Schimel, D.S., Cole, C.V. and Ojma, D.S. 1987. Analysis of factors controlling soil organic matter levels in Great Plains grasslands. *Soil Science Society of America J.* 51: 1173-79.

Paul, E.A, and Juma, N.G. 1981. Mineralisation and immobilisation of soil nitrogen by microorganisms. In Clark, F.E. and Rosswall, T. (eds) *Terrestrial Nitrogen Cycles. Ecological Bulletin (Stockholm)* 33:179-95.

Reitzer, L.J. and Magasanik, B. 1987. Ammonia, assimilation and the biosynthesis of glutamine, glutamate, aspartate, asparagine, L-alanine, and D-alanine. In Neidhardt, F.C., Ingham, J.H., Low, K.B., Magasanik, B., Schaecher, M. and Umbarger, H.E. (eds) Escherichia coli *and* Salmonella typhimurium: *Cellular and Molecular Biology*. Washington DC, USA: American Society of Microbiology.

Sayler, G.S., Nikbakht, K, Fleming, J.T. and Packard, J. 1992. Applications of molecular techniques to soil biochemistry. In Stotzky, G. and Bollag, J.-M. (ed) *Soil Biochemistry* (Vol. 7). New York, USA: Marcel Dekker.

Schimel, J.P. and Firestone, M.K. 1989. Inorganic N incorporation by coniferous forest floor material. *Soil Biology and Biochemistry* 21: 41-46.

Stanford, G. and Smith, J.J. 1972. Nitrogen mineralisation potentials of soils. *Soil Science Society of America Proc.* 36: 465-72.

Stevenson, F.J. 1982. *Humus Chemistry Genesis, Composition and Reactions*. New York, USA: Wiley.

Stevenson, F.J. 1986. *Cycles of Soil Organic Carbon, Nitrogen, Phosphorus, Sulfur, Micronutrients*. New York, USA: Wiley-Interscience.

Stevenson, F.J., Elliott, E.T., Cole C.V., Ingram, J., Oades, J.M., Preston, C. and Sollins, P.S. 1989. Methodologies for assessing the quantity and quality of soil organic matter. In Coleman, D.J., Oades, J.M. and Uehara, G. (ed) *Dynamics of Soil Organic Matter in Tropical Ecosystems*. Hawaii, USA: NifTAL.

Trevors, J.T. 1992. DNA extraction from soil. *Microbial Releases* 1: 3-9.

Tunlid, A. and White, D.C. 1992. Biochemical analysis of biomass. Community structure nutritional status and metabolic activity of microbial communities in soil. In Stotzky, G. and Bollag, J.-M. (eds) *Soil Biochemistry* (Vol. 7). New York, USA: Marcel Dekker.

Beyond the Biomass
Edited by K. Ritz, J. Dighton and K.E. Giller
© 1994 British Society of Soil Science (BSSS)
A Wiley-Sayce Publication

CHAPTER 24

Consequences of the spatial distribution of microbial communities in soil

P.J. HARRIS

Terms such as 'microbial communities' and 'spatial distribution' have the disadvantage of being understood by almost everyone, and yet precise definitions are invariably the subject of dispute. 'Microbial community' can be regarded, at its simplest, as a term used to cover a collection of different organisms which happen to be detectable in a particular environment or subsection of an environment. The mere fact that they co-exist at the moment of sampling is often taken as evidence that they are part of a community. At this level of investigation, the 'community' may be no more than a disparate collection of organisms that happen to be found within a sample of a particular size taken from the environment (for example, a soil). However, the term 'community' implies that there is some degree of aggregation of individuals, some of which may be the same, while others are different but live together at a level that involves a degree of interaction. If two communities of whatever kind (human, plant or microbial) are sufficiently separated from one another, then interaction becomes almost impossible. This introduces the concept of 'spatial distribution' and the significance that this may have in biological function.

This chapter investigates the relationship of microbial size and colonial form, microbial distribution and location to the concept of 'community' in microbial ecology and the influence of spatial distribution on microbial function. Where possible, an attempt is made to draw parallels with larger and perhaps more easily recognised 'communities'.

THE IMPORTANCE OF SCALE

Microorganisms or microbial communities in soil are regarded as inhabiting microhabitats, but microhabitats are poorly defined. How big is a microhabitat? Is it the same for bacteria, actinomycetes, fungi and protozoa? It is probable that it will be necessary to distinguish between different groups of soil microorganisms if community structure is to be investigated. Bacteria and fungi differ in size by one or two orders of magnitude and most prudent plant and animal ecologists would consider it necessary to make some distinctions between organisms if they ranged from 2 cm to 2 m in length or height. Just because microorganisms are generally very small, it does not follow that relative size is unimportant. Clay particles and fine sand particles are of a size that can be compared with soil bacteria and fungi, respectively, and yet the difference in size is regarded as fundamental in understanding the physical properties of soils.

There are other reasons for the distinction. Fungi, because of their filamentous nature, can explore the soil in a manner which is not available to the more sedentary bacteria. A bacterial development is probably influenced by the soil conditions within a radius of a few score microns at best. A developing fungus has the advantage of being able to extend beyond immediate conditions, even when these conditions are unfavourable, and does not encounter soil conditions in the same restricted way. A fungus experiences the conditions in the soil in a similar way to a plant root system which, by being in contact with different parts of the same soil, experiences a degree of averaging in soil conditions. It is no coincidence that when we measure chemical and physical properties of soils, we tend to measure average properties at a scale that is relevant to a growing plant. For investigations of plant nutrition or water availability, this has proved quite successful and may prove, on a slightly reduced scale, to be equally successful for filamentous fungi. For bacteria, a much smaller scale of analysis may be required if the heterogeneity of microhabitats is to be revealed.

For the above reasons, most of the following discussion applies to bacteria, where the problems of defining the microhabitat are more difficult.

BACTERIAL COMMUNITIES IN SOIL

The formation of bacterial communities in soil cannot be divorced from bacterial colonisation, since it is from colonisation that communities develop. The same is true of plant, human and other communities. For colonisation to occur there are a number of prerequisites. There must obviously be colonisers, there must be access to the area being colonised and there must be resources to support colonisation (Elliott et al., 1980). It is also true that there must be an absence of conditions which may seriously harm the developing colony.

There are a few parallels to be drawn between the requirements for bacterial colonisation and those for human colonisation. Colonisation by single human beings has been a rare event in human history, whereas, in theory at least, only a single bacterium should be necessary (the role of the single bacterial cell will be discussed again later in this chapter). Access to a new site for colonisation was limited in human terms until new forms of transport were developed. For the bacterial colonist there are more options. The organism can move actively to a new site, it can be moved passively by other agents, such as soil animals or even soil water, or the site may move to the coloniser in the form of a plant root. The importance to a bacterial coloniser of motility is still uncertain. Zvyagintsev (1962) observed that relatively few soil bacteria ($< 1\%$) are motile at any one time and it remains to be determined whether motility extends the size of a microhabitat or simply allows movement from one discrete habitat to another, the organisms needing only to survive passage through the intervening soil. It must be said that this isolated, though much quoted, observation deserves much wider substantiation using modern techniques.

The question of the availability of resources is temptingly simple. Human colonisation was originally based on the easy availability of food and water, and most concepts of microbial colonisation follow the same basic idea. Microorganisms colonise nutrient sources, which in soils usually means organic debris. Certainly, most studies of localised development of organisms in soil agree that the highest local concentrations of organisms are associated with organic matter rather than mineral material (Brian, 1957; Gray et al., 1968). Other common sites of human colonisation have developed around specialised resources, such as mineral deposits, but a third and very common reason for human colonisation has been related to sites through which resources pass, with or without processing as they do so; cross-roads, river crossings and ports are examples. Animal and plant communities such as corals, aquatic epiphytes and notorious river weeds, such as the water hyacynth, which rely on 'passing trade' rather than specifically located resources are also known.

Soil bacteria could benefit from such a strategy, which would overcome their relative immobility. Soil solution has a low but not insignificant content of dissolved organic carbon (DOC) which moves by both diffusion and mass flow. Water in soil moves continuously, especially within the larger pores of the soil, and this solution will bathe the organisms colonising pore surfaces with a continuous supply of nutrients. If the bacteria are near a transpiring plant, the inflow of water to the plant will carry the dissolved organic nutrients through an ever-decreasing volume of soil as it approaches the root surface. This will result in an increase in substrate supply per unit surface area; calculations based on conservative estimates of DOC suggest that this inflow may be of a similar order of magnitude to the amounts of soluble carbon compounds suggested by various authors for root exudates over the course of a growing season (Whipps, 1990). Evidence for the importance of DOC in the soil solution is provided by the correlation found between DOC and denitrification rates in soil reported by Burford and Bremner (1975). If soil DOC in the water flow to plant roots does contribute to microbial growth, the effect will be most pronounced (on the basis of simple geometry) in the radial zone extending approximately 2 mm from the root surface.

Relatively little information is available to those who wish to review spatial aspects of the bacterial colonisation of soil compared with the abundance of studies on the colonisation of such diverse surfaces as teeth (Kolenbrander, 1991) and sewage trickle filter beds. Perhaps it is worth remarking that the advent of biomass studies coincided with the cessation of some promising developments in microbial ecology based on observation. The classic studies at this time include those by Zvyagintsev (1962), Jones and Griffiths (1964), Gray et al. (1968) and Hattori (1973), all of whom were presenting evidence which emphasised the need to discriminate between ecological zones within soils, or more appropriately, within and upon soil aggregates. The results reported by these authors confirmed that bacteria were not evenly distributed in soils and that colonies were a characteristic feature of bacterial communities. It is to be hoped that recent developments in microscopy and image analysis of surfaces, such as those reported by Caldwell and Lawrence (1989), will revive interest in direct observation.

SOIL ECOLOGY AND THE SINGLE CELL

All examination methods, even non-destructive ones, reveal the presence in soils of isolated cells. The origin of these cells and their significance is unclear. A single cell could arise in a number of ways: by movement as outlined earlier, either active (Zvyagintsev, 1962) or passive (Opperman et al., 1987), or possibly as relicts from previous colonisation stages which have since died or been subject to predation. The important question which follows is the extent to which an isolated cell can develop further. Presumably, if conditions are suitable, the cell can develop and provide the seed for further colonisation. The existence of a single cell might indicate that suitable conditions have not yet occurred. If this is so, then the cell might be in a resting stage, awaiting suitable conditions. The fact that the soil can be a good medium in which to survive has been exploited in the use of sterile soil for storing cultures for decades, and certain organisms such as *Bacillus anthracis* can survive under field conditions for similar periods. However, these studies often rely on large starting populations and do not address the possibility that survival as an isolated cell might be different from survival as part of a colony. The biochemical potential of an isolated bacterial cell, in terms of quantities of enzymes produced, is very limited, and the crucial stage in colonisation by bacteria is likely to be the period during which the organism achieves the first few divisions, leading to a condition where the combined enzymic potential of the young colony is large enough to make a significant impact on a substrate. This early stage of colonisation is one in which the new coloniser might stand to benefit from the support of other organisms, perhaps already in a colonial form. At this level, the concept of the community might have real meaning.

BACTERIAL COLONIES IN SOIL

Evidence of the existence of colonies on soil aggregate surfaces (Harris, 1972; Waid, 1973) suggests that these may be larger than previously suspected from the much earlier work of Jones and Mollison (1948). They exist as fragile sheets of cells that are generally of a single cell thickness, in contrast to other biofilms which have been reported as being 80 μm thick and more (Caldwell et al., 1992). They may cover many hundreds of square micro-metres and usually appear as large rafts of a single cell type; more rarely, they appear to have a more mixed constitution; occasionally, distinct morphological types of cells may overlap. Investigations of their activity by autoradiographic (Waid et al., 1971) or histological (MacDonald, 1980) means could determine whether the whole colony is fully functional or is metabolically active only at the periphery. A major feature of colony formation is the creation of a concentration of enzymic activity which will be mutually beneficial to all the organisms in the colony and possibly confer benefits in terms of antagonistic effects on other would-be colonisers. The danger of this strategy is that the organisms present an easy target to grazing predators which might decimate them with ease.

The existence of large colonies of more than 10 000 morphologically similar cells, separated by relatively uncolonised soil surfaces, can be compared with the spatial separation of human colonisation. Even in densely populated areas such as the Netherlands or southern England, there are still large tracts of countryside with little evidence of human habitation. The extent of the non-uniformity shown by bacterial colonisation can be assessed by calculating the degree to which the distribution differs from a uniform distribution, a statistic known as the Index of Dispersion (Fisher, 1970). This is calculated from the chi squared value (Equation 1) for the number of fields counted for each slide or set of slides:

$$\text{Chi squared } \chi^2 = \sum \frac{(x - \bar{x})^2}{\bar{x}} \tag{1}$$

The Index of Dispersion is then calculated using Equation 2, which takes account of the number of microscopic fields examined. The index value exceeds about 2 (or 1.65 if p = 0.05) when the cells show aggregation rather than uniform distribution:

$$\text{Index of Dispersion} = \sqrt{(2\chi^2)} - \sqrt{(2n - 1)} \tag{2}$$

Experiments conducted using the surface film removal technique on 12 soils (Al-Khadeer, 1975) showed that in all soil films the bacterial cells were distributed in a highly non-uniform manner. It was not possible to find linear correlations between the irregularity of microbial colonisation and any single soil property (see Table 24.1). However, the soils with the most irregular distribution of cells were all soils with considerable quantities of free calcium carbonate. In the remaining soils, the high degree of non-uniformity was associated with high coarse sand content. Gray et al. (1968) commented on the marked localisation of bacterial colonisation in the sand dune soils which they examined. It is possible that sand grains present problems in terms of the ease with which cells can become, and remain, attached, or it could also be a function of the heterogeneity of organic matter distribution in coarse sandy soils.

Table 24.1　Index of Dispersion of bacterial cells in removed surface films of individual aggregates of 12 surface soils[a]

Soil number	Coarse sand (%)	Organic matter (%)[b]	Calcium carbonate (%)	pH	Index of Dispersion[c]	
					Mean	(SD)
1	11.3	7.5	21.9	7.8	162	(113)
2	3.0	9.2	15.2	8.0	147	(65)
3	31.7	5.6	14.0	8.0	121	(66)
4	11.3	2.8	28.9	8.4	108	(82)
5	15.0	6.4	24.1	7.9	106	(74)
6	51.6	1.6	0.0	5.4	96	(42)
7	41.6	6.0	0.0	4.8	92	(40)
8	44.6	2.3	0.5	6.5	88	(42)
9	43.8	7.2	0.0	4.8	74	(28)
10	37.0	1.9	0.4	7.4	60	(23)
11	8.2	5.2	0.0	6.2	48	(21)
12	3.5	6.3	0.0	6.6	48	(24)

Note: a 20 aggregates per soil, 1.5-2.5 mm diameter
　　　 b Organic matter determined using the Walkley and Black (1934) method
　　　 c 20 microscopic fields were examined for each aggregate film. An Index of Dispersion of 2 would indicate a perfectly uniform distribution of cells in the film. Higher values indicate a greater degree of non-uniform distribution. The SD values provide a guide to the differences between individual aggregates.

Despite the large size of the colonies observed in surface films from soil aggregates, there is currently no way, using conventional methods such as the dilution plate technique, to determine how much the members of such colonies contribute to the organisms isolated from soils. They could be over-represented because they are likely to give rise to colony-forming units (CFUs) of many cells with a good chance of viability in laboratory culture, or they could be under-represented for much the same reason — that is, that the size of the CFU is large and hence a large colony of many thousands of cells makes only a small numerical contribution to dilution plates or most probable number (MPN) methods.

Work at the University of Reading using the agar film method described by Jones and Mollison (1948) on suspensions prepared for dilution pour plates confirmed the original authors' findings that the bacteria in a well-dispersed suspension have an average cell clump size (or potential CFU) of between three and seven cells with reasonable uniformity between different soils (*see* Table 24.2 *overleaf*). These figures were obtained from soils that had been well dispersed and hence included bacteria from both the surface and the interior of aggregates. No evidence was found of the large colonies described earlier, suggesting that they do not survive vigorous dispersion.

Table 24.2   **Average size of bacterial clumps found in soil suspension after ultrasonic dispersion for 2 minutes at 18 Khz using an MSE Soniprep 150 machine**

Soil type	Sample number	Average number of bacteria in clump
Sandy loam	1	6.9
	2	6.6
	3	5.6
	4	5.7
Clay loam	1	4.7
	2	4.3
	3	4.5
	4	3.9

The organisms which have been observed in the centres of soil aggregates (Jones and Griffiths, 1964) present an interesting spatial problem. There is little opportunity for them to grow into large sheets; indeed, when they have exhausted the available nutrients in their immediate vicinity there may be little opportunity for them to grow at all. How did they come to be in the centre of an aggregate? For how long can they be active? Are they culturable? It is possible that some of the intra-aggregate bacteria have penetrated fine pores to reach their present location. To do so would imply good motility and/or very small size. The sites within an aggregate may offer good protection from predation by protozoa and nematodes. Their strategy would seem to be one of living frugally but safely. These intra-aggregate bacteria might be expected to include the dwarf cells described by Bae et al. (1972). An alternative explanation for their location is that they originally developed on an aggregate surface but were transferred to their present position by soil perturbation and reworking. This raises the question of the average age of soil aggregates.

CONCLUSION

Bacterial cells in soil are unevenly distributed and in some locations may show colonisation at high densities. These areas of intense colonisation may often, but not necessarily, be related to high levels of adjacent substrate. Other locations in the soil may favour different strategies involving low nutrient use and prolonged survival (van Veen et al., 1985). It is tempting to suggest that the most important location for intense microbial activity is on the accessible surfaces of soil aggregates as this favours large-scale colony development. Substrates are likely to be more easily available in the form of DOC moving in the soil pore system and root proximity is more likely. It would be interesting to investigate the extent to which this surface population contributes to the overall microbial activity of soils, since the organisms are probably more susceptible to manipulation than those buried within aggregates. The latter group may be only 'sleepers' or 'dwarf cells' awaiting a redistribution of soil particles (or perhaps the attentions of a friendly soil microbiologist who will apply soluble substrates at such excessive levels that some may eventually diffuse into their lonely fastness).

Microbial ecology — that is, microbial ecology at a scale that is relevant to soil bacteria — must warrant further investigation if we are to understand the biological nature of soil. Human physiology might have progressed slowly if we had remained content with knowing what a pancreas looked like but had no idea of where it was located or how it interacted with other organs. There has been an explosion of new techniques during the era of the biomass, and it is the responsibility of soil microbiologists to assess which of these new methods will help improve our understanding of how the soil works at a detailed level.

References

Al-Khadeer, M.A. 1975. Studies of the bacterial population on the surface of soil aggregates. MPhil. thesis, University of Reading, UK.

Bae, H.C., Cota-Robles, E.H. and Casida, L.E. 1972. Microflora of soil as viewed by transmission electron microscopy. *Applied Microbiology* 23: 637-48.

Brian, P.W. 1957. Ecological significance of antibiotic production. In Williams, R.E.D and Spicer, C.C. (eds) *Microbial Ecology*. Cambridge, UK: Cambridge University Press.

Burford, J.R. and Bremner, J.M. 1975. Relationships between the denitrification capacities of soils and total, water soluble and readily decomposable soil organic matter. *Soil Biology and Biochemistry* 7: 389-94.

Caldwell, D.E. and Lawrence, J.R. 1989. Study of attached cells in continuous-flow slide culture. In Wimpenny, J.W.T. (ed). *A Handbook of Model Systems for Microbial Ecosystem Research*. Boca Raton, Florida, USA: CRC Press.

Caldwell, D.E., Korber, D.R. and Lawrence, J.R. 1992. Imaging of bacterial cells by fluorescence exclusion using confocal laser microscopy. *J. Microbiological Methods* 15: 249-61.

Elliott, E.T., Anderson, R.V., Coleman, D.C. and Cole, C.V. 1980. Habitable pore space and microbial trophic interactions. *Oikos* 35: 327-35.

Fisher, R.A. 1970. *Statistical Methods for Research Workers*. Edinburgh, UK: Oliver and Boyd.

Gray, T.R.G., Baxby, P., Hill, I.R. and Goodfellow, M. 1968. Direct observation of bacteria in soil. In Gray, T.R.G. and Parkinson, D. (eds) *The Ecology of Soil Bacteria*. Liverpool, UK: Liverpool University Press.

Harris, P.J. 1972. Micro-organisms in surface films from soil crumbs. *Soil Biology and Biochemistry* 4: 105-06.

Hattori, T. 1973. *Microbial Life in the Soil*. New York, USA: Marcel Dekker.

Jones, D. and Griffiths, E. 1964. The use of thin soil sections for the study of soil microorganisms. *Plant and Soil* 20: 232-40.

Jones, P.C.T. and Mollison, J.E. 1948. A technique for the quantitative estimation of soil microorganisms. *J. General Microbiology* 2: 54-69.

Kolenbrander, P.E. 1991. Coaggregation: Adherence in the human oral microbial ecosystem. In Dworkin, M. (ed) *Microbial Cell Interactions*. Washington DC, USA: American Society for Microbiology.

MacDonald, R.M. 1980. Cytochemical demonstration of catabolism in soil microorganisms. *Soil Biology and Biochemistry* 12: 419-23.

Opperman, M.H., McBain, L. and Wood, M. 1987. Movement of cattle slurry through soil by *Eisenia foetida* (Savigny). *Soil Biology and Biochemistry* 19 : 741-45.

van Veen, J.A., Ladd, J.N. and Amato, M. 1985. Turnover of carbon and nitrogen through the microbial biomass in a sandy loam and a clay soil incubated with [^{14}C(U)] glucose and [^{15}N] $(NH_4)_2SO_4$ under different moisture regimes. *Soil Biology and Biochemistry* 17: 747-56.

Waid, J.S. 1973. A method to study microorganisms on surface films from soil particles with the aid of the transmission electron microscope. *Bulletin of Ecological Research Communications (Stockholm)* 17: 103-08.

Waid, J.S., Preston, K.J. and Harris, P.J. 1971. A method to detect metabolically active microorganisms in leaf litter habitats. *Soil Biology and Biochemistry* 3: 235-41.

Walkley, A. and Black, I. A. 1934. An examination of the Degtjareff method for determining soil organic matter and a proposed modification of the chromic acid titration method. *Soil Science* 37: 29-38.

Whipps, J.M. 1990. Carbon economy. In Lynch, J.M. (ed) *The Rhizosphere*. Chichester, UK: Wiley Interscience.

Zvyagintsev, D.G. 1962. Adsorption of microorganisms by soil particles. *Soviet Soil Science* 2: 19-25.

Beyond the Biomass
Edited by K. Ritz, J. Dighton and K.E. Giller
© 1994 British Society of Soil Science (BSSS)
A Wiley-Sayce Publication

CHAPTER 25

Pattern-generating processes in fungal communities

A.D.M. RAYNER

The functioning of natural fungal communities cannot be fully understood without knowledge of the distributional patterns of individual mycelial systems and the processes that create those patterns. However, a vast amount of fungal ecological work has been done, and continues to be done, in the absence of such knowledge. This seems to be due to a reluctance to recognise the significance, from methodological and conceptual points of view, of an indeterminate body form. Efforts to quantify the natural occurrence of mycelial fungi have therefore focused on ways to enumerate or weigh these organisms, assuming that they can, at least approximately, be treated as fully particulate or homogeneous entities.

Such assumptions are invalid. Mycelia are structurally and functionally heterogeneous, potentially limitless (indeterminate) systems that thrive in locally unpredictable niches by dint of their ability to modify their developmental pattern as circumstances change. How, then, can mycelial fungi be counted meaningfully when, depending partly upon methodology and partly upon concept, a countable entity may be anything from a viable spore to a system occupying areas measurable in hectares (Rayner, 1989; Smith et al., 1992)? What is the meaning of a biomass estimate if it is not known how that biomass is distributed asymmetrically through space and time? How, in any case, is biomass related to the ecological functioning of a mycelium as an energy-distributing system?

These questions introduce the need not so much to count or weigh fungi as to be able to construct maps that configure both the regional boundaries of mycelial systems and the heterogenous disposition of structure within those boundaries. Only then can the task of relating structure to ecological functioning really begin. However, two widespread fallacies have hindered attempts to produce mycelial distribution maps. The first is that mycelia in natural habitats are invariably too hidden, too microscopic and too unidentifiable on the basis of morphology to enable their domains to be directly visualised. Consequently, their presence is usually detected indirectly (for example, by isolation onto culture media, often sampling at scales that are either too coarse or too fine to be appropriate to actual distributional patterns). The second fallacy is that it is not possible to distinguish meaningfully between functionally discrete mycelial systems of the same species in adjacent domains.

PRODUCING MYCELIAL DISTRIBUTION MAPS

Mapping regional boundaries

Providing that a suitable scale of observation is chosen, it is often easy to visualise mycelial domains directly by examining either the hyphal systems themselves or their effects on the living or non-living materials that they inhabit. These domains are commonly macroscopic, as in fairy rings, lichen communities, decay columns in wood, cankers in bark, rhizomorphic and mycelial cord systems and forest die-backs caused by certain root-rot fungi. The latter may indeed may be so large as to be most easily observed from aircraft! Even where the domains and/or materials inhabited are very small, and indirect methods are used, resolution can be greatly improved by selecting an appropriate particle size for transfer to isolation media (for example, Kirby et al., 1990). Moreover, because of characteristic physiological and morphological changes that occur at their interfaces, adjacent mycelial systems, both of the same and of different species, commonly map their own regional boundaries (Rayner and Todd, 1979). The boundaries are most evident in communities of crustose lichens (Smith, 1921) and in bulky substrata such as decaying wood (*see* Figure 25.1). However, they can also be found within

Figure 25.1   Transverse and longitudinal slices through a beech log placed upright in woodland soil and allowed to become colonised by wood decay fungi

The upper surface has become colonised by numerous genets of *Coriolus versicolor* and *Stereum hirsutum*, whereas much of the lower surface has been invaded by a single genet of the mycelial cord-former, *Phanerochaete velutina*.

Source: Coates and Rayner (1985)

smaller substrata such as individual decomposing leaves and twigs if these are examined carefully (Boddy and Rayner, 1984; Cooke and Rayner, 1984).

The formation of boundary zones between adjacent mycelia of the same species in higher fungi is commonly attributable to a 'rejection' or 'somatic incompatibility' response following hyphal fusions between systems that are sufficiently distinct genetically to recognise one another as 'non-self'. This response prevents physiological integration and so maintains the individual integrity of each system (for example, Rayner, 1991a). It also provides a convenient means of distinguishing between mycelial systems of different genetic origin (genets) (Brasier and Rayner, 1987), both in the field and in laboratory culture (Adams and Roth, 1967, 1969; Rayner and Todd, 1979; Rayner et al., 1984). Genets may also be distinguished using a variety of other genetic as well as molecular markers (Rayner, 1992; Smith et al., 1992), and represent the most meaningful application of the concept of 'individuals' within the context of fungal populations. They differ from animal individuals in being indeterminate and hence occupying variably sized spatio-temporal domains or 'genetic territories' (Rayner, 1991a, 1992), but exhibit fundamentally similar dynamic properties to social collectives of animals, such as army ant swarms (Rayner and Franks, 1987). Ultimately, mapping regional boundaries in natural communities of higher fungi means mapping genets.

Mapping regional topography

Mapping the disposition of structure within mycelial domains may present somewhat greater difficulties than mapping regional boundaries between genets. However, it is certainly easily achievable in the case of macroscopic systems, such as those formed by fairy rings and mycelial cord-forming fungi (Thompson, 1984; Dowson et al., 1989a) and there is no particular reason why suitable observational techniques cannot be devised for use at smaller spatial scales.

In quantifying the distribution patterns, it is particularly important to be aware of the heterogeneity of mycelia, which is implicit in their branching pattern and capacity to combine explorative, assimilative, conservative and redistributive modes of operation (Rayner et al., 1987a, 1993a, b). The problem of how best to characterise the heterogeneity of indeterminate systems is of general significance in biology. The most fundamental property of such systems is that they fill space incompletely, such that their structural density varies with the scale of observation. The most suitable measure of their distribution is therefore probably to be found in their fractal dimension (Ritz and Crawford, 1990; Bolton and Boddy, 1993), which is larger (approaching 2 in a plane, 3 in a volume) the more branched (tangentially rather than radially oriented) the structure. Image analysis techniques (Bolton et al., 1991; Bolton and Boddy, 1993) may prove to be of particular value in making such quantifications.

EXTERNAL DETERMINANTS OF FUNGAL FRONTIERS

Once suitable maps have been prepared at scales appropriate to habitat parameters influencing the size of individual mycelial domains, the challenge then is to interpret their meaning in terms of the dynamic processes underlying the organisation and functioning of fungal communities. There are two basic approaches to this issue. The first is to examine adaptive responses to extrinsic biotic and abiotic constraints and opportunities, and to relate these to concepts of evolutionary niche and ecological strategy (Cooke and Rayner, 1984; Stenlid and Rayner, 1989; Andrews, 1992). The second is to examine the *intrinsic* generative processes that cause mycelial systems to self-organise and reproduce and study how these processes cause particular developmental patterns to be organisationally impelled under given environmental circumstances (Rayner et al., 1993a, b; cf. Gould and Lewontin, 1979).

With respect to extrinsic factors, the distribution and accessibility of nutritional resources is critical to the determination of colonisation patterns. These resources characteristically occur naturally in discrete 'resource units', so that potential fungal colonists are required to have some means of arriving at them. Two means of arrival are possible: propagules (typically spores) that enable dispersal in space and/or survival over time, and the migratory mycelium that is capable of development across nutritionally barren domains between resource units.

Genets able to proliferate only as propagules are limited by the physical boundaries of resource units (that is, they are 'resource unit-restricted'), whereas those able to produce a migratory mycelium are 'non-unit-restricted' (Cooke and Rayner, 1984; Rayner et al., 1985a, b, 1987a). This distinction is crucial to the structure and functioning of fungal communities, particularly in soil systems.

Colonisation patterns of resource unit-restricted fungi

The mycelial domains of resource unit-restricted fungi are bounded not only by the physical dimensions of the resource units that they occupy, but also by the extent to which those dimensions become occupied by competitive organisms of the same or different species. Moreover, the incidence of competitors in a community has obvious implications for its ecological functioning, both in terms of its overall activity and the heterogeneous partitioning of activity amongst individual community members (Rayner and Todd, 1979; Rayner and Boddy, 1987).

According to ecological strategy theory (Grime, 1979; Cooke and Rayner, 1984), the potential incidence of competitors in a fungal community is determined by the extent to which colonisation occurs under ecologically disturbed or stressful conditions. Disturbance imposes R-selection, favouring so-called 'ruderal' strategists that are quick to arrive at newly available sites for colonisation and to deplete easily assimilable resources. Such organisms will be expected to reproduce both early (before or as competitors establish) and clonally (because of the relative uniformity of initial conditions and short individual life spans). They will therefore tend to have a limited developmental repertoire, commensurate with reduced heterogeneity, and, whilst not being subject to intraspecific competition, will tend to be replaced readily by other species.

Under undisturbed conditions, the incidence of competitors may be limited by environmental stress, abiotic conditions that inhibit growth of most organisms in question (Cooke and Rayner, 1984). The minority that are able to grow effectively under these conditions are S-selected, having an S-tolerant or S-adapted strategy. In the relative absence of competitors they may establish extensive mycelial domains and have individually prolonged life spans. Many of the fungi which cause decay of intact sapwood or heartwood in standing trees provide a good example (Rayner and Boddy, 1988).

A prolonged life span enhances the risk to evolutionary fitness of the spread of disease between genetically identical organisms (for example, Brasier, 1988), and stressful environments are often spatio-temporally heterogeneous. S-selection may therefore often favour sexual outcrossing and a high degree of developmental versatility. The latter applies particulary to those organisms, such as some endophytic fungi (for example, Boddy and Griffith, 1989), which establish under stressful conditions but become active and persist once these conditions have been alleviated (Cooke and Rayner, 1984). On the other hand, the type of S-selection that operates, for example, on a population at the edge of its geographical range is liable to favour clonal reproduction of resilient genotypes

(Rayner, 1992). In the basidiomycete, *Stereum hirsutum*, for example, northern European populations are non-outcrossing, whereas more southern populations reproduce by sexual outcrossing (Ainsworth et al., 1990a). On the same lines, the population structure of the ascomycete, *Ophiostoma novo-ulmi*, the causal agent of Dutch elm disease (Brasier, 1991), is clonal at epidemic fronts but genetically diverse because of sexual outcrossing behind these fronts (Brasier, 1988).

Undisturbed and relatively unstressful conditions favour sexually outcrossing organisms with long individual life spans and 'combative' strategies that enable them either to 'defend' or 'capture' resources from others. Such strategies depend upon both the possession of and resistance to antagonistic mechanisms (Rayner and Webber, 1984). Where defence predominates, as often applies amongst resource unit-restricted fungi, then it is common for numerous genets of the same and of different species to exist alongside one another in relatively small, mutually exclusive domains (*see* Figure 25.1). Such small domains limit the reproductive output possible from individual genets, enhancing the genetic heterogeneity of populations. However, under particularly favourable conditions for colonisation, the incidence of competitors may be so high as to impair the ability of individual genets to fruit and to exploit resources, so exerting a form of population control applicable even to indeterminate organisms (Rayner and Todd, 1979; Coates and Rayner, 1985).

Generally, the incidence of competitors will be expected to be highest near the surface of a resource unit; within the interior, individual domains of positionally or otherwise advantaged genets will both expand and be fewer in number, simplifying population and community structure (Adams and Roth, 1969; Cooke and Rayner, 1984; Rayner, 1992). An interesting consequence, since the surface area of an object relative to its volume increases as the scale of linear measurement is reduced, is that similar numbers of genets may be expected to occur in the interior of resource units differing vastly in size (such as a twig and a tree trunk). This is important in studies of population and community organisation with respect to designing a sampling pattern appropriate to the scale of the resource unit being studied.

There are some instructive exceptions to the above generalisations amongst wood-inhabiting fungi. For example, certain host-selective species of *Hypoxylon*, such as *H. fragiforme* on *Fagus* and *H. fuscum* on *Corylus*, produce populations consisting of numerous genets occupying small domains alongside one another (Chapela and Boddy, 1988; A. Inman and A.D.M. Rayner, unpubl.). However, these fungi exhibit latency, becoming active only when the sapwood becomes dysfunctional but having colonised earlier under highly selective conditions. Fungi with apparently similar behaviour, such as *Piptoporus betulinus* on *Betula* and *Daldinia concentrica* on *Fraxinus*, form relatively few, spatially extensive genets in individual trunks (Rayner and Boddy, 1986). The difference may be related to the role of ascospores in establishing colonisation foci in *H. fuscum* and *H. fragiforme*. In the latter species, ascospore germination has been shown to be stimulated by specific, host-mediated recognition responses (Chapela et al., 1991, 1993).

Another exception is provided by the basidiomycetes *Pseudotrametes gibbosa* and *Lenzites betulina*, which produce large mycelial domains and fruit bodies in spite of colonising detached wood under circumstances that would normally lead to a high incidence of competitors. They do this by 'temporary parasitism', a colonisation strategy analogous to those of certain ant species and strangler figs, which entails initial parasitism and then replacement of resident populations of specific pioneer colonists, *Bjerkandera* species with *P. gibbosa* and *Coriolus* species with *L. betulina* (Rayner et al., 1987b).

Colonisation patterns of non-unit-restricted fungi

Fungal genets capable of mycelial extension beyond the physical boundaries of individual resource units have the potential to form enormous genetic territories, measurable in hectares and centuries old in the case of some wood-inhabiting forms (for example, Smith et al., 1992). Such genets rarely encounter one another within individual resource units, where they commonly replace resident unit-restricted fungi (for example, Coates and Rayner, 1985; *see* Figure 25.1), but they may do so outside these units (Dowson et al., 1988a). Non-unit-restricted systems have the ecologically important property of interconnecting activities in different microenvironmental sites, whether these be plant or animal remains or, in the case of mycorrhizal fungi, living roots (Read, 1984, 1992). The manner in which such interconnections are established is of fundamental biological and ecological importance, and may be interpreted in terms of 'foraging strategies', paralleling those, for example, of army ant raids or stoloniferous plants. Foraging can be studied in artificial microcosms (*see* Figure 25.2), in semi-natural microcosms (Dowson et al., 1986, 1988b, 1989b) and in the field (Dowson et al., 1988c, d, 1989a).

Foraging patterns vary from species to species, indicating a genetic component to their regulation, and with the size, proximity and frequency of resource units. Where resource units are small, but frequent and close to one another, as in a layer of leaf litter, close-range foraging is apposite. This is illustrated by the fairy ring-forming basidiomycete, *Clitocybe nebularis*, which drives short ex-plorative mycelial cords ahead of an annulus of a diffuse exploitative mycelium with a degenerative trailing edge (Dowson et al., 1989a). Short-range foraging is apposite when resource units are relatively small and frequent but spatially separated, and is similarly redistributive. It is exemplified

Figure 25.2   Networking of *Coprinus picaceus* mycelium grown in a matrix of interconnecting chambers alternately containing high (H) and low (L) nutrient media (2% malt agar and tap water agar)

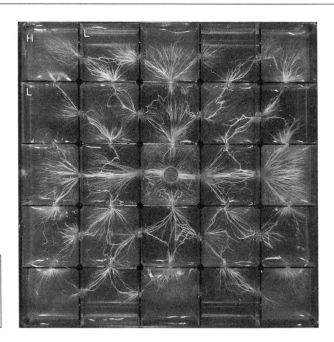

It is worth noting the prolifera-tion of interconnecting, anastomosed mycelial cords across the low nutrient media

Source: L. Owen and E. Bower

by such wood decay fungi as *Hypholoma fasciculare* and *Steccherinum fimbriatum* (Dowson et al., 1986, 1988b, 1989b). Outgrowth of a dense mycelial cord system (high fractal dimension) from a food base ceases when it somes into contact with previously uncolonised wood, after which non-connective cords degenerate as renewed outgrowth occurs from the new base. Long-range foraging, exemplified by *Phanerochaete velutina* and *Armillaria* species, involves outgrowth of more sparsely distributed (lower fractal dimension) mycelial cords or rhizomorphs, which cease extension and degenerate less readily after contact with a new food base. It is most effective when resource units are large but widely scattered.

The formation of anastomoses in foraging systems increases the persistence and distributive capacity of initially explorative mycelia. Moreover, the production of persistent networks adds the ability to forage in time to the ability to forage in space because any resource that comes into contact with the network can become quickly colonised. Networks can therefore conserve energy until the opportunity for assimilation arises, rather than actively locating resources. The stinkhorn, *Phallus impudicus*, and the agaric, *Tricholomopsis platyphylla*, both wood decomposers, exemplify this strategy (Dowson et al., 1988b).

INTERNAL DRIVES

From the above discussion it should be clear that fungal mycelia are capable of producing patterns which, by maximising the efficiency of resource capture and distribution, have adaptive value in the varied and heterogeneous conditions of natural environments. For many, recognition of such adaptive qualities provides, in itself, sufficient explanation for the existence of these patterns, and identification of their genetic specification becomes a primary research objective. However, following the same kind of logic, a river system might be perceived as an adaptation to environmental topography such as to achieve maximum drainage to the sea (Rayner et al., 1993b). The latter point draws attention to the need to think systemically about mycelia in order to understand the processes that determine their varied developmental patterns.

'Self-plumbing'

According to a recent hypothesis (Rayner, 1991b; Rayner et al., 1993a, b) the patterns produced by a mycelium can be viewed as the automatic consequence of the way that it is organised as an assimilative *and* distributive system. Feedback *processes* cause this system to respond to changing circumstances, using material *mechanisms* whose *exact* specification is encoded in genes.

In broad outline, this hypothesis envisages the mycelium as an intrinsically unstable, fluid-dynamical system in which expansive processes resulting from input of energy are counteracted by constraints that prevent absolute dispersion. When it experiences a drop in internal energy charge, the system activates metabolic pathways that lead either to the rigidification and sealing off (insulation) of external boundaries (hyphal walls) or to the onset of degeneration. These pathways are of also of fundamental importance in the implementation of and resistance to antagonistic mechanisms within and between species (Rayner et al., 1993a, b). They probably involve the production of hydrophobic aromatic and terpenoid compounds, certain polypeptides known as 'hydrophobins' (Wessels, 1991, 1992) and the operation of phenol-oxidising enzymes.

By such means, mycelia can vary the deformability and penetrability of their external boundaries and the partitioning of their interior to accord with fluctuations in assimilable energy supplies. In so doing, they generate variable amounts of hydraulic thrust and, thereby, diverse developmental patterns (*see* Figure 25.3). The exact pattern depends upon the relative rates of uptake of water and nutrients from the environment and throughput of energy following such uptake to existing sites of deformation on the system's boundary.

Where uptake exceeds throughput capacity (maximum thrust), as would be likely to occur in energy-rich sites where the boundary is largely permeable, the system is prone to branch. By contrast, the sealing of lateral boundaries in nutrient-poor sites suppresses branching and converts hyphae from assimilative to distributive structures (conduits). The angle of branching may depend on whether thrust is developed proximally or distally to the point of origin. Anastomosis has the effect of amplifying the throughput capacity of the system, so enhancing the thrust that can be applied to sites of deformation, lowering the fractal dimension and allowing the emergence of fast-growing sectors, mycelial cords and rhizomorphs.

Figure 25.3   **Regulation of patterns of uptake and development of thrust in axially elongated metabolic systems by internal partitioning (membrane discontinuity) and boundary properties**

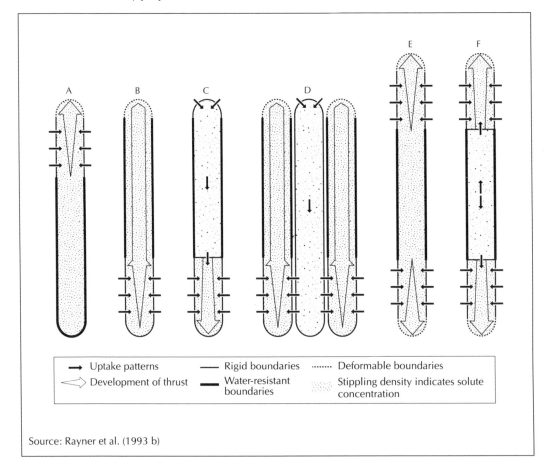

| → Uptake patterns | —— Rigid boundaries | ······· Deformable boundaries |
| ◁ Development of thrust | ▬ Water-resistant boundaries | ░ Stippling density indicates solute concentration |

Source: Rayner et al. (1993 b)

Essentially, this hypothesis implies that mycelial systems are *impelled* to develop heterogeneity by creating and following paths of least resistance. Whereas the pattern-generating processes are fundamentally the same in all systems, the parameters governing their operation in individual cases are genetically specific. Adaptation therefore lies not so much in the processes themselves, but in their pre-set or variable tuning to meet the predictable and unpredictable requirements of diverse niches, definable in terms of ecological strategies (Stenlid and Rayner, 1989).

Genomic distribution

Mixed populations of genomic organelles, nuclei and mitochondria, may result from hyphal fusion between genetically different mycelia. Questions therefore arise as to how the immigration and emigration of these organelles from fusion sites may influence, and be influenced by, mycelial

---

Notes to Figure 25.3

External water potential is assumed to be uniformly high, but evapotranspirative or other mechanisms leading to extraction or discharge of water will enhance or suppress the fluxes according to circumstances, as will the formation and mobilisation of 'storage' compounds such as glycogen.

A   A single protoplasmic continuum, where the deformable region encounters an energy-rich domain, and thrust builds as it extends, increasing the surface through which net uptake occurs (for example, a germinating spore or a migratory mycelium encountering a food base). When the uptaking surface exceeds the critical value (which may be prevented if boundary-sealing keeps pace with extension) supporting maximum thrust (equivalent to linear extension), the system will be prone to branching.

B   A continuum, where the deformable region enters non-nutritive domains, such that lateral boundaries are insulated, as in a hypha or hyphal system emerging from a food base. Thrust, developed in the assimilative region, is sustained to an extent which depends on the efficacy of lateral sealing.

C   A protoplasmic discontinuum, in which the assimilative region is separated from the non-assimilative region by a complete membrane, as would occur at a sealed septum. Here, water and solutes would be drawn from and through the less metabolically active non-assimilative component into the more competitive (in terms of active transport) assimilative component. This mechanism would allow redistribution from degenerative regions.

D   A coupled system, in which a central, freely permeable, metabolically inactive system is surrounded by highly insulated hyphae. The central system provides a channel through which water and solutes absorbed passively at the mycelial margin are distributed to assimilative components, as could apply to foraging mycelial cords emerging from a food base.

E   A linked system, with assimilative sites at each end. The connection will remain intact as long as it does not become isolated (for example, a system bridging two assimilative sites as a result of foraging or self-anastomosis) (cf. Figure 25.2).

F   A linked system, in which the connection becomes isolated, degenerates, and loses resources to assimilative phases, as between somatically incompatible mycelia.

instability. That certain combinations of organelles are able to co-exist stably is evident in sexually outcrossing higher basidiomycetes, where vigorous, independently growing heterokaryons containing nuclei with complementary mating-type alleles are formed. However, how these heterokaryons remain stable is not known, and they do not in any case seem able persistently to contain combinations of more than two genetically disparate kinds of nuclei and more than one mitochondrial type (for example, Rayner, 1991a; Rayner and Ross, 1991). On the other hand, the existence of somatic incompatibility responses that demarcate individual genotypes indicates that the co-existence of disparate genomes within the same protoplasm can lead to instability. Such responses may be interpreted as the degenerative consequence of conflicting genetic information, perhaps resulting from mitochondrial dysfunction (Rayner, 1991a; Rayner and Ross, 1991). A dynamic relationship exists between these rejection responses and somatic acceptance responses that allow genetic non-self access, both in ascomycetes and basidiomycetes (Rayner, 1991a). This relationship is critical to understanding how boundaries within and between populations are configured. Amongst basidiomycetes, the failure of somatic acceptance to override rejection responses results in reproductive isolation, and hence in speciation and the evolution of non-outcrossing breeding strategies (Rayner et al., 1984; Rayner, 1991a). On the other hand, allowing access between strains which for one reason or another are not fully competent to mate, can lead to extensive degeneracy, genomic replacement and complex patterns of phenotypic expression (Ainsworth et al., 1990a, b, 1992). Such effects link the pattern-generating processes within fungal individuals to the isolation and speciation processes that bring diversity in fungal populations.

References

Adams, D.H. and Roth, L.F. 1967. Demarcation lines in paired cultures of *Fomes cajanderi* as a basis for detecting genetically distinct mycelia. *Canadian J. Botany* 45: 1583-89.

Adams, D.H. and Roth, L.F. 1969. Intraspecific competition among genotypes of *Fomes cajanderi* decaying young growth Douglas fir. *Forest Science* 15: 327-31.

Ainsworth, A.M., Rayner, A.D.M., Broxholme, S.J. and Beeching, J.R. 1990a. Occurrence of unilateral genetic transfer and genomic replacement between strains of *Stereum hirsutum* from non-outcrossing and outcrossing populations. *New Phytologist* 115: 119-28.

Ainsworth, A.M., Rayner, A.D.M., Broxholme, S.J., Beeching, J.R., Pryke, J.A., Scard, P.T., Berriman, J., Powell, K.A., Floyd, A.J. and Branch, S.K. 1990b. Production and properties of the sesquiterpene, (+)-torreyol, in degenerative mycelial interactions between strains of *Stereum*. *Mycological Research* 94: 799-809.

Ainsworth, A.M., Beeching, J.R., Broxholme, S.J., Hunt, B.A., Rayner, A.D.M. and Scard, P.T. 1992. Complex outcome of reciprocal exchange of nuclear DNA between two members of the basidiomycete genus *Stereum*. *J. General Microbiology* 138: 1147-57.

Andrews, J.H. 1992. Fungal life-history strategies. In Carroll, G.C. and Wicklow, D.T. (eds) *The Fungal Community*. New York, USA: Marcel Dekker.

Boddy, L. and Rayner, A.D.M. 1984. Fungi inhabiting oak twigs before and at fall. *Transactions of the British Mycological Society* 82: 501-05.

Boddy, L. and Griffith, G.S. 1989. Role of endophytes and latent invasion in the development of decay communities in sapwood of angiospermous trees. *Sydowia* 41: 41-73.

Bolton, R.G. and Boddy, L. 1993. Characterisation of the spatial aspects of foraging mycelial cord systems using fractal geometry. *Mycological Research* 97: 762-65.

Bolton, R.G., Morris, C.W. and Boddy, L. 1991. Non-destructive quantification of growth and regression of mycelial cords using image analysis. *Binary* 3: 127-132.

Brasier, C.M. 1988. Rapid changes in genetic structure of epidemic populations of *Ophiostoma ulmi*. *Nature* 332: 538-41.

Brasier, C.M. 1991. *Ophiostoma novo-ulmi* sp.nov., causative agent of current Dutch elm disease pandemics. *Mycopathologia* 115: 151-61.

Brasier, C.M. and Rayner, A.D.M. 1987. Whither terminology below the species level in fungi? In Rayner, A.D.M., Brasier, C.M. and Moore, D. (eds) *Evolutionary Biology of the Fungi*. Cambridge, UK: Cambridge University Press.

Chapela, I.H. and Boddy, L. 1988. Fungal colonisation of attached beech branches. II. Spatial and temporal organisation of communities arising from latent invaders in bark and functional sapwood under different moisture regimes. *New Phytologist* 110: 47-57.

Chapela, I.H., Petrini, O. and Hagmann, L. 1991. Monolignol glucosides as specific recognition messengers in fungus-plant symbioses. *Physiological and Molecular Plant Pathology* 39: 289-98.

Chapela, I.H., Petrini, O. and Bielser, G. 1993. The physiology of ascospore eclosion in *Hypoxylon fragiforme*: Mechanisms in the early recognition and establishment of endophytic symbiosis. *Mycological Research* 97: 157-62.

Coates, D. and Rayner, A.D.M. 1985. Fungal population and community development in cut beech logs. I. Spatial dynamics, interactions and strategies. *New Phytologist* 101: 183-98.

Cooke, R.C. and Rayner, A.D.M. 1984. *Ecology of Saprotrophic Fungi*. London, UK: Longman.

Dowson, C.G., Rayner, A.D.M. and Boddy, L. 1986. Outgrowth patterns of mycelial cord-forming basidiomycetes from and between woody resource units in soil. *J. General Microbiology* 132: 203-11.

Dowson, C.G., Rayner, A.D.M. and Boddy, L. 1988a. The form and outcome of mycelial interactions involving cord-forming decomposer basidiomycetes in homogeneous and heterogeneous environments. *New Phytologist* 109: 423-32.

Dowson, C.G., Rayner, A.D.M. and Boddy, L. 1988b. Foraging patterns of *Phallus impudicus*, *Phanerochaete laevis* and *Steccherinum fimbriatum* between discontinuous resource units in soil. *FEMS Microbiology Ecology* 53: 291-98.

Dowson, C.G., Rayner, A.D.M. and Boddy, L. 1988c. Inoculation of mycelial cord-forming basidiomycetes into woodland soil and litter. I. Initial establishment. *New Phytologist* 109: 335-41.

Dowson, C.G., Rayner, A.D.M. and Boddy, L. 1988d. Inoculation of mycelial cord-forming basidiomycetes into woodland soil and litter. II. Resource capture and persistence. *New Phytologist*:109: 343-49.

Dowson, C.G., Rayner, A.D.M. and Boddy, L. 1989a. Spatial dynamics and interactions of the woodland fairy ring fungus, *Clitocybe nebularis*. *New Phytologist* 111: 699-705.

Dowson, C.G., Springham, P., Rayner, A.D.M. and Boddy, L. 1989b. Resource relationships of foraging mycelial systems of *Phanerochaete velutina* and *Hypholoma fasciculare*. *New Phytologist* 111: 501-09.

Gould, S.J. and Lewontin, R.C. 1979. The spandrels of San Marco and the Panglossian paradigm: A critique of the adaptionist programme. *Proc. Royal Society, London (Series B)* 205: 581-98.

Grime, J.P. 1979. *Plant Strategies and Vegetation Processes*. Chichester, UK/New York, USA: John Wiley.

Kirby, J.J.H., Webster, J. and Baker, J.H. 1990. A particle plating method for analysis of fungal community composition and structure. *Mycological Research* 94: 621-26.

Rayner, A.D.M. 1989. Developmental and genecological strategies of wood-inhabiting fungi. In Hattori, T., Ishida, Y, Maruyama, Y., Morita, R.Y. and Uchida, A. (eds) *Recent Advances in Microbial Ecology*. Tokyo, Japan: Scientific Press.

Rayner, A.D.M. 1991a. The challenge of the individualistic mycelium. *Mycologia* 83: 48-71.

Rayner, A.D.M. 1991b. Conflicting flows — the dynamics of mycelial territoriality. *McIlvainea* 10: 24-35.

Rayner, A.D.M. 1992. Monitoring genetic interactions between fungi in terrestrial habitats. In Wellington, E.M.H. and Van Elsas, J.D. (eds) *Genetic Interactions among Microorganisms in the Natural Environment*. Oxford, UK: Pergamon Press.

Rayner, A.D.M. and Todd, N.K. 1979. Population and community structure and dynamics of fungi in decaying wood. *Advances in Botanical Research* 7: 333-420.

Rayner, A.D.M. and Webber, J.F. 1984. Interspecific mycelial interactions — an overview. In Jennings, D.H. and Rayner, A.D.M. (eds) *The Ecology and Physiology of the Fungal Mycelium*. Cambridge, UK: Cambridge University Press.

Rayner, A.D.M. and Boddy, L. 1986. Population structure and the infection biology of wood-decay fungi in living trees. *Advances in Plant Pathology* 5: 119-60.

Rayner, A.D.M. and Boddy, L. 1987. Fungal communities in the decay of wood. *Advances in Microbial Ecology* 10: 115-66.

Rayner, A.D.M. and Franks, N.R. 1987. Evolutionary and ecological parallels between ants and fungi. *Trends in Ecology and Evolution* 2: 127-33.

Rayner, A.D.M. and Boddy, L. 1988. *Fungal Decomposition of Wood*. Chichester, UK: John Wiley.

Rayner, A.D.M. and Ross, I.K. 1991. Sexual politics in the cell. *New Scientist* 129: 30-33.

Rayner, A.D.M., Coates, D., Ainsworth, A.M., Adams, T.J.H., Williams, E.N.D. and Todd, N.K. 1984. The biological consequences of the individualistic mycelium. In Jennings, D.H. and Rayner, A.D.M. (eds) *The Ecology and Physiology of the Fungal Mycelium*. Cambridge, UK: Cambridge University Press.

Rayner, A.D.M., Powell, K.A., Thompson, W. and Jennings, D.H. 1985a. Morphogenesis of vegetative organs. In Moore, D., Casselton, L.A., Wood, D.A. and Frankland, J.C. (eds) *Developmental Biology of Higher Fungi*. Cambridge, UK: Cambridge University Press.

Rayner, A.D.M., Watling, R. and Frankland, J.C. 1985b. Resource relationships — an overview. In Moore, D., Casselton, L.A., Wood, D.A. and Frankland, J.C. (eds) *Developmental Biology of Higher Fungi*. Cambridge, UK: Cambridge University Press.

Rayner, A.D.M., Boddy, L. and Dowson, C.G. 1987a. Genetic interactions and developmental versatility during establishment of decomposer basidiomycetes in wood and tree litter. *Symposia of the Society for General Microbiology* 41: 83-123.

Rayner, A.D.M., Boddy, L. and Dowson, C.G. 1987b. Temporary parasitism of *Coriolus* spp. by *Lenzites betulina*: A strategy for domain capture in wood decay fungi. *FEMS Microbiology Ecology* 45: 53-58.

Rayner, A.D.M., Griffith, G.S. and Ainsworth, A.M. 1993a. Mycelial interconnectedness and the dynamic life styles of filamentous fungi. In Gow, N.A.R. and Gadd, G.M. (eds) *The Growing Fungus*. London, UK: Chapman and Hall.

Rayner, A.D.M., Griffith, G.S. and Wildman, H.G. 1993b. Differential insulation and the generation of my–celial patterns. In Ingram, D.S. (ed) *Shape and Form in Plants and Fungi*. London, UK: Academic Press.

Read, D.J. 1984. The structure and function of the vegetative mycelium of mycorrhizal roots. In Jennings, D.H. and Rayner, A.D.M. (eds) *The Ecology and Physiology of the Fungal Mycelium*. Cambridge, UK: Cambridge University Press.

Read, D.J. 1992. The mycorrhizal fungal community with special reference to nutrient mobilisation. In Carroll, G.C. and Wicklow, D.T. (eds) *The Fungal Community* (2nd edn). New York, USA: Marcel Dekker.

Ritz, K. and Crawford, J. 1990. Quantification of the fractal nature of colonies of *Trichoderma viride*. *Mycological Research* 94: 1138-41.

Smith, A.L. 1921. *Lichens*. Cambridge, UK: Cambridge University Press.

Smith, M.L., Bruhn, J.N. and Anderson, J.B. 1992. The fungus *Armillaria bulbosa* is among the largest and oldest living organisms. *Nature* 356: 428-31.

Stenlid, J. and Rayner, A.D.M. 1989. Environmental and endogenous controls of developmental pathways: Variation and its significance in the forest pathogen, *Heterobasidion annosum*. *New Phytologist* 113: 245-58.

Thompson, W. 1984. Distribution, development and functioning of mycelial cord systems of decomposer basidiomycetes of the deciduous woodland floor. In Jennings, D.H. and Rayner, A.D.M. (eds) *The Ecology and Physiology of the Fungal Mycelium*. Cambridge, UK: Cambridge University Press.

Wessels, J.G.H. 1991. Fungal growth and development: A molecular perspective. In Hawksworth, D.L. (ed) *Frontiers in Mycology*. Wallingford, UK: CAB International.

Wessels, J.G.H. 1992. Gene expression during fruiting in *Schizophyllum commune*. *Mycological Research* 96: 609-20.

PART V

Overview

Beyond the Biomass
Edited by K. Ritz, J. Dighton and K.E. Giller
© 1994 British Society of Soil Science (BSSS)
A Wiley-Sayce Publication

CHAPTER 26

Perspectives on the compositional and functional analysis of soil communities

D.C. Coleman, J. Dighton, K. Ritz and K.E. Giller

This chapter summarises some of the main issues which arose during the symposium, particularly in the discussion sessions, and provides further perspectives on the subject addressed by the symposium. We consider it useful at this stage to go a little beyond the 'Beyond'!

The symposium demonstrated clearly that progress is now being made in the compositional and functional analysis of soil communities. There was much emphasis on the development and refinement of methodology, which is to be expected given the way soils-related research is constrained by the inherent opacity, complexity and heterogeneity of the system it studies. An important aspect of the symposium was that it enabled molecular biologists and ecologists to exchange views on various approaches and problems. A major obstacle in applying methods from molecular biology to the study of microbial biomass has been the inability to isolate DNA from the soil in sufficient amounts and with sufficient purity. Soil factors such as humic acids are persistent contaminants of extracted DNA which interfere with many molecular biology protocols. The methods pioneered by MacDonald (1986) went some way towards this goal by using dispersion and elutriation techniques to separate microbial cells from other soil components, but it has been difficult to scale up such methods. The innovative application of aqueous two-phase partitioning (A2PP) to this problem by Smith and Stribley (*Chapter 5, this volume*) provides a method which can be used to process sufficient quantities of soil to obtain hitherto unattainably large quantities of microbial cells indigenous to the soil. This method excited considerable interest during the symposium and it will be interesting to see whether its application will lead to advances in the study of soil microbial DNA.

BIODIVERSITY

One of the principal concerns for biologists in the final decade of this century is: what are the fates of the diverse array of organisms in all ecosystems of the world? Do we know enough about the full species richness even to make educated guesses about the extent to which organisms are, or may be, endangered (Hawksworth, 1991a)? The diversity of soil organisms is apparently vast. This had been suspected, on the basis of observation (comparative discrepancies between number of culturable versus observable cells in soil preparations), as well as inference and extrapolation. For example, approximately 70 000 species of fungi are currently known (*see* Table 26.1). By analogy with a factor used to estimate the ratio of the species of vascular plants already known to those yet to be described, a total of 1.5 million fungal species is mooted as a conservative estimate (Hawksworth, 1991b). If the vast number of insects yet to be described, and the fungi associated with them, are considered, the actual number of fungal species will be even higher. As noted by Price (1988, 1992), mutualism facilitates adaptive radiation, so many more fungi associated with the as yet undescribed species of plants and insects would be expected.

Table 26.1   **Comparison of the numbers of known and estimated total species globally of selected groups or organisms**

Group	Known species	Estimated total species	Percentage known
Vascular plants	220 000	270 000	81
Bryophytes	17 000	25 000	68
Algae	40 000	60 000	67
Fungi	69 000	1 500 000	5
Bacteria	3 000	30 000	10
Viruses	5 000	130 000	4

Source: Hawksworth (1991b)

To give one example from the soil fauna, the oribatid mites constitute a representative group of soil animals which have been fairly well studied in the northern hemisphere. Yet, only 30-35% of oribatids in North America have been well described (Behan-Pelletier and Bissett, 1993), despite the extensive research carried out on them over the past 20-30 years. If one adds undescribed oribatids from the species-rich tropical environments, there may be more than 100 000 undescribed species within this group alone. Whilst it is acknowledged that there are many unknown faunal species below ground, the situation is complicated by the fact that only a tiny fraction of *immature* stages of soil fauna have been conclusively identified (*see* Figure 26.1). And, as Torsvik et al. (*Chapter 4*) demonstrate, molecular techniques are further reinforcing the evidence for startling below-ground diversity even among the prokaryotes.

These aspects of biodiversity present major challenges in attempting to analyse soil communities. We need to decide whether such a comprehensive inventory of species is actually necessary, especially since a complete catalogue may be impossible to achieve. Why does such diversity exist? Is it a consequence of the enormous range of niches over many scales that the soil architecture provides, coupled with the wide spectrum of substrates, both of which are arranged in a highly heterogeneous manner?

In order to answer some of these problems, Behan-Pelletier and Bissett (1993) suggest that 'advances in systematics and ecology must progress in tandem: systematics providing both the basis and predictions for ecological studies, and ecology providing information on community structure and explanations for recent evolution and adaptation.'

Figure 26.1   Percentage of adults and immatures described for selected arthropod groups in North American fauna

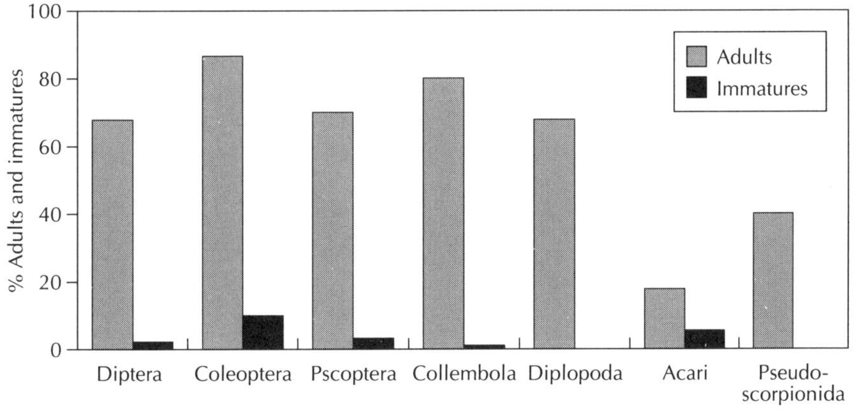

Source: Behan-Pelletier and Bisset (1993)

CONSEQUENCES OF BIODIVERSITY

The argument for the maintenance of biodiversity is akin to that of conservation in general. In the light of evolutionary trends, where species and higher taxonomic groups have waxed, waned and become extinct, or in seral successions where one species replaces another over time, we have to ask what purpose conservation serves. By maintaining artificially large biodiversity and genetic pools, are we suppressing, rather than encouraging, the evolution of new taxa which may be more adapted to life in, for example, environments modified by man? Is this why there appears to be an excess of diversity in the soil microflora, where many species exist at low frequency in an inactive state. Are they 'waiting in the wings' for changes to occur which will enable them to fulfil equivalent functions when some of the more dominant taxa are lost from the community? This would imply a degree of inbuilt redundancy in the microbial community, and that some organisms are vestigal relics of past conditions. Alternatively, could this part of the community be performing some vital, but as yet poorly understood, function? Wener (1992) cites an example where gnotobiotically grown sheep inoculated with 11 of the most common gut bacteria grew normally for 4 months. After that time, most animals died. It was postulated that further bacterial species were essential, but that they occurred in such small numbers that their origin and function were obscure. Only by studying both prevalent and rare organisms can the question raised by Dighton and Kooistra (1993) — 'who is who in the zoo and what do they do?' — be answered. However, as Torsvik et al. (*Chapter 4*) suggest, even 'rare' species may have fairly high population sizes in numerical terms. In many ways, the less common organisms pose the greater research challenges; perhaps such trace organisms may become a new theme in soil biology. Community diversity in the sense of a range across several trophic groups (vertical as opposed to horizontal diversity) has been shown to affect nutrient cycling (Coûteaux and Bottner, *Chapter 17*), but there appears to be a further interaction here with ecosystem type.

Recent studies have been conducted which link the biodiversity of organisms found to particular environmental stresses; for example, at the level of changes in microbial DNA in response to waterlogging, as described by Harris (*Chapter 12*), or the reduction in diversity of rhizobial populations in polluted soils as a result of long-term heavy metal contamination (Giller et al., 1989; Hirsch et al., 1993). Reductions in the diversity of rhizobia were noticed initially with the loss of function of a single specific group of microorganisms which led to yield loss in white clover. The potential danger is that in polluted soils other beneficial organisms could be lost without an immediately obvious loss of function. The diversity of organisms capable of carrying out a specific function is thus reduced, and the overall resilience of the biomass may be compromised. Evidence from ^{3}H-labelling experiments and fractionation of community DNA indicates that replication rates of different microorganisms are uniform and that activity and, most importantly, growth are not confined to a small fraction of the microbial biomass (Harris, *Chapter 12*).

The question of the relationship between diversity and function leads us to ask whether diversity necessarily leads to stability (or the corollary, that a loss of diversity will result in a trend towards instability). Diversity has been assumed to confer greater stability on ecological systems, although there is little experimental evidence to demonstrate such a link (Walker, 1989). Gross measurements of microbial diversity have been used to assess environmental stress, but such studies have been hampered by problems of sampling and extraction, resulting in bias towards certain 'species' types within mixed populations (Atlas, 1984). However, in a review of progress in research on microbial ecology, Brock (1987) challenged the purpose of studying species diversity. He suggested that the species diversity of microbial communities was not related to stability because microbial populations respond to environmental fluctuations by changing and not by resisting environmental stress. By contrast, it is likely that greater genetic diversity within a species will confer greater resilience in the face of further environmental perturbations. The spatial distribution of microorganisms, or rather, the provision of a matrix for such distribution, may also confer stability.

Clearly, the biomass is functionally diverse, as demonstrated so elegantly by the Biolog® plate method (Garland and Mills, *Chapter 8*; Winding, *Chapter 9*). However, as postulated by Young (*Chapter 11*), most processes are carried out by a diverse range of species and any one species may carry out a diverse range of processes. The use of reporter genes linked to gene promoters in order to measure the *in situ* activity of specific enzymes related to particular processes (Wilson, *Chapter 16*) may have great utility in monitoring the activity of processes in soil. We need to know how much loss of diversity can be borne without significant loss of function, and this question can now be realistically addressed.

Interactions

Interactions result both from the diversity of organisms in soil and the heterogeneous nature of the resources available to them. In the discussions at the symposium, fungi were considered an important but poorly understood component of the soil microflora.

The fungal/bacterial ratio, a difficult parameter to define given the indeterminate growth modes of eucarpic fungi, varies greatly in soils. In some systems, fungi may predominate (Zvyaginstev, *Chapter 3*), while in others they may constitute less than 5% of the biomass (Brussaard et al., 1990; J. Bloem, pers. comm.), with a complete range between these two extremes (for example, Jenkinson et al., 1976; Schnürer et al., 1986; Beare et al., 1992). Molecular data were presented at the meeting (Harris, *Chapter 12*) which suggested the eukaryotic fraction formed a very small proportion of the total DNA pool. It can be argued that the size of microbial pools *per se* are of less importance than how active they are (Clark and Paul, 1970; Scheu, 1992); the proportion of FDA-active hyphae is usually a small, and presumably variable, fraction of the total hyphal biomass (Schnürer et al., 1986; Ingham et al., 1989). In terms of ecological roles, quantity may not relate to quality in any case and, more importantly, where an interdependent system is being considered it may be irrelevant to ask which is the most prevalent. In arable soils subject to frequent physical disturbance, it is possible that continued disruption of hyphal networks may erode fungal populations. In undisturbed soils, such as those of forests and other natural ecosystems, fungi appear to play more important roles. Unlike determinate organisms, fungal hyphae are able to colonise new resources whilst still attached to old resource units. The direction and rate of hyphal growth is influenced by feedback signals (Rayner, *Chapter 25*) and the community structure which develops on a resource unit is dependent upon the physical and chemical dimension of the resource (Swift, 1976). By connecting resource islands, hyphae are able to transport nutrients from one to the other (Dighton and Boddy, 1989). Where competition between fungi occurs, the respiration of mixed species assemblages is greater than that expected by the combined respiration of the individuals (*see* Figure 26.2). Given no increase in mineralisation rate with mixed fungal species, the suggestion is that the increase in respiration results from the activity of maintaining competitive interactions (Robinson et al., 1993). If combative strategies are energetically demanding, how much less 'efficient' is an organism in competition rather than as the sole user of a resource?

Figure 26.2   Respiration from single and mixed species of fungi decomposing straw

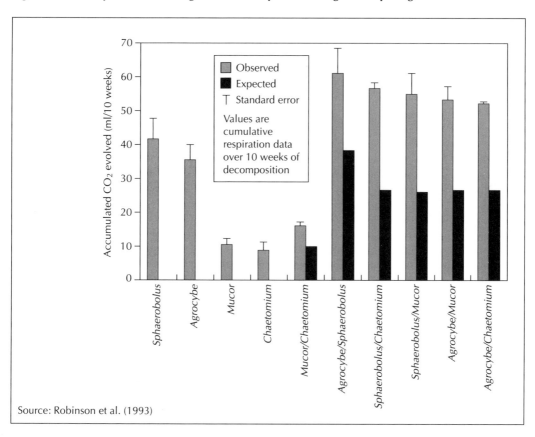

Source: Robinson et al. (1993)

Interdependencies were shown to exist between fungi and bacteria (Clarholm, *Chapter 22*) where bacterial rods (indicative of well-nourished cells) occurred in close association with fungal hyphae, whereas cocci (indicative of starved cells) occurred in the bulk soil. Similarly, roots supply resources which nourish microbial populations. The rhizosphere and symbiotic associations have been given attention by Lynch (1990) and Wener (1992), but in the case of mycorrhizae, little attention has been focused on the distal parts of hyphae in contact with nutrient resources. Here again, the marriage of molecular tools with ecology could enhance our understanding of interactions and processes.

Another persistent theme at the symposium concerned the interactions between soil microflora and fauna (for example, Coûteaux and Bottner, *Chapter 17*; Kandeler et al., *Chapter 19*). Such interactions have a number of effects:

- negative effects of grazing on population size of a given organism; if the system is conservative, then there will be a consequential loss of function

- positive effects of grazing on population size by increased resource availability in animal faeces (Parle, 1963; Shaw and Pawluk, 1986), stimulating the growth of fungi by, for example, low levels of grazing intensity (Parkinson et al., 1970) and influencing community structure by preferential grazing (Newell, 1984a, b)

- movement of microorganisms within the soil, allowing access to new resources (Moody, 1993)

To complete our understanding of interactions in heterogeneous environments, we need to advance into the third (spatial) and fourth (time) dimensions and avoid the two dimensions so beloved of scientific papers, computer screens and linear administrators. A notable example of this is the paper by Foster and Dormaar (1991) which showed that naked amoebae can extend pseudopodia into very small pore necks and consume the bacteria contained within the seemingly inaccessible pores. A meso-scale version of 3-D visualisation was presented by Joschko et al. (1991), where X-ray computed tomography permitted non-destructive assessment of earthworm burrows. Further, more detailed observation of the spatial relationship of organisms in small soil volumes (Ponge, 1988, 1990) may help improve our understanding of microbial interactions. Methods for *in situ* detection of micro–organisms in soil have been enhanced for some groups of microorganisms through the use of specific gene probes (Hahn, *Chapter 15*) and conventional microscopy. It is now feasible to couple such specific methods with soil thin-section techniques designed to preserve biological features in soils intact and *in situ* (Tippkötter et al., 1986; Postma and Altemüller, 1990; *see* Figure 26.3). The linking of such advances to automated image analysis procedures (Morgan et al., 1991; Bloem et al., 1992) offers tremendous possibilities in the analysis of the spatial distribution of microbes, both generally and specifically, in relation to their environment. Contemporary research in fractal mathematics (Crawford et al., 1993) also suggests that there are real possibilities for the development of a comprehensive, quantifiable and mechanistic framework for studying the relationships between soil architecture and microbial functioning below ground.

COMMUNITY FUNCTIONING

Much of the discussion at the symposium concerning the functional aspects of soil communities has been covered above. Here, we consider the functional aspects of different species of organisms and the modification of these processes by interactions.

Figure 26.3   Microbial features observed in thin-sections of undisturbed soil, prepared to show biological material *in situ*

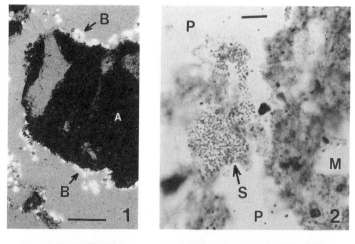

1.
Bacterial colonies (B) associated with aggregate (A) of mineral soil. Combined epifluorescence and bright field illumination. Bacteria fluoresce strongly due to staining with Calcofluor W® and are very distinct in original colour image. Scale bar = 50 μm.

2.
Spore cluster (S) associated with organic matter in vicinity of aggregate; M = mineral component, P = pore space. Spores stained with basic fuchsin, bright field transmitted illuminated. Scale bar = 50 μm.

3.
Fungal network (F, arrows) in pore (P) of forest soil. Hyphae stained with MgANS, UV epifluorescence illumination. Soil matrix also fluoresces due to high organic matter content. Scale bar = 100 μm.

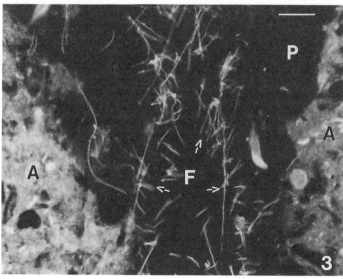

Source: R. Tippkötter and K. Ritz

As suggested earlier, there may be no foundation for the belief that there is a one-to-one congruence between biodiversity, expressed by species richness, and ecosystem function. If there were such a relationship, then one would expect the system to come into some form of 'steady state' at high levels of diversity. Physiologists and ecologists seem to have different opinions on the meaning of 'steady state'. Can we consider such a diverse collective as the biomass as a single entity, analagous to an organism? Soils, or rather parts of them, such as rhizospheres and 'hot spots', are very dynamic systems, and some argue that it may be irrelevant to consider such systems as being in steady state. This also provides an example of the problems inherent in extrapolating concepts devised for chemostats to a ped. Perhaps the approaches proposed by Kunc (*Chapter 2*), coupled with the concepts described by Anderson (*Chapter 7*) may provide further insight.

By encouraging molecular biologists to avoid concentrating on a few 'type' species, and to integrate their techniques with the information of process rates gained from the ecologists, further strides could be made in understanding the functional aspects of soil communities. For example, extension of the Biolog plate method (Garland and Mills, *Chapter 8*; Winding, *Chapter 9*) could demonstrate functional diversity, analysis of which could be 'fine tuned' using reporter genes attached to sites of enzyme activity (Wilson, *Chapter 16*) to locate active microsites. Analysis of profiles of phospholipid-linked fatty acid methyl esters (PL-FAMES) extracted from the biomass will also have a role to play here (Tunlid and White, 1992; Zelles et al., 1992). Specific signature lipids can be used to quantify subsets of the biomass, and changes in such profiles may be attributable to alterations in the physiological status of extant populations or to actual shifts in community structure. Selective removal of components of the ecosystem is a further approach in identifying where gaps in processes occur (Kandeler et al., *Chapter 19*; Ingham et al., 1989).

The application of approaches using techniques in molecular biology, together with 'classical' methods, may provide insights or even raise new questions for research. For example, in the enumeration of *Frankia* in soil, plant infection methods gave numbers of 'infection units' some 1000 times below the numbers of 'genomic units' estimated to be present by an approach based on polymerase chain reaction (PCR) amplification of *Frankia*-specific primers (Myrold, *Chapter 14*). It seems unlikely that 1000 genomic units are required to form each plant infection but it is unclear whether the widely used plant-infection method has given a misleading underestimation of numbers in the past, or whether the PCR-based method gives an overestimation of the population size.

Attention must always be paid to the functioning of the biota in interaction with the organic and inorganic components of soil. It has been hypothesised that soil architecture may exert a dominant influence over soil microbial processes; for example, in relation to decomposition (van Veen and Kuikman, 1990) or nitrogen mineralisation (Hassink, 1992). The notion of bacteria being protected by occupying pore spaces inaccessible to their grazing predators is often vaunted (for example, Vargas and Hattori, 1986; Postma and van Veen, 1990); however, as Foster and Dormaar (1991) have shown, the situation is not so clear-cut where amoebal pseudopodia are involved! There may also be direct consequences of soil mineralogy on soil organic matter sequestration and turnover. Does clay mineralogy have a significant impact on the size classes and turnover rates of the micro- and macro-aggregates which form in soil? The studies conducted by Oades and Waters (1991), which included comparing Oxisols with less-weathered Mollisols, need to be repeated in soils developed from several parent materials and of various ages. Many simulation models are unable to model long-term levels of organic matter, possibly because of poor understanding of these features; for example, current versions of the extensively used CENTURY model are able to generate narrow C/N ratio material, but do not mimic the carbon allocation patterns of most forest soil profiles (W. Pulliam, pers. comm.).

CONCLUSION

A conclusion so often discussed at symposia, the present one included, may be referred to as the 'question syndrome': what are the questions that we really need to ask, and at what scale of resolution do we wish to ask them? We must tailor our methodologies to these scales and think seriously about the consequences of extrapolating results up, or down, between them (Coleman, *Chapter 21*; Harris,

Chapter 24). How to scale from individuals to colonies to communities, and hence to ecosystems, in a realistic and meaningful manner, promises to be a major research theme in the future.

The symposium and this volume have highlighted a number of successes and raised some problems to be investigated in the future. It is clear that the initiation of dialogue between molecular biologists and ecologists has the potential to advance the science of soil ecology significantly. By using molecular tools in conjunction with traditional methods, making precise observations in four dimensions and acknowledging heterogeneity and scaling, soil biology will certainly progress in the quest to assign functionality to the entity of the microbial biomass which researchers have so avidly measured over recent years.

References

Atlas, R.M. 1984. Use of microbial diversity measurements to assess environmental stress. In Klug, M.J. and Reddy, A. (eds.) *Current Perspectives in Microbial Ecology*. Washington DC, USA: American Society for Microbiology.

Beare, M.H., Parmelee, R.W., Hendrix, P.F., Cheng, W., Coleman, D.C. and Crossley, D.A. 1992. Microbial and faunal interactions and effects on litter nitrogen and decomposition in agroecosystems. *Ecological Monographs* 62: 569-91.

Behan-Pelletier, V.M. and Bissett, B. 1993. Biodiversity of nearctic soil arthropods. *Canadian Biodiversity* 2: 5-14.

Bloem, J., de Ruiter, P.C., Koopman, G., Lebbink, G. and Brussaard, L. 1992. Microbial numbers and activity in dried and rewetted arable soil under conventional and integrated management. *Soil Biology and Biochemistry* 24: 655-65.

Brock, T.D. 1987. The study of microorganisms *in situ*: Progress and problems. In Fletcher, M., Gray, T.R.G. and Jones, J.G. (eds) *Ecology of Microbial Communities*. Cambridge, UK: Cambridge University Press.

Brussaard, L., Bouwman, L.A., Geurs, M., Hassink, J. and Zwart, K.B. 1990. Biomass, composition and temporal dynamics of soil organisms of a silt loam under conventional and integrated management. *Netherlands J. Agricultural Science 38*: 283-302.

Clark, F.E. and Paul, E.A. 1970. The microflora of grassland. *Advances in Agronomy* 22: 375-435.

Crawford, J.W., Ritz, K. and Young, I.M. 1993. Quantification of fungal morphology, gaseous transport and microbial dynamics in soil: An integrated framework utilising fractal geometry. *Geoderma* 56: 157-72.

Dighton, J. and Boddy, L. 1989. Role of fungi in nitrogen, phosphorus and sulphur cycling in temperate forest ecosystems. In Boddy, L., Marchant, R. and Read, D.J. (eds) *Nitrogen, Phosphorus and Sulphur Cycling in Temperate Forest Ecosystems*. Cambridge, UK: Cambridge University Press.

Dighton, J. and Kooistra, M. 1993. Measurement of proliferation and biomass of fungal hyphae and roots. *Geoderma* 56: 317-30.

Foster, R.C. and Dormaar, J.F 1991. Bacteria-grazing amoebae *in situ* in the rhizosphere. *Biology and Fertility of Soils* 11: 83-87.

Giller, K.E., McGrath, S.P and Hirsch, P.R. 1989. Absence of nitrogen fxation in clover grown on soil subject to long term contamination with heavy metals is due to survival of only ineffective *Rhizobium*. *Soil Biology and Biochemistry* 21: 841-48.

Hassink, J. 1992. Effects of soil texture and structure on carbon and nitrogen mineralisation in grassland soils. *Biology and Fertility of Soils* 14: 126-34.

Hawksworth, D.L. 1991a. *The Biodiversity of Microorganisms and Invertebrates*. Wallingford, UK: CAB International.

Hawksworth, D.L. 1991b. The fungal dimension of biodiversity: Magnitude, significance and conservation. *Mycological Research* 95: 641-55.

Hirsch, P.R., Jones, M.J., McGrath, S.P. and Giller, K.E. 1993. Heavy metals from past applications of sewage sludge decrease the genetic diversity of *Rhizobium leguminosarum* biovar *trifolii* populations in field soil. *Soil Biology and Biochemistry* (in press).

Ingham, E.R., Coleman, D.C. and Moore, J.C. 1989. An analysis of food-web structure and function in a shortgrass prairie, a mountain meadow, and a lodgepole pine forest. *Biology and Fertility of Soils* 8: 29-37.

Jenkinson, D.S., Powlson, D.S. and Wedderburn, R.W.M. 1976. The effects of biocidal treatments on metabolism in soil. III. The relationship between soil biovolume, measured by optical microscopy, and the flush of decomposition caused by fumigation. *Soil Biology and Biochemistry* 8: 189-202.

Joschko, M., Müller, P.C., Kotzke, K., Linder, P., Pretschner, D.P. and Larink, O. 1991. A non-destructive method for the morphological assessment of earthworm burrow systems in three dimensions by X-ray computed tomography. *Biology and Fertility of Soils* 11: 88-92.

Lynch, J.M. 1990. *The Rhizosphere*. London, UK: John Wiley.

Macdonald, R.M. 1986. Sampling soil microfloras: Dispersion of soil by ion exchange and extraction of specific microorganisms from suspension by elutriation. *Soil Biology and Biochemistry* 18: 399-406.

Moody, S.A. 1993. Aspects of dispersal of wheat straw fungi by earthworms and springtails. PhD thesis, University of Lancaster, UK.

Morgan, P., Cooper, C.J., Battersby, N.S., Lee, S.A., Lewis, S.T., Machin, T.M., Graham, S.C. and Watkinson, R.J. 1991. Automated image analysis method to determine fungal biomass in soils and on solid matrices. *Soil Biology and Biochemistry* 23: 609-16.

Newell, K. 1984a. Interactions between two decomposer basidiomycetes and a collembolan under *Sitka* spruce: Distribution, abundance and selective grazing. *Soil Biology and Biochemistry* 16: 227-33.

Newell, K. 1984b. Interactions between two decomposer basidiomycetes and a collembolan under Sitka spruce: Grazing and its potential effects on fungal distribution and litter decomposition. *Soil Biology and Biochemistry* 16: 235-39.

Oades, J.M. and Waters, A.G. 1991. Aggregate hierarchy in soils. *Australian J. Soil Research* 29: 815-28.

Parle, J.N. 1963. A microbiological study of earthworm casts. *J. General Microbiology* 31: 13-22.

Parkinson, D., Visser, S. and Whittaker, J.B. 1970. Effects of collembollan grazing on fungal colonisation of leaf letter. *Soil Biology and Biochemistry* 18: 583-88.

Ponge, J.F. 1988. Etude ecologique d'un humus forestière par l'observation d'un petit volume. III. La couche Fl d'un moder sous *Pinus sylvestris*. *Pedobiologia* 31: 1-64.

Ponge, J.F. 1990. Ecological study of a forest humus by observing a small volume. I. Penetration of pine litter by mycorrhizal fungi. *European J. Forest Pathology* 20: 290-303.

Postma, J. and Altemüller, H.-J. 1990. Bacteria in thin soil sections stained with the fluorescnce brightener Calcofluor M2R. *Soil Biology and Biochemistry* 22: 89-96.

Postma, J. and van Veen, J.A. 1990. Habitable pore space and survival of *Rhizobium leguminosarum* biovar *trifolii* introduced into soil. *Microbial Ecology* 19: 149-61.

Price, P.W. 1988. An overview of organismal interaction in ecosystems in evolutionary and ecological time. *Agriculture Ecosystems and Environment* 24: 369-77.

Price, P.W. 1992. The resource-based organisation of communities. *Biotropica* 24: 273-82.

Robinson, C.H., Dighton, J., Frankland, J.C. and Coward, P.A. 1993. Nutrient and carbon dioxide release by interacting species of straw-decomposing fungi. *Plant and Soil* 151: 139-42.

Scheu, S. 1992. Automated measurement of the respiratory response of soil microcompoartments: Active microbial biomass in earthworm faeces. *Soil Biology and Biochemistry* 24: 1113-18.

Schnürer, J., Clarholm, M. and Rosswall, T. 1986. Fungi, bacteria and protozoa in soil from four arable cropping systems. *Biology and Fertility of Soils* 2: 119-26.

Shaw, C. and Pawluk, S. 1986. Faecal microbiology of *Octolasion tyrtaeum*, *Aporrectodea turgida* and *Lumbricus terrestris* and its relation to carbon budgets of three artificial soils. *Pedobiologia* 29: 377-89.

Swift, M.J. 1976. Species diversity and the structure of microbial communities in terrestrial habitats. In Anderson, J.M. and Macfadyen, A. (eds) *The Role of Aquatic and Terrestrial Organisms in Decomposition Processes*. Oxford, UK: Blackwell.

Tippkötter, R., Ritz, K. and Darbyshire, J.F. 1986. The preparation of soil thin sections for biological studies. *J. Soil Science* 37: 681-90.

Tunlid, A. and White, D.C. 1992. Biochemical analysis of biomass, community structure, nutritional status and metabolic activity of microbial communities in soil. In Stotzky, G. and Bollag, J.M. (eds) *Soil Biochemistry* (Vol. 7). New York,USA: Marcel Dekker.

van Veen, J.A. and Kuikman, P.J. 1990. Soil structural aspects of decomposition of organic matter. *Biogeochemistry* 11: 213-33.

Vargas, R. and Hattori, T. 1986. Protozoan predation of bacterial cells in soil aggregates. *FEMS Microbiology Ecology* 38: 233-42.

Walker, D. 1989. Diversity and stability. In Cherrett, J.M. (ed) *Ecological Concepts*. Oxford, UK: Blackwell.

Wener, D. 1992. *Symbiosis of Plants and Microbes*. London, UK: Chapman and Hall.

Zelles, L., Bai, Q.Y. and Beese, F. 1992. Signature fatty acids in phospholipids and lipopolysaccharides as indicators of microbial biomass and community structure in agricultural soils. *Soil Biology and Biochemistry* 24: 317-23.

Index